Contributions to Management Science

For further volumes:
http://www.springer.com/series/1505

Thomas Ehrmann • Josef Windsperger
Gérard Cliquet • George Hendrikse
Editors

Network Governance

Alliances, Cooperatives and Franchise Chains

Physica-Verlag
A Springer Company

Editors
Thomas Ehrmann
Institut für Strategisches Management
Universität Münster
Münster
Germany

Josef Windsperger
Department of Management
University of Vienna
Wien
Austria

Gérard Cliquet
Institute of Management of Rennes (IAE)
Université de Rennes
Rennes Cédex
France

George Hendrikse
Rotterdam School of Management
Erasmus University Rotterdam
Rotterdam
The Netherlands

ISSN 1431-1941
ISBN 978-3-642-43091-6 ISBN 978-3-7908-2867-2 (eBook)
DOI 10.1007/978-3-7908-2867-2
Springer Heidelberg New York Dordrecht London

Printed on acid-free paper

Physica is a brand of Springer
Springer is part of Springer Science+Business Media (www.springer.com)

Acknowledgements

Funding from the Austrian Ministry for Education, Science and Culture is gratefully acknowledged.

Contents

Part B: Cooperatives

Part C: Franchise Chains

Contributors

Ana Aleksić Mirić University of Belgrade, Belgrade, Serbia

Nieves Arranz Faculty of Economics and Business Administration (UNED), Madrid, Spain

Guy Basset Graduate School of Management – University of Rennes, Center for Research in Economics and Management (CREM UMR CNRS 6211), Rennes Cedex 7, France

Maryse Brand University of Groningen, Groningen, The Netherlands

Danièle Briand-Meledo Faculty of Law and Political Science – University of Rennes 1, Institute for Business Law, Estate and Torts (CDA-PR UPRES EA 3195), Rennes Cedex, France

Jenny Buchan School of Taxation and Business Law, Australian School of Business, University of New South Wales, Kensington, NSW, Australia

Richard Burton Duke University, Fuqua School of Business, Durham, NC, USA

Clarietta Chagwiza Centre for International Development Issues Nijmegen, Radboud University Nijmegen, Nijmegen, The Netherlands

Evelien Croonen University of Groningen, Groningen, The Netherlands

Gérard Cliquet Institute of Management of Rennes (IAE), Université de Rennes 1, Rennes Cedex, France

Wendong Deng Rotterdam School of Management, Erasmus University, Rotterdam, The Netherlands

Carlos Fdez. de Arroyabe ESIC Business and Marketing School, Pozuelo de Alarcón, Spain

Cary Di Lernia Discipline of Business Law, The University of Sydney Business School, University of Sydney, Sydney, NSW, Australia

Thomas Ehrmann Institute of Strategic Management, University of Muenster, Muenster, Germany

Hanane El Zeiny University of Alexandria, Alexandria, Egypt

Li Feng Department of Economics, Swedish University of Agricultural Sciences, Uppsala, Sweden

Taras Gagalyuk Ukrainian Agribusiness Club, Kiev, Ukraine

Karin Hakelius Department of Economics, Swedish University of Agricultural Sciences, Uppsala, Sweden

John Hanf University of Applied Sciences RheinMain, Geisenheim, Germany

George Hendrikse Rotterdam School of Management, Erasmus University Rotterdam, Rotterdam, The Netherlands

Kostas Karantininis Department of Economics, Swedish University of Agricultural Sciences, Uppsala, Sweden

Roldan Muradian Centre for International Development Issues Nijmegen, Radboud University Nijmegen, Nijmegen, The Netherlands

Rozenn Perrigot Graduate School of Management – University of Rennes 1 & ESC Rennes, Center for Research in Economics and Management (CREM UMR CNRS 6211) & Institute for Business Law, Estate and Torts (CDA-PR UPRES EA 3195), Rennes Cedex 7, France

Mirjana Petkovic University of Belgrade, Belgrade, Serbia

Ruerd Ruben Centre for International Development Issues Nijmegen, Radboud University Nijmegen, Nijmegen, The Netherlands

Marijana Srećković Chair of Real Estate Development, Faculty of Architecture and Planning, Vienna, University of Technology, Vienna, Austria

Odile Streed Concordia College, Moorhead, MN, USA

Andrew Terry Discipline of Business Law, The University of Sydney Business School, University of Sydney, Sydney, NSW, Australia

Worku Tessema Department of Agricultural Economics, Hawassa University, Awassa, Ethiopia

Jenni Torikka Jyväskylä University School of Business and Economics, University of Jyväskylä, Jyväskylä, Finland

Josef Windsperger Center for Business Studies, University of Vienna, Vienna, Austria

Muhammad Zafar Yaqub Center for Business Studies, University of Vienna, Vienna, Austria

Introducing "Network Governance: Alliances, Cooperatives and Franchise Chains"

Josef Windsperger, Gérard Cliquet, Thomas Ehrmann, and George Hendrikse

Network governance through alliances, cooperatives and franchise chains aims to explore and exploit knowledge in interfirm relationships in order to realize competitive advantage by creating relational rents and/or reducing coordination and agency costs (e.g. Blair and Lafontaine 2005; Cliquet and Penard 2012; Combs et al. 2011; Dant et al. 2011; Dyer 1997; Dyer and Singh 1998; Grandori 2010; Gulati 2007; Hendrikse and Feng 2013; Kale and Singh 2007, 2009; Madhok and Tallman 1998; March 1991; Mayer and Salomon 2006; Meiseberg and Ehrmann 2013; Poppo and Zenger 2002; Schilke and Goerzen 2010; Tuunanen et al. 2011; Williamson 1991; Windsperger 2013). The current book emphasizes network research by offering new perspectives on formal and informal network governance. In particular, issues on interorganizational learning, relational norms (e.g. trust), knowledge transfer in alliances, governance and incentives in cooperatives, governance of international retail chains, goal achievement in supply chain networks, network uniformity and intercultural aspects in franchising, development of new franchise governance forms, determinants of the decision-making process regarding franchising versus self-employment as well as efficacy of pre-purchase disclosure in franchise relationships are discussed.

J. Windsperger (✉)
Center for Business Studies, University of Vienna, Bruenner Str. 72, 1210 Vienna, Austria
e-mail: josef.windsperger@univie.ac.at

G. Cliquet
IGR-IAE, University of Rennes 1, 11, rue Jean Macé, CS 70803, 35708 Rennes Cedex, France

T. Ehrmann
Institute of Strategic Management, University of Muenster, Leonardo-Campus 18, 48149 Muenster, Germany

G. Hendrikse
Rotterdam School of Management, Erasmus University Rotterdam, PO Box 1738
Office T08-56, 3000 DR Rotterdam, The Netherlands

T. Ehrmann et al. (eds.), *Network Governance*, Contributions to Management Science,
DOI 10.1007/978-3-7908-2867-2_1,

A first version of these papers from different areas in economics and management of networks (alliances, cooperatives and franchising) were initially presented at the fifth international conference on Economics and Management of Networks (*EMNet*) that took place at the Frederick University in Limassol, Cyprus, from December 1 to December 3, 2011.

The book is structured in three parts:

Alliances
Cooperatives
Franchise chains

1 Alliances

Aleksić Mirić, Burton and Petkovic investigate the issue of inter-organizational learning coordination throughout the evolution of the strategic alliance. They argue that alliances should be designed to learn, and alliance partners' choices about mechanisms of coordination should be made depending on the stage of alliance evolution. They operationalize alliance evolution through two dimensions—alliance age and alliance maturity. Alliance age is simply the time from birth of the alliance; alliance maturity is the degree of past involvement and experience of the two alliance partners. The study suggests that alliance age is positively associated with the application of formal mechanisms of coordination, while alliance maturity is positively associated with the application of informal mechanisms of coordination. Application of both formal and informal mechanisms of coordination is important for learning and knowledge transfer in strategic alliances; thus, their importance is moderated by alliance maturity. The results contribute to the knowledge on strategic alliance dynamics and to the organizational design theory.

Arranz and Fdez. de Arroyabe examine the role of network position in the performance of joint R&D projects using data on European networks of excellence. The empirical work asks the question whether centrality (structural embeddedness) and connectivity (junctional embeddedness) increase the project performance. Their results show that while structural embeddedness exerts a clear influence on the performance of exploitation projects, the effect of centrality degree has a lower impact on exploration projects due to the redundant information that this location implies. Similarly, while junctional embeddedness positively affects the performance of exploration projects, the heterogeneity of partners involved in a betweenness centrality position does not favor the performance of exploitation projects. This study contributes to social capital theory by offering new empirical evidence of the effect of network position on performance, which depends on the technological objectives of joint R&D projects.

According to *Yaqub*, a number of studies have shown that the relational norms positively affect the relational outcomes like satisfaction, trust and/or commitment that eventually lead to an enhancement in the value co-created by the exchange partners. However, most of these studies have theoretically and/or empirically treated relational norms and/or the intermediate relational outcomes as complex higher-order constructs. *Yaqub* argues that such a treatment of these relational constructs may lead to the loss of useful information about the interaction of individual norms with the various facets of these relational outcomes. The results from this empirical study conducted in the context of supplier-intermediate buyer dyadic relationships offer some interesting insights into the dynamics of relational exchanges in the downstream structural arrangements. The study specifically contributes to the relational exchange literature by empirically demonstrating the phenomena like unilateral-relationalism and the relationality-paradox over and above presenting an unprecedented discussion on the association among the individual relational norms and the key relational outcomes.

Srećković and Windsperger examine the impact of trust on the use of knowledge transfer mechanisms of cluster firms by deriving hypotheses from a relational governance perspective. Specifically, they analyse the influence of trust on the use of face-to-face knowledge transfer mechanisms in cluster relationships. Based on the relational view of governance, it is argued that trust may influence the choice of knowledge transfer mechanisms of the cluster companies in the following way: First, if trust reduces relational risk, an increase in trust reduces the firms' use of face-to-face knowledge transfer mechanisms. Second, if trust increases knowledge sharing between the cluster partners, it increases the firms' use of face-to-face knowledge transfer mechanisms. The hypotheses are tested by using data from 118 companies in the Italian textile and fashion sectors. The data supports the hypothesis that experience-based trust increases knowledge sharing between the cluster partners by increasing the use of face-to-face knowledge transfer mechanisms. It also supports the knowledge-based hypothesis that tacitness influences the choice of knowledge transfer mechanisms. The study extends the knowledge-based view of the choice of knowledge transfer mechanisms by showing that trust is an additional determinant of the knowledge transfer strategy.

Streed and Cliquet investigate the determinants of market entry failures of international retail networks. Specifically, they ask the question what causes an international retailer to divest from a specific country or region. Based on a sample of 112 cases, exploratory results show that time of entry, brand penetration, entropy level and local store density are strongly correlated to failure or success of an international retailer in emerging countries. Additionally, preliminary results in terms of market entry mode choice tend to show that governance modes with low level of control such as franchising, licensing or minority joint ventures may be the best market entry choice for international retailers expanding into emerging countries.

Based on the relational view of governance, *Gagalyuk and Hanf* develop and test a model of goal achievement in supply chain networks. They conceptualise goals of a whole supply chain network as those set by members at the firm and network levels. Moreover, they relate the achievement of network-level and firm-level goals of network members to the achievement of goals of the network management. The latter include the alignment of interests and the alignment of actions which, if fulfilled simultaneously, pave the way for the achievement of both network-level and firm-level goals of the network participants. Furthermore, they hypothesize that the interest and action alignment are contingent upon a number of network characteristics. They test the model in the context of supply chain networks in the food industry.

2 Cooperatives

Hakelius, Karantininis and Feng address the phenomenon of cooperative beehiving. Members de-associate themselves from large cooperatives and form smaller entities, just as bees swarm out of the old crowded beehive in search for a new one. They show in the framework of transaction cost theory that the exiting farmers are those who have experience and advantages in organizing cooperatives and are willing to take risks as entrepreneurs. The new beehives, organized also as cooperatives, rely heavily on outsourcing and start-up assistance plans. Two cases from the Swedish agrifood industry illustrate their claims.

The study of *Chagwiza, Muradian, Ruben and Tessema* deals with the comparison of two types of honey producers' enterprises in the Masha district, western Ethiopia. Cooperatives and private limited companies (PLCs) are both collectively owned by a group of farmers, but the former do not face a legal restriction regarding the number of members, are strongly regulated by the government, and their shares are not tradable. They argue that the collective entrepreneurial capacity varies significantly among the two types of organizations. They found that members of PLCs have higher productivity and income derived from honey, are more prone to adopt new technologies, as well as receive higher dividends and price per kilo of honey. Additionally, the incentive scheme exercised by the PLCs was more market-oriented. Furthermore, as compared to cooperatives, PLC members market a higher proportion of honey through their organizations. These results are relevant for the design of development interventions aiming at enhancement of market integration of small farmers in Ethiopia.

Feng and Hendrikse develop a multi-task principal-agent model to capture the effect of membership size and heterogeneity on the incentive provision of the CEO in a cooperative. The study shows that an increase in membership size as well as an increase in member heterogeneity decreases the optimal incentive intensity of the CEO.

The work of *Deng and Hendrikse* compares the product quality provision of cooperatives and investor-owned firms (IOFs) by highlighting the impacts of uncertainties in agricultural production and marketing, and farmers' risk aversion. In a principal-agent model, they show that the linear contract can shift the risk of market uncertainty from farmers to processors, and pooling can share the risk of production uncertainty among cooperative members. Complete pooling places the cooperative at a disadvantage relative to the IOF in a quality-differentiated market due to the loss of free-riding dominating the gain of risk-sharing. Product quality of cooperatives decreases when the membership size increases. Cooperatives can overcome this disadvantage by partial pooling. Product quality of cooperatives will be equivalent to that of IOFs when an optimal income rights structure with partial pooling is adopted.

3 Franchise Chains

Croonen and Brand's study develops a theoretical framework on antecedents of franchisees' trust in their franchisors and franchise systems. They integrate franchising literature with literature on trust in other organizational contexts to develop the framework. They argue that a franchisee's general propensity to trust combined with its perception of trustworthiness of its franchisor and franchise system determines this franchisee's level of organizational trust. *Croonen and Brand* distinguish three franchise system components that each entail a set of determinants by which franchisees evaluate a franchisor's and franchise system's trustworthiness; the system's strategic positioning in the market, the system's operational management, and franchisee management.

The study of *Perrigot, Basset, Briand-Meledo and Cliquet* aims to highlight the challenges associated with network uniformity and brand image for franchisors, more specifically when their franchisees set up and manage their own website. This practice has some impact on network uniformity which is a key concept in franchising. They analyze the presence of franchisees on the Internet of the 471 networks, both in retailing and services, described in the 2011 franchise directory. They find that only 38 franchise networks are concerned about this practice. They use a qualitative approach based on multiple cases studies of these 38 franchise networks. It points out the different aspects of franchisees' websites that can damage concept uniformity. Maintaining network uniformity when there are various websites set up and run by franchisees entails challenges to franchisors that are presented in this paper within a managerial perspective linked to technical and organizational know-how. Some insights from the legal perspective are also provided. If the concept of uniformity can be of great interest in managing franchise networks, the question is to know if this concept in the field of franchising can be identified as a legal concept and which would be its legal status.

El Zeiny and Cliquet investigate service quality variation among McDonald's fast food franchise chain outlets. Findings from the data collected from 162 customers indicate that McDonald's fast food chain restaurant is able to ensure service quality standardization across its franchised outlets located in Egypt, while it fails to ensure this standardization across franchised units in Egypt on the one hand and franchised units abroad on the other hand. The study concludes that, although standardization is expected from the franchised outlets, some outlets are not able to follow the same standards, especially, when they are functioning in different conditions.

According to *Terry and Di Lernia,* franchising's capacity for reinventing itself is a matter of record. This study suggests a role for a form of franchising which incorporates only back-of-house elements—the tried, tested and proven systems and procedures which are not directly visible to the customer—and eschews brand and other visible manifestations of a standardized "one-size-fits-all" approach to service provision. It proposes a form of quasi-franchising where brand and related front-of-house features are removed or, at least, significantly reduced. The franchisee acquires the right, and the obligation, to use the franchisor's back-of-house system while retaining flexibility for entrepreneurial endeavour in building an idiosyncratic, eclectic and individualized business.

Torrika's study reports the results of the third and final phase of the longitudinal study on trainees of the Finnish franchisee training program. The study targeted 46 respondents who in the previous phases indicated that they became either franchisees or self-employed in stand-alone businesses after graduation. The aim of this study is to find out what factors influence the respondents' entrepreneurial decision-making processes and what role the training program plays in these processes. The results show that trainees' entrepreneurial decision-making processes are dissimilar. Some are pushed while others are pulled to entrepreneurship. Regression analysis was performed to discover what factors could be used to predict respondents' likelihood of becoming franchisees or self-employed. Findings of the study provide implications for both franchisors and potential franchisees as well as for organizations' planning and offering entrepreneurship training.

Buchan argues that pre-purchase disclosure is an important aspect of the due diligence process for business format franchisees. It focuses on the financial fitness of the franchisor entity, and on the specific franchise opportunity the franchisee is evaluating. Equipped with disclosed information a diligent franchisee theoretically has the opportunity, *ex ante*, to identify and evaluate risks and protect itself from the consequences of making a bad investment decision. This study examines the efficacy of disclosure for franchisees whose insolvent franchisor enters administration. Problems arise out of the content and timing of disclosure, the difficulty of verifying the disclosed information and the conflicting requirements of the legislation protecting franchisees and that regulating administrators. Pre-purchase disclosure cannot empower franchisees to anticipate or address the consequences of franchisor administration. It fails to protect franchisees. Suggested avenues for redressing this situation are identified.

References

Blair R, Lafontaine F (2005) The economics of franchising. Cambridge University Press, Cambridge

Cliquet G, Penard T (2012) Plural form franchise networks: a test of Bradach's model. J Retailing Consum Serv 19:159–167

Combs JG, Ketchen DJ, Shook CL, Short JC (2011) Antecedents and consequences of franchising: past accomplishments and future challenges. J Manage 37:99–126

Dant RP, Grünhagen M, Windsperger J (2011) Franchising research frontiers for the twenty-first century. J Retailing 87:253–268

Dyer JH (1997) Effective interim collaboration: how firms minimize transaction costs and maximize transaction value. Strateg Manage J 18:535–556

Dyer JH, Singh H (1998) The relational view: cooperative strategy and sources of interorganizational competitive advantage. Acad Manage Rev 23:660–679

Grandori A (2010) Asset commitment, constitutional governance and the nature of the firm. J Inst Econ 6:351–375

Gulati R (2007) Managing network resources: alliances, affiliations, and other relational assets. Oxford University Press, Oxford

Hendrikse G, Feng L (2013) Interfirm cooperatives. In: Grandori A (ed) Handbook of economic organization. Integrating economic and organization theory. Edward Elgar Publishing, Cheltenham

Kale P, Singh H (2007) Building firm capabilities through learning: the role of the alliance learning process in alliance capability and firm-level alliance success. Strateg Manage J 28:981–1000

Kale P, Singh H (2009) Managing strategic alliances: what do we know now, and where do we go from here. Acad Manage Perspect 23:45–62

Madhok A, Tallman S (1998) Resources, transactions and rents: managing value through interfirm collaborative relationships. Organ Sci 9:326–339

March JG (1991) Exploration and exploitation. Organ Sci 2:71–87

Mayer KJ, Salomon RM (2006) Capabilities, contractual hazards, and governance: integrating resource-based and transaction cost perspective. Acad Manage J 49:942–959

Meiseberg B, Ehrmann T (2013) Tendency to networks of SMEs: combining organizational economics and resource-based perspectives. Managerial and Decision Economics (forthcoming)

Poppo L, Zenger T (2002) Do formal contracts and relational governance function as substitutes or complements? Strateg Manage J 23:707–725

Schilke O, Goerzen A (2010) Alliance management capability: an investigation of the construct and its measurements. J Manage 36:1192–1219

Tuunanen M, Windsperger J, Cliquet G, Hendrikse G (eds) (2011) New developments in the theory of networks: franchising, alliances and cooperatives. Springer, Heidelberg

Williamson OE (1991) Comparative economic organization: the analysis of discrete structural alternatives. Adm Sci 36:269–296

Windsperger J (2013) Governance of franchising networks. In: Grandori A (ed) Handbook of economic organization. Integrating economic and organization theory. Edward Elgar Publishing, Cheltenham

Part A
Alliances

Coordinating Inter-organizational Learning Throughout Alliance Evolution

Ana Aleksić Mirić, Richard M. Burton, and Mirjana Petković

Abstract Based on three case studies we investigate the issue of inter-organizational learning coordination throughout the evolution of the strategic alliance. We imply that alliances should be designed to learn, and alliance partners' choices about mechanisms of coordination should be made depending on the stage of alliance evolution. We operationalize alliance evolution through two dimensions:—alliance age and alliance maturity. Alliance age is simply the time from birth of the alliance; alliance maturity is the degree of past involvement and experience of the two alliance partners. The study suggests that alliance age is positively associated with the application of formal mechanisms of coordination, while alliance maturity is positively associated with the application of informal mechanisms of coordination. Application of both formal and informal mechanisms of coordination is important for learning and knowledge transfer in strategic alliances; thus, their importance is moderated by alliance maturity. Alliance maturity, unlike alliance age, can decrease. Our results contribute to the knowledge on strategic alliance dynamics, and to the organizational design theory.

Keywords Alliance evolution • Alliance maturity • Coordination mechanisms • Interorganizational learning

A. Aleksić Mirić (✉) • M. Petković
Faculty of Economics, University of Belgrade, Kamenicka 6 Street, Belgrade 11000, Serbia
e-mail: anaa@ekof.bg.ac.rs

R.M. Burton
Duke University, Fuqua School of Business, 100 Fuqua Drive, Durham, NC 27708-0120, USA

T. Ehrmann et al. (eds.), *Network Governance*, Contributions to Management Science, 11
DOI 10.1007/978-3-7908-2867-2_2, © Springer-Verlag Berlin Heidelberg 2013

1 Introduction

Organizational learning and transfer of knowledge are among the main reasons why companies enter strategic alliances (Inkpen and Ramaswamy 2006; Child 2001; Lyles 2001). As structured systems of established relationships, alliances develop gradually over time. Learning as an inter-organizational process between allies depends on time as well: time is essential for bringing people together, for developing mutual trust, shared cultural values, and joint ideas. However, the interplay between these two evolutionary processes, learning and aging, is not extensively researched. Existing literature recognizes that alliances are a fertile ground for organizational learning and knowledge transfer, and several explanations of the relationship between alliance evolution and learning have been offered. One stream of the research says that organizational learning in strategic alliances will come as a result of alliance evolution, and partners that work together will eventually come to new knowledge and learn from each other, even in cases when alliances are not created with learning intentions (Muthusamy and White 2005; Grant and Baden-Fuller 2004; Inkpen 1998a, b; Mowery et al. 1996). The underlying logic is that the relationship between knowledge and age is positively related; the older a partnership is, the more knowledge will be transferred between the alliance partners, the more a firm will know, and more it will be able to learn (Faulkner 1995). Others state that a direct connection between learning and a life cycle of a partnership can be studied only in those partnerships in which there is a genuine devotion of both partners to mutual learning and cooperation. If a partnership is seen as a short-term opportunity, the possibilities of such a result are quite limited (Child 2001; Khanna et al. 1994). Researchers have tried to explain various aspects of learning and knowledge transfer in strategic alliances [see Meier (2011) for the review of empirical studies]; yet few have discussed it in the context of organizational design properties and associated concepts. Existing research does not reveal if organizational design properties contribute to organizational and inter-organizational learning in strategic alliances.

In this paper we explain further the connection between the evolutionary processes in alliances, operationalized through two dimensions: age (longevity) and maturity, and the importance of active management of knowledge and learning through the application of adequate mechanisms of coordination. Alliance age is simply the time from birth of the alliance; alliance maturity is the degree of past involvement and experience of the two alliance partners. Our beginning assumption is that learning processes and knowledge transfer occur all the time through alliance evolution; it is an emergent organizational process. Learning and knowledge transfer may be intended, but they exist even when not intended. However, inter-organizational learning and knowledge transfer of that kind are difficult to identify and control: we do not know when they happen, how they happen, who is involved, and what the effects are, so it actually represents a passive approach to management of organizational learning and knowledge in alliance settings. Active approach to management of knowledge and organizational learning in alliances suggests that they should be designed in a way to transfer knowledge and learn.

The contribution is to explain that differences in knowledge stocks and learning flows between alliance partners can arise as a result of the application of coordination mechanisms, depending on the stage of alliance age and maturity.

The paper is organized as follows. Firstly, we explain the conceptual framework for the paper and the case study method we apply. Second, we focus on the evaluation of important organizational properties with respect to their changes as organizations entered partnerships and analyze learning and knowledge outcomes of the partnerships. Finally, we discuss major finding and give implications for both theory and practice.

2 Research Design

2.1 Conceptual Framework

We investigate the influence of organizational design in the context of organizational and alliance ability to receive and accept knowledge, as well as to promote learning process, which is known as "absorptive capacity" (AC). The concept of absorptive capacity was recognized by Kedia and Bhagat (1988) in their explanation on cultural constraints on technology transfer across nations, but established and further developed by Cohen and Levinthal (1990). They define absorptive capacity as "a firm's general ability to value, assimilate, and commercialize new, external knowledge."

Throughout the 1990s the cognition about the influence of organizational characteristics on organizational ability to give and receive knowledge evolved further. Hamel (1991, p. 87) recognizes the importance of organizational ability to receive knowledge by defining receptivity as a determinant of learning. In his work, receptivity is a function of skills and absorptiveness of receptors and exposure of position and parallelism in facilities, while asymmetry in receptivity pre-ordains asymmetric learning. Szulanski (1996) notes that knowledge transfer is less difficult if a recipient is prepared to receive that knowledge. Lane and Lubatkin (1998) point to the importance of understanding absorptive capacity as relative phenomenon, and affirm the term "relative absorptive capacity." They relate the idea of absorptive capacity to receptivity and investigate how absorptive capacity of a firm depends on the relevance of the student firm's basic knowledge to the teacher's firm knowledge base, similarities in organizational structures of teaching and learning firms, and similarity of the student and teacher firms' compensation practices.

Subsequent work through 2000s increased knowledge in the anteceding role of organizational characteristics for learning and knowledge transfer. Child (2001, p. 659) states that companies must have "the experience or capacity to acquire and absorb knowledge available from the alliance partner." Lyles (2001, p. 681) draws on the importance of ability to absorb knowledge by defining organizational learning as "embedded know-how resulting from absorptive capacity, receptivity of the firm to new knowledge, and the firm's ability to develop knowledge utilization skills." Jansen et al. (2005) argue that there is a lack of research regarding organizational antecedents of absorptive capacity. Units need to develop "combinative capabilities" (Kogut and Zander 1992) that enable them to synthesize and apply current and newly acquired external knowledge. Jansen et al. (2005) focus on the explication of: (1) coordination capabilities that enhance knowledge exchange across disciplinary and hierarchical boundaries, (2) system capabilities that program behaviors in advance of their execution and provide a memory for handling routine situations (formalization, routinization), and (3) socialization capabilities contribute to common codes of communication and dominant values (connectedness and socialization tactics). Inkpen and Ramaswamy (2006) have discussed the importance of strategic fit and organizational fit in the context of strategic alliances. Previously, Lorange and Roos (1991) recognized the importance of mutual adjustment of partnering organizations in the strategic and organizational sense by identifying three fundamental reasons why strategic alliances are so difficult to manage: according to them, (1) shared decision making, (2) separate corporate cultures, and (3) different (conflicting) strategic interests are the main reasons for difficulties in managing strategic alliances. Volberda et al. (2010) gave thorough analysis of the concept of absorptive capacity, concluding that research on absorptive capacity should examine the relationship between intra-organizational and inter-organizational antecedents (p. 947). They systematize intra-organizational antecedents of absorptive capacity as organizational form, incentive structure, informal networks and internal communication, and inter-organizational configuration as knowledge creation and sharing, alliance management system, dyad and network knowledge management, and transfer and relatedness of organizations. Lewin et al. (2011) made a significant contribution toward explicit operationalization of the absorptive capacity construct, proposing a routine-based model of absorptive capacity. They suggest decomposition of the AC construct into its internal and external components and identify underlying metaroutines.

Though results of these studies give us grounds for thinking about how learning should be coordinated in alliances, they do not reflect to the dynamic nature of this type of inter-organizational linking.

In this paper we explain further the connection between evolutionary processes in alliances and the importance of active management of knowledge and learning through the application of adequate mechanisms of coordination.[1] We argue that

[1] From the beginning of 2000s management literature turns its focus from knowledge transfer to knowledge management. Understood as a conscious coordination and monitoring of knowledge processes (Inkpen 2000), knowledge management becomes an organization design issue to improve the efficiency and effectiveness of an organization and its people by sharing knowledge and information (Burton and Obel 2004, p. 10). Coordination is one of the most important organizational design properties (Burton et al. 2011).

designing alliances to learn is a complex managerial job which should be guided by the identification of stocks and flows involved in the process of knowledge transfer and interorganizational learning between alliance partners. This whole process is influenced by the phase in alliance evolution. This general goal can be further explained through the research questions: (1) is there interdependence between alliance evolution, coordination of knowledge transfer and learning processes happening between alliance partners, and the learning outcomes the alliance achieves? and (2) should management approaches toward learning and knowledge be different depending on the differences in alliance age vs. alliance maturity?

In their perspective paper on absorptive capacity, Volberda et al. (2010, p. 937) note that most empirical studies apply unidimensional operationalization of absorptive capacity, and fail to recognize "internal mechanisms that can influence firm's level of AC...." We propose a bidirectional approach to interorganizational learning and knowledge transfer and explain the intra- and inter-organizational antecedent of organizational absorptive capacity and their influence on organizational ability to receive and give knowledge and to learn. We consider the interplay between knowledge, as a stock category, and learning, as a flow category. We define knowledge as "information that corresponds to a particular context" (Burton et al. 2011), and learning as "a capacity of organization to gain insight from its own experience, the experience of others, and to modify the way it functions according to such insight, which leads to the development of knowledge base" (Shaw and Perkins 1991; Shrivastava 1981).

We use the term "coordination" to explain active learning management practices and active knowledge management, which need to be carried out as to enable knowledge transfer and learning processes in alliances. We differentiate between formal (institutional) and informal (behavioral) mechanisms of coordination (Fig. 1). Formal coordination includes the way decisions are made and the way they are shared in an alliance, and formalization of relationships between alliance partners. Informal coordination is based on the use of mechanisms of organizational culture and trust.

Alliances are dynamic systems that evolve gradually over time. Alliance dynamics is recognized as an important aspect of various processes happening in alliances (Khanna et al. 1994; Ring and Van de Ven 1994; Faulkner 1995; Doz 1996; Ariño and de la Torre 1998; Iyer 2002; Child 2001). We define alliance evolution as an independent variable, operationalized through the two variables—longevity (age), which is a managerial uncontrollable variable, and maturity, which is a managerial controllable variable.

Alliances of the same age can differ in the level of their maturity. Alliance maturity depends on the pre-existing experience in working together (Aleksić Mirić 2011). Child (2001, p. 669) discussed the importance of previous experience for organizational learning in strategic alliances. He argued that two aspects of experience facilitate learning in alliances: experience in working in alliance context and experience of having collaborated with the same partner. Both of these previous experiences are important and relevant for building organizational capacities to teach and learn. In addition to the two experiences recognized by Child, we add one

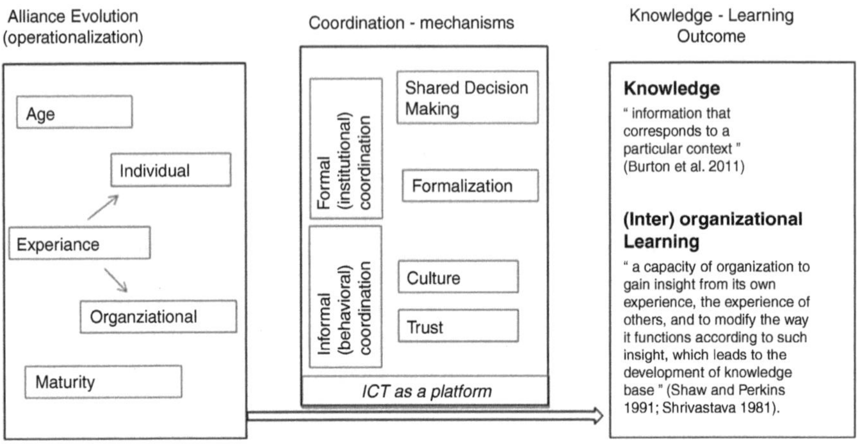

Fig. 1 The model. Source: authors

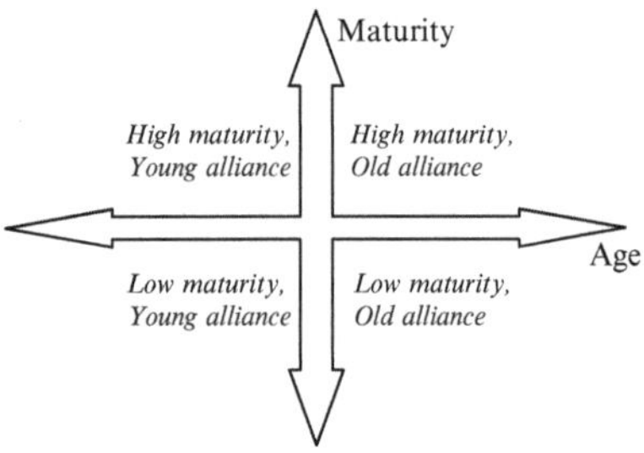

Fig. 2 Age-maturity relationship. Source: authors

more type of experience: pre-existing ties between employees. Organizations might have never cooperated before, but employees might have; they might have been colleagues or classmates, they might have cooperated on some other projects in other organizations. This kind of micro-level generated interorganizational ties can have strong influences on alliance evolution and performance.

We develop a two-dimensional model and four different kinds of age-maturity relationships: (1) young alliances with low maturity, (2) young alliances with high maturity, (3) old alliances with low maturity, and (4) old alliances with high maturity (Fig. 2).

2.2 *Case Study Method*

Empirical part of the research was carried out through multiple exploratory case study research and started in 2007. Multiple exploratory case studies are advised to be used when one wants to examine the phenomena of interest in their real settings (Yin 1984). The purpose of the case study in this research is not only to describe the phenomena under research, but to explore and to explain them in a more detailed manner.

We built our arguments on case studies of three alliances created by three Serbian organizations with foreign partners from EU countries.[2] All partnerships are created in media (electronic and print) business. This allowed more precise control and clearly defined frames for generalization of conclusions (as suggested by Eisenhardt 1989, p. 537). Above all, it enabled control of factors that come from inter-industrial differences. The research applied the following techniques: interview, observation, analysis of historical data, and quantitative social network analysis. We applied these methods simultaneously in order to enable triangulation of the validity of our findings (Eisenhardt 1989). The interviews were conducted in semi-structuralized form.[3]

Alliances as a form of interfirm collaboration cover a wide range of interfirm linkings. In order to capture different forms of alliances, our three cases under investigation differ in ownership criteria, including 50–50 joint venture (Case 1), alliance with one partner's majority in ownership (Case 2), and non-ownership based partnership (Case 3). Argote and Miron-Spektor (2011) state "...because organizational learning occurs over time, studying organizational learning requires time series or longitudinal data." Now, in order to capture the influence of pre-existing experience on learning in alliances, we analyze cases with different longevity and different maturity.

The first partnership—Case 1—was created by the two renowned organizations, A_1 and B_1, both national leaders with respectable traditions in their businesses. However, during their long individual histories they had never cooperated before. At the time of the research, the partnership was almost 7 years old. Entering the Serbian market, B_1 created an alliance with a local publishing company, A_2 (Case 2). Like the partnership with A_1, the A_2 partnership with B_1 projected economic recovery and taking back seriously disturbed market positions. When this research was undertaken, the partnership was in its fourth year of existence. As in A_1's case, A_2 did not have any previous experience in cooperating with B_1, but its relative comparative advantage came from the fact that it had A_1's case to look upon to

[2] Partnerships also in depth described in Aleksić Mirić (2011).

[3] Data collection included interviews with managers in key positions in both alliance partners, and the investigation of archival data. The first author conducted the interviews using a semi-structured format. In order to ensure validity of the content of the interview and to enable systematic data gathering during the interview procedure, we developed a questionnaire which in content met the research questions. All the interviews were conducted face-to-face. In total, 20 interviews were conducted. The interviews lasted from 45 to 210 min. Most of the interviews were conducted once, but in some cases it was necessary to go over the research questions once more. Archival data included various historical data about the companies that created the network—contracts, manuals, and bylaws, minutes from managerial meetings, press releases, and so forth. The method of observation was also applied where appropriate.

when deciding on whether to enter this venture or not. The Case 3 partnership was created by the two companies (A_3 and B_2) that did not match in ownership case (public–private partnerships) but did strategically recognize mutual interest in cooperation. At the time of the research, this partnership was the youngest—only one year old. Still, pre-existing experience in working together, driven both from individual and organizational levels, this partnership could be categorized as the one with highest maturity among the three investigated.

3 Alliance Evolution and the Learning Coordination: Experience from the Three Cases

We evaluated important organizational properties with respect to their changes as organizations entered partnerships. We analyzed the following properties: (1) formal (institutional) mechanisms of coordination: shared decision making and formalization, and (2) informal (behavioral) mechanisms of coordination: culture and trust. We also included the analysis of the role of information-communication systems, as they are platforms for both institutional and behavioral coordination. Then, as presented by the model (Figs. 1 and 2), we analyze learning and knowledge outcomes.

3.1 Coordination

Shared Decision Making: As cooperation between firms for the purpose of improving ability to achieve strategic goals (Child 2001), alliance creation brings challenges to the way in which decisions are made. Partners voluntarily agree to exchange, share, or co-develop products, technologies, or services (Inkpen and Ramaswamy 2006, p. 81; Inkpen and Tsang 2005, p. 148; Gulati 1998, p. 293) and to constantly contribute to the accomplishment of one or more strategic goals. One of the main characteristics of alliances is that partners share the benefits that come as a result of an alliance creation, as well as control over the responsibilities related to the functioning of an alliance. In order to achieve that, they need to balance centralization and decentralization of decision making. Shared decision making might be the most important characteristic of alliances and, at the same time, the most difficult task to realize.

Our three cases differ in the handling of shared decision making. In Case 1, the awareness of how and where in a value-chain the profit is made in the newspaper-publishing industry determined the model of cooperation. Respecting the logic that the individual price of a newspaper in the time of declining circulations cannot provide a positive financial result, it becomes evident that, following the economic logic, revenue was to be searched for elsewhere. A_1 and B_1 divided responsibilities and authority over decision making in a way that created two basically independent parts within alliance. The B_1 CEO remained formally the leading authority; he had two main associates—his deputy (B_1's representative), in charge of logistics, advertising, marketing and printing, and the editor in chief (A_1's representative), in charge of the editorials. The part of the alliance controlled by B_1 underwent radical changes. The advertising unit was centralized, the organization was transformed significantly, and the learning within this unit became very intensive. As the B_1 head of advertisements explained,

> ... when deciding on how to organize Advertisement Sector we have been searching for the most suitable model. All of us who were involved in this struggled to realize what is good in our existing practices and should not be changed, and what is that we should change instantly. A number of consultation meetings were held with an aim to find out what we should implement from B_1.

The printing plant also transformed where necessary—in the supply department, which was centralized in order to enable more economical supply of materials.

In contrast, the part controlled by the A_1 remained unchanged. The operationalization of this strategic plan over the shared decision making as explained was much harder: everyday practice within the Case 1 shows that there is considerable overlapping between the economic and non-economic issues to be decided upon, as well as that the economic and editorial interest are difficult to separate one from the other, and that it was not easy to decide with certainty which domain specific decisions belonged to.

In its other Serbian partnership, Case 2, B_1 started the establishment of a new company with a proposal of a new organizational scheme. The initial proposal of the organizational scheme was set as an ideal pattern to be followed, while in reality the organizational scheme changed slightly. The reasons for the discrepancy between the real and the ideal organizational schemes were primarily technical and technological in nature. The part that radically changed was the advertising unit. Previously, in A_2, advertising was perceived as a non-core business, a unit of lower importance than the editorials. Nevertheless, after partnership formation, the A_2 advertising unit experienced tremendous changes.

Although B_1 held the majority in ownership, a part of the editorial office remained intact, just as in Case 1. Just like Case 1, the reasons for forming the Case 2 partnership were dominantly economic (financial) in nature. However, unlike Case 1, the Case 2 partnership possessed a certain level of stability regarding its management structure, so that the group that participated in the negotiations in the process of the foundation of the partnership was very similar in its structure to the managerial structure at the time of the research. The only difference is in the position of the CEO deputy, who was initially a B_1 representative. As time passed and the partnership started functioning regularly, this principle was abandoned; the position of the CEO deputy was abolished and substituted for a staff position of the CEO's associate, taken over by a foreign partner's representative. This way, the associate became the only direct B_1 representative but without managerial duties and responsibilities, which is obviously an unfavorable situation in comparison to that in Case 1.

The Case 3 partnership constantly balanced centralization and decentralization over the shared decision making. The fact that the product of cooperation was a radio show broadcasted live implied the need for decentralized, on-the-spot decision making. However, the partnership creation was followed by strict rulers, and though operated in Serbia, by legal standards was under the British law, which meant centralization over decisions on higher managerial levels.

Formalization: In terms of formalization we can draw a distinction between formalization of the relationship through the official agreement between the partners, and formalization of behavior. Strategic alliances can be classified according to whether they are formed on contract or ownership basis, with or without forming a new entity. Ariño and Reuer (2004) point to the importance of the criterion of ownership and the distinction between ownership-based and non-ownership based partnerships, relying on the fact that distinguishing between contract-based and ownership-based strategic alliances makes no real sense, because every partnership implies forming some kind of contract. All of the analyzed partnerships in our investigated cases have been announced as official through the contract, while the Case 1 and Case 2 partnerships are ownership-involved as well.

The Case 1 and 2 partnerships did not experience significant changes in the level of formalization compared to the pre-alliance stage. The bureaucratic nature of their organizations is revealed in the partnerships they created. The Case 3 partnership did include changes in the level of formalization, moving A_3 toward more formalized behavior. For instance, the way of reporting in B_2 was a novelty for the A_3 reporters because the duration of a particular story being analyzed is rather long for the standards of a radio show. Such duration of a report contributes to the analysis of a piece of news from different angles. For A_3, the cooperation with B_2 involved other sources of useful input in the very domains that represent B_2's main strengths: the code of behavior, standardization and formalization, unbiased advertising, and training on legal matters. For an example, the code of behavior is reflected through the rule that each sound-assistant should get the schedule of each broadcast in advance. Unbiased reporting is reflected through the rule that in each piece of news, all the sides involved in the topics and the issue must be heard. Consequently, no news which could be characterized as biased can be broadcasted. Finally, legal training involves all the situations in which the company can be sued.

Culture: Organizational culture represents shared beliefs among employees. Organizational culture is a very important factor of organizational learning and knowledge transfer in alliances. Child (2001, p. 669) argues that internal differentiation within and external differentiation between organizations introduces barriers to organizational learning within alliances, identifying social identities, typically represented by different organizational and national cultures.

Case 1 and Case 2 report similar pattern in behavior: Serbian partners were positive about the improvement they could get in finance and technology through the partnerships they created, but strong organizational cultures of A_1 and A_2, corporate pride, and the sense of strong professional standards did not create positive attitudes toward the potential to learn from partner B_1. How important it was for the employees and the public that A_1 did not lose its "national identity," that it was not sold but ventured into an arrangement to bring betterment for all stakeholders, can be seen from the fact that A_1's leading daily newspapers published an interview with one of the sprouts of the founder and the first CEO of the Case 1 partnership, which was titled—"*The* A_1 *is not sold*, and further—*don't be concerned*."

> We did not sell the A_1. We basically invested our editorials, our good-will and our offset printing-office in into joint venture with German media giant. In return, our partner invests financial resources and logistics. All the issues, and all the employees will be moved into the new company, which preserved the name – the A_1 (...) the X A_1 will not change editorial policy.

In the interview given by the German representative, he stated:

> With the German partner entering (into the Serbian market, AAM) the existing standards in media are going to change. In addition to new business philosophy, the German partner brings into the A_1 rich experience in the field of publishing, design and marketing.

Driven from the contract-based delegation of authorities and decision making, we can clearly see that cooperation is restricted to operative publishing, design of the products, and marketing, while other parts of organization are not included. This strategic management orientation was recognized by the employees as well. One of the employees said:

> ...Germans do know it very well that the market is a source of survival. So why is that we apprehend potential threat to the Company's basic values – they do know which market our products are bound for...

These statements can also help us understand internal system of beliefs and values of the companies entering partnership. The Serbian partner perceived itself, and was perceived in public, as a media house which established professional standards. This considerably changed during 1990s, so the creation of this partnership was seen as a way to take back lost market positions. One of the employees (from the Serbian side) explains this as following:

> ... somehow I do believe that we will succeed to implant the seed that we have been carefully growing for more than one century into our partner, who comes with new ideas and financial resources. And somehow I believe that they will nurture this seed successfully...

In Case 1 partners report similar cultural values—strong respect of own history and emphasized corporate pride. Both partners are characterized with strong culture. Thus, as operationalized, there were no true initiatives for building joint, shared culture. Actually, both partners demonstrated intention toward cultural dominance over the other. The Case 2 partnership is characterized with a weaker position of a Serbian partner and the dominance of the "German-teacher" (Table 1).

The Case 3 partnership was operationalized by people who used to be co-workers at some point in the past, who shared the same "rebellion spirit of the 1990s in Serbia" and who were positive toward experiments they jointly created. The fact of cooperation was seen as a "value-added" element of both cultures.

Table 1 The use of mechanisms of coordination and knowledge transfer

	Case 1	Case 2	Case 3
Type of the relationship	Teacher–learner	Teacher–learner	Learner–learner
The use of knowledge transfer mechanisms lower in information richness (as intranet, chat, fax, emails, newsgroups...)	Moderate, strictly centralized	Very low, strictly centralized	Moderate to high, decentralized
The use of knowledge transfer mechanisms higher in information richness (as seminars, workshops, meetings, media conferences)	Low	Very low	Moderate to high
Trust	Weak	Weak	Strong
Culture	Similar cultural values; strong culture within both partners	Dominance of a "teacher"; weak culture within "learner" partner	Similar cultural values; strong culture within both partners
Ties between the partners	Weak	Weak	Strong

Furthermore, partnerships report similarity in cultural values: strong respect of own history and emphasized corporate pride. Strong culture within both partners was not, however, an obstacle for the initiatives for building joint, shared culture, which was further supported with no intention toward cultural dominance (Table 1).

Trust: Extensive research has been carried out about the influence that trust can have on strategic alliances (Ariño and de la Torre 1998; Child and Faulkner 1998; Inkpen and Currall 1998; Inkpen and Ramaswamy 2006). Lorange and Roos (1991) recognize the importance of good climate (which trust can be considered a part of) within organizations toward strategic alliance as one of the things that might significantly influence the success of partnership. They say that "...during the more intense formation phases [management should ensure] that a broad range of people within the organization is committed to and enthusiastic about the venture." Some of the findings on the influence that trust has on organizational learning in an alliance context are quite contradictory. In their research on trust and organizational learning in inter-organizational joint ventures in Hungary, Lane and associates (2001) did not report statistically relevant dependence between trust and learning, but they did report statistically relevant dependence between trust and performance. Contrarily, Muthusamy and White (2005) reported that trust between partners has strong influence on the process of learning between the partners. The mutual devotion depends on the moral responsibility of the partner, and heavily reflects learning in strategic alliances.

The role of trust in managing Case 1 and Case 2 partnerships is low. There was no previous experience in business involving these two partners based on which trust could be built. The group that brought the contract was not the group that worked on its implementation. In contrast, the role of trust in managing the Case 3 partnership is extensive and essential. Mutual trust is a result of the previous cooperation and social network that exists among employees working in these two organizations (Table 1).

Information and Communication Systems: Information and communication systems represent a platform for coordination of joint activities between alliance partners. Information technology was not extensively used to support partnership implementation, so the role of IT as a mechanism of coordination and active knowledge management and learning was not significant. Partnerships mostly relied on the use of top-management meetings and exchange of official documents. Meetings between representatives of both partners are exclusively held on the top-management level, while visits of one partner's representatives to the locations of other partner, joint training sessions, employees' rotations, mixed teams, and coordination mechanisms are not used.

Communication technology (via telephone) played an important role in the Case 3 partnership. Although it would be expected that partners would rely more extensively on contemporary communication software based on the Internet, they did not; rather, they basically used telephone communication. This channel enabled virtual meetings before, during, and after the radio show they produced (Table 1).

3.2 Learning and Knowledge Outcomes

The goal of forming the Case 1 partnership was primarily economic in its nature. In the process of partnership creation, both sides used their organizational knowledge as a strong instrument for negotiation in order to estimate the relative values of their individual investments into the partnership. Both A_1 and B_1 showed the value of their organizational knowledge they were to invest in the partnership by means of turning their non-material investments into material ones worth 24 million euros. With regard to the investment structure, both partners recognized organizational knowledge as the most valuable investment. However, the contract between the partners did not define precisely what that investment involved, nor was there an action plan to specify what exactly knowledge transfer covered. Furthermore, organizational learning through the new business venture was not explicitly recognized as an incentive for the partnership creation, at least not in this stage of its development. The organizational learning in the partnership was mainly characterized by the exploitative learning within one circle. When new products were created, the process of learning was exploitative in nature and resulted from taking over the existing practice and the ways of doing business in other organizations within B_1's system. Up to a certain extent, individual learning occurred; however, there were no mechanisms that would enable the integration within the organization as a whole, which would enable transfer of individual, group, and organizational learning and knowledge. Apart from the hardware and the SAP software investments financed by the B_1, there was no significant exchange of technology, and the attempts of joint learning of management and marketing skills were very few. There was no attempt at creating more opportunities for the partners to try joint accomplishment of these tasks. We could also argue that what happened in this partnership is a kind of a precise transfer of B_1's explicit knowledge, or, to interpret it in Nonaka's (1994) terms, knowledge combination. Products that were very successful in other markets were taken over, and their life cycle was extended through the implementation on the new market. When saying that precise knowledge transfer occurred we assume that authority, delegation, and strict division of responsibilities between the partners prevented free knowledge diffusion, which could have been good for explorative learning promotion.

The Case 2 analysis shows the following: both sides recognized organizational knowledge as an important element of investment when negotiating alliance creation; organizational learning and knowledge transfer were not explicitly recognized as the aims for forming the partnership; at the very beginning of the partnership, a new organizational scheme was established and connections between the partners defined; organizational learning in the partnership had characteristics of exploitative, single-loop learning. A low level of organizational adaptation to partnership settings was also followed by a relatively low level of learning, which showed the characteristics of exploitative, single-loop learning (Argyris and Schön 1985, 1996). Understandably, even such a limited form of learning was precious to the company, which was trying to regain its seriously shaken position in the market.

The Case 3 partnership was built on the need of both organizations to think of the new output that would bring improvement in market positions and listening rates to both of them. They created a show which represents a completely innovative product requiring a high level of explorative orientation and interaction between partners. The interactive concept of the show and the intensive contacts during its realization determined to a large extent the interaction between individual and organizational learning. This case also illustrates two ways of enabling the spread of learning and knowledge within an organization. One of the employees reflected:

The execution of this show initiated changes in existing operating procedures, and caused introductions of some new standards in the radio shows of A_3, so now technicians ask for show-outline plan in other informative shows as well.

Everyday virtual contacts, telephone communication, and interactive adjustments of partners during the show enabled continuous exchange of implicit knowledge. Although this partnership was the youngest measured in terms of time, it was also the most mature.

Knowledge transfer was the focus of the Case 1 and Case 2 partnerships. The Serbian partners were interested in gaining new knowledge, which was new to them, but basically relied on the direct knowledge transfer from B_1's existing knowledge portfolio. This kind of direct knowledge transfer was a good solution for all the sides involved. The parties involved in the Case 1 and Case 2 partnerships did not evaluate learning as an important alliance goal. On the other hand, the explorative concept that joined B_2 and A_3 called for an experimental learning, and no knowledge transfer, as the repetition of the known patterns in behavior could jeopardize the idea. They searched for innovation, radically different output, and inspiration that could be packed in an hour-long radio show.

The Case 3 partnership (A_3 and B_2), for several important reasons, represents a case different from the two we analyzed earlier. First, the partners achieved a high level of strategic fit, on the basis of strategic complementarily between partners and recognition of the shared goals as important ones. The basic aim of partnership creation was exploration: how to create new product (new radio-show) which will enrich the program and increase listening rates. Convergence of strategic goals alike was additionally supported with pre-existing experience and strong friendship and collaborative ties among the employees, which was, as we see it, the secret formula for success of this partnership. Second, the partnership between the B_2 and A_3 is the youngest in origin and only one year old at the time of the research. Third, the project that initiated the cooperation between A_3 and B_2 represents an innovative solution which was to bridge the problem both organizations faced: the decrease in listening rates. On the scale of organizational learning, creating a completely new show (i.e., new product) represents a very high level of innovation. The program in Case 3 brought benefits to both sides; A_3 was definitely motivated by its basic principles: explorative orientation and innovative program. For A_3 this cooperation created an opportunity to learn from B_2, which is characteristic for the affirmation of analytical way of reporting and a higher level of the reporters' independence. Namely, B_2's priority is not to provide brief information but rather precise and complete information. This principle gains in importance with knowledge that legally the business of this partnership was to undergo the British legal system. Through the joint production with A_3, B_2 gained in speed and working dynamics, atypical for their standard working conditions, which could be labeled as slow and bureaucratic. The link to B_2's website was incorporated into A_3's website, and B_2 also got an opportunity to increase the listening rates of its program through the innovative approach toward a radio show. B_2's previous experience with non-standard programs had not been successful. Unexpectedly B_2 gained new idea about how to improve the listening rates in some other countries. The concept of the alliance with A_3, the dynamics, and the success of Case 3 initiated B_2 to rethink the ways of doing business with partners, and to try to transfer knowledge from Case 3 to their other alliances in Ukraine, Russia, and Turkey.

3.3 *Age, Maturity, and Alliance Evolution*

In order to explain further the importance of alliance evolution through two parallel dimensions—longevity and maturity—we expand our research on the analysis of the application of mechanisms of coordination, as a way to enable active management of knowledge and learning. Comparative analysis is given in Fig. 3.

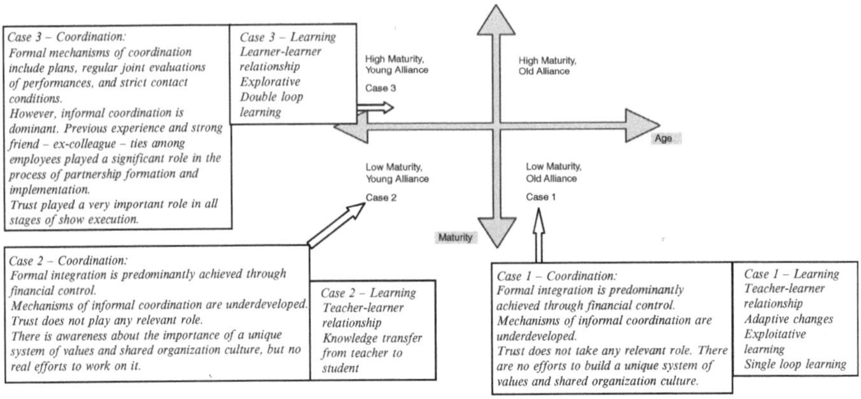

Fig. 3 Age-maturity-coordination: case studies' findings. Source: authors

4 Discussion and Implications

4.1 Discussion

Age–Knowledge–Learning: Age is positively connected to knowledge, but not necessarily to learning (Aleksić Mirić 2011). Older alliances are characterized with significant knowledge depositories, so older alliances are certainly richer in knowledge than younger ones. Organizational processes in older alliances support knowledge utilization and circles within the same knowledge framework, while explorative processes of knowledge generation are neglected and rejected by internal barriers (Child 2001), which are built in over the years.

Maturity – Knowledge – Learning: Maturity is positively connected to learning, but does not necessarily imply high knowledge stocks (Aleksić Mirić 2011). In young partnerships partners may not have enough knowledge about each other, and their knowledge bases are likely to be disconnected and separated, but what counts more in the context of inter-organizational maturity, as defined, is their commitment to acquire and use existing knowledge and to create new knowledge—that is, to learn. This learning might be less connected to time dimension per se, but rather to the dynamics of the development of social connections among individuals and groups, information channels, and shared experience.

Mature partnerships are characterized by a learner–learner relationship, while those low in maturity are characterized with teacher–learner relationship (as on Fig. 3).

- Teacher–learner relationships are followed by the use of "hard" (institutional, formal) methods of coordination intended to enable knowledge transfer from one partner to the other, and not suitable for inter-organizational learning purposes. This kind of relationship occurred in the two cases of ours: Case 1 and Case 2. Both cases report that formal integration is predominantly achieved through financial control, while mechanisms of informal coordination are underdeveloped. Trust did not take any relevant role in functioning of the partnerships. There were practically no efforts to build a unique system of values and shared organizational culture.

- Learner–learner relationships are followed by the use of "soft" (behavioral, informal) methods of coordination, suitable for inter-organizational learning without being a barrier for direct knowledge transfer. On the contrary, soft methods of coordination enhance transfer of tacit knowledge. This kind of relationship occurred in our Case 3 partnership. Formal mechanisms of coordination included plans, regular joint evaluations of performances, and strict contact conditions. However, informal coordination was dominant. Previous experience and strong friend—ex-colleague—ties among employees played a significant role in the process of partnership formation and implementation. Trust played a very important role in all stages of show execution.

From the comparative analysis of the three cases we derive an explanation of how the alliance moves within the age-maturity space and how inter-organizational learning depends on the application of the mechanisms of coordination.

The path of organizational moving within the defined 2 × 2 space depends not only on the time passing per se, but on the application of the mechanisms of organizational integration and coordination. From the point of inter-organizational coordination, alliance age and alliance knowledge are static components of inter-organizational relationships, while alliance maturity and alliance learning are dynamic components.

Finding 1: Age-Coordination	Alliance age is positively associated with the application of formal (institutional) mechanisms of coordination.
Finding 2: Maturity-Coordination	Alliance maturity is positively associated with the application of informal (behavioral) mechanisms of coordination.
Finding 3: Coordination-Learning	Application of both formal and informal mechanisms of coordination is important for learning and knowledge transfer in strategic alliances; thus, their importance is moderated by alliance maturity.
Finding 4: Maturity – Coordination-Learning	Alliance maturity, unlike alliance age, can decrease. Some alliances can start from the position of high maturity, but inadequate application of the mechanisms of coordination can negatively influence the alliance maturity, which will consequently decrease learning in alliance.

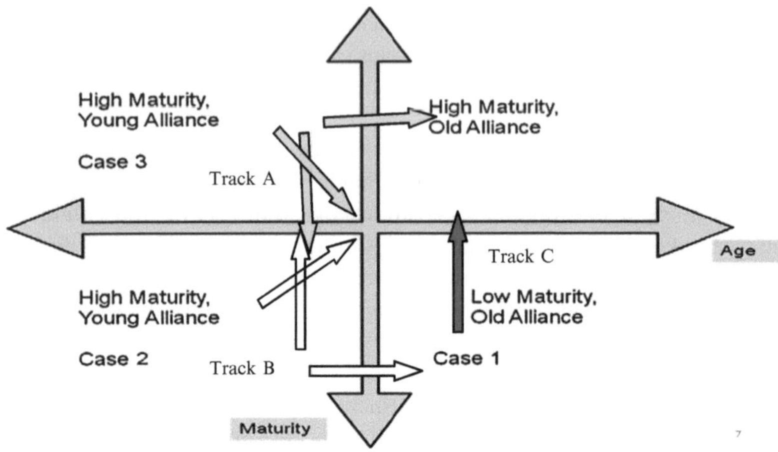

Fig. 4 Potential implications. Source: authors

4.2 Implications

Implications for Theory: Conceptualizing the difference between alliance longevity (age) and maturity, and the corresponding implications on the use of organizational design properties and the effects on interorganizational learning, contributes to the understanding of the dynamic nature of alliances and organizational and inter-organizational processes related to it. We explain the connection between the evolutionary processes in alliances, and the importance of active management of knowledge and learning through the adequate approach to organizational and interorganizational design. With this framework we contribute both to the research in strategic alliances and to the theory of organizational design.

Implications for Practice: We identify two potential starting points: (1) young alliances with low maturity—start-up alliances without previous individual or organizational experience between alliance partners, and (2) young alliances with high maturity—young alliances with previous individual or organizational experience between alliance partners.

Option 1 (Fig. 4—track A)—young alliances with low maturity—if learning is a goal, start-up alliances without previous individual or organizational experience between alliance partners are advised to work on building mechanisms of informal coordination in order to increase their maturity. Focusing on mechanisms of formal coordination will bureaucratize their immature alliance and build learning barriers.

Option 2 (Fig. 4—track B)—young alliances with high maturity—if learning is a goal, young alliances with previous individual or organizational experience between alliance partners must focus on institutionalization of informal practices and controlled formalization. This formalization should be oriented toward increasing knowledge base. Maturity gets positively connected to knowledge if partners succeed in nurturing explorative orientation together with incorporating mechanisms of formal coordination that will enable stability, formalization, alliance longevity, and progress over the time.

Old alliances are in a different situation: (3) old alliances with low maturity are characterized with unused knowledge depositories (Fig. 4—track C). These stocks should be moved through the application of behavioral mechanisms of coordination, while existing mechanisms of formal coordination should be evaluated from the point of their contributions to the creation of new knowledge through learning. Finally, (4) old alliances with high maturity are in a potential winning position.

5 Conclusion

This research had three important goals to further build theoretical understandings of organizational learning and knowledge transfer in strategic alliances. Firstly, our goal was to explicitly express the role of organizational design in alliance knowledge management. Secondly, our intention was to capture dynamics of all the constructs under investigation: organizational design, learning, and knowledge transfer and alliances. As structured systems of established relationships, alliances develop gradually over time. Learning as an inter-organizational process between alliance partners depends on time as well: time is essential for bringing people together and developing mutual trust, shared cultural values, and joint ideas. However, the interplay between these two evolutionary processes happening in alliances, learning and aging, and the connected role of (inter-)organizational design, is not extensively researched. We explain the connection between the evolutionary processes in alliances, operationalized through the two dimensions: age (longevity) and maturity, and the importance of management of knowledge and learning through the adequate approach to (inter-) organizational design. Thirdly, it was our intention to contribute to organizational learning theory. We build upon the concept of absorptive capacity (Cohen and Levinthal 1990), which examines inter-organizational learning as a unidirectional issue—from the outside environment to inside the organization—and argue that learning, as a mutual or two-way phenomenon, in the context of dynamic nature of strategic alliances, yields more in-depth understanding.

From the aspect of management practice, this research intends to explain that designing alliances to learn is a complex managerial job which should be guided by the identification of stocks and flows involved in the process of knowledge transfer and interorganizational learning between alliance partners, and aligned with the stage in the achieved level of alliance development.

Based on three case studies we investigated the issue of inter-organizational learning coordination throughout the evolution of strategic alliance. Through alliance evolution, various learning processes and knowledge transfer happen. If they are not actively managed, they are uncontrolled, hard to identify and follow, while their effects are hard to predict and measure. Active knowledge management is an organization design issue to improve the efficiency and effectiveness of an organization and its people by sharing knowledge and information (Burton and Obel 2004, p. 10). Particularly, we focused on one organizational design property, coordination, exploring its mediating effect on the relationship between alliance evolution (operationalized through the two dimensions: age and maturity), and the learning effects alliances achieve. Perceiving alliance age as an uncontrollable variable, and alliance maturity as a controllable variable, we show how managers can moderate maturity through the application of mechanisms of coordination, and consequently influence learning processes. We imply that alliances should be designed to learn, and the choices about mechanisms of coordination alliance partners apply should be made depending on the alliance's age and maturity. We develop a two-dimensional model and four different kinds of age-maturity relationships and propose that:

1. Alliance movements within the age-maturity space depend significantly on the application of the mechanisms of coordination.
   (a) Alliance age is positively associated with the application of formal (institutional) mechanisms of coordination, while alliance maturity is positively associated with the application of informal (behavioral) mechanisms of coordination.
   (b) Alliance maturity, contrary to the alliance age, can decrease. Some alliances can start from the position of high maturity, but inadequate application of the mechanisms of coordination can negatively influence alliance maturity, which will consequently decrease learning in alliance.
2. Inter-organizational learning that follows alliance moves within the age-maturity space depends significantly on the application of the mechanisms of coordination. Application of both formal and informal mechanisms of coordination is important for learning and knowledge transfer in strategic alliances, thus their importance is moderated by alliance maturity.
   (a) Young alliances with low maturity are advised to work on building mechanisms of informal coordination in order to increase their maturity. Focusing on mechanisms of formal coordination will bureaucratize their immature alliance and build learning barriers.
   (b) Young alliances with high maturity should focus on institutionalization of informal practices and controlled formalization. This formalization should be oriented toward increasing the knowledge base.
   (c) Old alliances with low maturity should focus on the activation of unused knowledge depositories through the application of behavioral mechanisms of coordination, while existing mechanisms of formal coordination should be evaluated from the point of their contributions to the creation of new knowledge through learning.
   (d) Old alliances with high maturity are in a potentially winning position in both learning and knowledge transfer terms. However, being an inherently unstable form of inter-organizational linking, a very small number of alliances will actually reach this position.

In this research we exclusively focused on the coordination as an organizational design property. Future research should expand our knowledge on the relationship between alliance age/maturity, organizational design, and inter-organizational learning through addressing organizational design properties other than coordination. Specification of formal and informal aspects of organizational configuration within this framework would further help us realize how to design alliances to learn through their life cycle (Aleksić Mirić and Burton 2012).

References

Aleksić Mirić A (2011) Influence of network maturity on organisational learning and knowledge transfer in strategic alliances. In: Tuunanen M, Windsperger J, Clinquet G, Hendrikse G (eds) New developments in the theory of networks: franchising, alliances and cooperatives. Springer, Berlin

Aleksić Mirić A, Burton RM (2012) Designing alliances to learn, Workshop on Information and Organizational Architecture, Brussels, Belgium

Argote L, Miron-Spektor E (2011) Organizational learning: from experience to knowledge. Organ Sci 22(5):1123–1137

Argyris C, Schön D (1985) Organizational learning: a theory of action perspective. Jossey-Bass, San Francisco

Argyris C, Schön D (1996) Organizational learning II: theory, method and practice. Addison-Wesley, Reading

Ariño A, de la Torre J (1998) Learning from failure: towards an evolutionary model of collaborative ventures. Organ Sci 9(3):306

Ariño A, Reuer JJ (2004) Designing and renegotiating strategic alliance contract. Acad Manage Exec 18(3):37–48

Burton RM, Obel B (2004) Strategic organizational diagnosis and design: the dynamics of fit. Springer, Berlin

Burton RM, Obel B, DeSanctis G (2011) Organizational design: a step by step approach, 2nd edn. Cambridge University Press, Cambridge

Child J (2001) Learning through strategic alliances. In: Dierkes M, Bethoin AA, Child J, Nonaka I (eds) Handbook of organizational learning and knowledge. Oxford University Press, Oxford

Child J, Faulkner D (1998) Strategies of cooperation: managing alliances, networks, and joint ventures. Oxford University Press, New York

Cohen W, Levinthal D (1990) Absorptive capacity: a new perspective on learning and innovation. Adm Sci Q 35:128–152

Doz YL (1996) The evolution of cooperation in strategic alliances: initial conditions or learning processes? Strateg Manage J 17:55–83

Eisenhardt KM (1989) Building theories from case study research. Acad Manage Rev 14 (4):532–550

Faulkner D (1995) International strategic alliances: cooperating to compete. McGraw Hill, London

Grant RM, Baden-Fuller C (2004) A knowledge accessing theory of strategic alliances. J Manage Stud 41(1):61–84

Gulati R (1998) Alliances and networks. Strateg Manage J 19:293–317

Hamel G (1991) Competition for competence and interpartner learning within international strategic alliances. Strateg Manage J 12:83–103

Inkpen AC (1998a) Learning and knowledge acquisition through international strategic alliances. Acad Manage Exec 12(4):69–80, 602

Inkpen AC (1998b) Learning, knowledge acquisition and strategic alliances. Eur Manage J (2):223–229(7)
Inkpen, A. (2000). Learning through joint ventures: a framework of knowledge acquisition. J Manage Stud, 37 (7):1019–1044
Inkpen A, Currall SC (1998) The nature, antecedents and consequences of joint venture trust. J Int Manage 4:1–20, 600
Inkpen A, Ramaswamy K (2006) Global strategy: creating and sustaining advantage across borders. Oxford University Press, Oxford
Inkpen AC, Tsang EWK (2005) Social capital, networks, and knowledge transfer. Acad Manage Rev 30(1):146–165
Iyer KNS (2002) Learning in strategic alliances: an evolutionary perspective. Acad Mark Sci Rev 10:1–16
Jansen J, Van Den Bosch FAJ, Volberda HW (2005) Managing potential and realized absorptive capacity: how do organizational antecedents matter? Acad Manage J 48(6):999–1015
Kedia BL, Bhagat RS (1988) Cultural constraints on transfer of technology across nations: implications for research in international and comparative management. Acad Manage Rev 13:558–571
Khanna T, Gulati R, Nohria N (1994) Alliances as learning races. Academy of Management Proceedings
Kogut B, Zander U (1992) Knowledge of the firm, combinative capabilities, and the replication of technology. Organ Sci 3:383–397
Lane P, Lubatkin M (1998) Relative absorptive capacity and inter-organizational learning. Strateg Manage J 19:461–477
Lane P, Salk J, Lyles MA (2001) Knowledge acquisition and performance in transitional economy international joint ventures. Strateg Manage J 22:1139–1162
Lewin AY, Massini S, Peeters C (2011) Microfoundations of internal and external absorptive capacity routines. Organ Sci 22(1):81–98
Lorange P, Roos J (1991) Why some strategic alliances succeed and others fail. J Bus Strategy, Jan-Feb, 25–30
Lyles M (2001) Organizational learning in international joint ventures: the case of Hungary. In: Dierkes M, Bethoin AA, Child J, Nonaka I (eds) Handbook of organizational learning and knowledge. Oxford University Press, Oxford
Meier MM (2011) Knowledge management in strategic alliances: a review of empirical evidence. Int J Manage Rev 13(1):1–23
Mowery DC, Oxley JE, Silverman BS (1996) Strategic alliances and interfirm knowledge transfer. Strateg Manage J 17:77–79 (winter special issue)
Muthusamy SK, White MA (2005) Learning and knowledge transfer in strategic alliances: a social exchange view. Organ Stud 26(3):415–441
Nonaka I (1994) A dynamic theory of organizational knowledge creation. Organ Sci 5(1):14–37
Ring PS, van de Ven AH (1994) Developmental processes of cooperative interorganizational relationships. Acad Manage Rev 19(1):90–118
Shaw RB, Perkins DNT (1991) The learning organization: teaching organizations to learn. Organ Dev J 9(4):1–12
Shrivastava P (1981) A typology of organizational learning systems. J Manage Stud 20(1):7–28
Szulanski G (1996) Exploring internal stickiness: impediments to the transfer of best practice within the firm. Strateg Manage J 17:27–43 (winter special issue)
Volberda HW, Foss NJ, Lyles MA (2010) Absorbing the concept of absorptive capacity: how to realize its potential in the organization field. Organ Sci 21(4):931–951. doi:10.1287/orsc.1090.0503
Yin RK (1984) Case study research-design and methods. Sage Publications, Thousand Oaks

Network Embeddedness and Performance of Joint R&D Projects

Nieves Arranz and J. Carlos Fdez. de Arroyabe

Abstract The purpose of this paper is to examine the role of network position in the performance of joint R&D projects using data on European networks of excellence. Our empirical work asks whether centrality (structural embeddedness) and connectivity (junctional embeddedness) increase the project performance. Results show that while structural embeddedness exerts a clear influence on the performance of exploitation projects, the effect of centrality degree has a lower impact on exploration projects due to the redundant information that this location implies. Similarly, while junctional embeddedness positively affects the performance of exploration projects, the heterogeneity of partners involved in a betweenness centrality position does not favor the performance of exploitation projects. This paper contributes to social capital theory by offering new empirical evidence of the effect of network position on performance, which depends on the technological objectives of joint R&D projects.

Keywords Exploitation • Exploration • Joint R&D projects • Network embeddedness

N. Arranz (✉)
Faculty of Economics and Business Administration (UNED), Senda del Rey, 11, 28040 Madrid, Spain
e-mail: narranz@cee.uned.es

J.C. Fdez. de Arroyabe
ESIC Business and Marketing School, Av Valdenigrales, s/n, 28223 Pozuelo de Alarcón, Spain
e-mail: juancarlos.fernandez@esic.es

T. Ehrmann et al. (eds.), *Network Governance*, Contributions to Management Science,
DOI 10.1007/978-3-7908-2867-2_3,

1 Introduction

Joint R&D projects are becoming a viable alternative to the traditional firm-based model, and the proliferation of cooperation has been one of the more enduring features of the organization environment over the last two decades. Increasingly, joint R&D projects have become the fundamental way to generate and to transfer innovations to useful technology, products, or services (Sctock and Tatikonda 2000; Thamhain 2003; Pek-Hooi and Roberts 2005). In fact, researchers have increasingly moved from the dyadic level to a network level analysis in order to understand the nature, effects and interdependencies of such networks (Gulati 1998; Ahuja 2000; Koka and Prescott 2002). Such research has focused on the notion of social capital to explain the nature and benefits to firms provided through these networks. Moran (2005) pointed the different levels in which has been studied the impact of social capital on performance, and emphasized that these analyses ranging from the individual and small groups to larger organizations and even communities and nations. Prior research on interfirm agreements has also highlighted that network effect (Ruef et al. 2003), or embeddedness (Granovetter 1985), affects economic and innovative performance. By their nature, network effect has been found to have a significant impact on the development of joint R&D projects as well as on their success (Grewal et al. 2006).

The evolving structure of the relationships between the heterogeneous partners involved in a joint project—the social capital involved in the system—provides a focus in which organizational researchers have highlighted the importance of embeddedness, or the architectural nature of partners' relationships (Granovetter 1985; Grewal et al. 2006). Thus, Ruef et al. (2003) define the "network effect" as the relations among partners and projects that provide partners with access to information and perhaps embedded resources (Portes 1998; Moran 2005). The impact of network effect on performance has been studied at multiple levels using different measures of performance. In general, it is accepted that social capital and the resulting network embeddedness have a significant impact on project performance. However, this is a complex relationship and it is not so clear how the network embeddedness of both projects and partners influences project performance. Thus, Burt (1992) pointed out the benefits of access to non-redundant contacts in order to obtain novel information, and Coleman (1988) argued that the location or form of embeddedness allows the actors to access different types of information which have a different impact on the performance of actors. Rosa et al. (1999) and Koka and Prescott (2002) showed that high network embeddedness implies that partners may be exposed to too much information, leading to cognitive overload and poorer work performance. More recently, Grewal et al. (2006) argued that social capital varies across projects and developers and that it plays a critical role in their success. As March (1991) points out, an R&D partnership comprises a great spectrum of projects, from those whose objectives are exploitation, that is, the refinement and extension of existing competencies, technologies and paradigms, to those whose objectives are exploration, that is, experimentation with new, uncertain alternatives. Thus, exploitation projects involve using existing information to improve efficiency and returns from present strategies, competencies, and procedures, while exploration projects entail searching and experimenting to find emerging innovations that will produce future profits. According to Gilsing et al. (2008) and Koka and Prescott (2002) those different objectives will imply different information needs; hence it is expected that not all positions in the network provide the same type of information. Therefore, our research question is how location or network embeddedness impact on project performance.

Several theoretical variants of embeddedness can be found in the literature. Following Grewal et al. (2006), we focus on two key dimension of embeddedness—structural and junctional—as a way to capture and integrate its cumulative effect on joint performance. These two forms of embeddedness bring different types of information and resources to both the project and the project manager. Structural embeddedness captures the centrality of the project and the project manager; hence a central position provides access to a large amount of information derived from the interconnection with many nodes of the social network. Junctional embeddedness assesses connectivity; consequently a central location provides the project and the project manager with access to information which some authors call quality information as a result of its bridge effect between groups of projects not directly interconnected (Koka and Prescott 2002). Accordingly, we argue that higher values of any of the two locations or forms imply greater embeddedness and social capital.

To understand how network embeddedness is associated with performance in joint R&D projects, we argue that the effect of embeddedness on performance is contingent upon network structure and project objectives. Thus, following Grewal et al. (2006), we examine the embeddedness-performance relationship by simultaneously examining the project/partner's embeddedness and network structure and objectives. By considering a project/partner's network structure, we are able to specifically answer our research questions: under what conditions will structural or junctional embeddedness provide a significant impact on project performance? How might the different objectives of the project (exploration/exploitation) influence this network position-performance relation?

The central statement of our argument is that social capital varies across projects and partners and that it plays a critical role in the performance of joint R&D projects. To examine this, we focused on the European networks of excellence as our research setting. The networks of excellence of the Sixth Framework Programme provide an ideal setting because these are designed to strengthen excellence in R&D by integrating a critical mass of resources and expertise which are networked around different joint R&D projects. A network of excellence is, therefore, an instrument for strengthening excellence by tackling the fragmentation of European research, aimed at creating a durable integration of the research capacities of the network participants while advancing knowledge on R&D topics. By simultaneously investigating the project/partner's position and network embeddedness in European networks of excellence, our paper advances a theory of competitive positioning in the management of networks and provides a more defined understanding of how European networks perform.

2 Theory and Research Hypotheses

2.1 Network Embeddedness

The concept of embeddedness, according to Zukin and DiMaggio (1990), comprises four broad categories: cognitive, cultural, political and structural. In this paper, we focus on what these authors refer to as structural embeddedness which they define as the conceptualization of economic exchange in the pattern of ongoing interpersonal relations. However, this concept has different theoretical variants (see Table 1). Thus, Uzzi (1996, p. 675) suggested that "structural embeddedness focuses on the relational quality of inter-actor exchanges and the architecture of network ties". Nahapiet and Ghoshal (1998) and Krause et al. (2007) pointed out that structural dimension results from the structural configuration, diversity, centrality and boundary-spanning roles of network participants. According to this perspective, Tsai and Ghosal (1998) highlighted that structural dimension includes social interaction, and recognized that the resulting interactions due to the location of an actor's contacts in a social structure provide certain advantages. Gulati (1995, 1998) explained that structural embeddedness or positional perspectives on networks go beyond the immediate ties of firms and emphasize the informational value of structural position that the nodes occupy in the network, using the terms structural and positional embeddedness interchangeably. To capture the architecture of network embeddedness, Grewal et al. (2006, p. 1045) proposed three subconstructs: structural, positional and junctional embeddedness. "Structural embeddedness captures the extent to which an actor is entrenched in a network of relationships"; "positional embeddedness appraises the extent to which an actor is connected with other structurally embedded entities", and "junctional embeddedness assesses the extent to which an actor connects with other actors". These authors explain that each form of network embeddedness allows the network nodes to access or spread information with different characteristics. Thus, structural embeddedness enables the node to access more/less information, i.e., a node very well connected with others which receive a great amount of information. Positional embeddedness makes reference to the information received or sent by the node, i.e., a node closely connected with a central nucleus of nodes, allows it to access/send important information. Finally, junctional embeddedness allows the node to access higher/lower quality information depending on its network position, i.e., a node which serves as bridge between two important clusters of the network. Our focus here is on structural and junctional embeddedness, that is, on quantity information vs. quality information as a way to capture and integrate the cumulative effect of embeddedness in joint project performance.

Table 1 Selected definitions of "embeddedness"

Concepts	Definitions	Authors
Embeddedness	Highlights the importance of location into a network (a) How central is the location (b) How strong are the ties that this location provides	Portes (1998) and Granovetter (1973)
Categories of embeddedness		
Cognitive	• Ways in which the structured regularities of mental processes limit the exercise of economic reasoning	Zukin and DiMaggio (1990)
Cultural	• Role of shared collective understandings in shaping economic strategies and goals	
Political	• Manner in which economic institutions and decisions are shaped by a struggle for power that involves economic actors and nonmarket institutions	
Structural	• Contextualization of economic exchange in the pattern of ongoing interpersonal relations	
Types of network embeddedness		
Relational embeddedness	• Highlights the effects of cohesive ties between social actors on subsequent cooperation between those actors	Gulati (1998) and Gulati and Gargiulo (1999)
Structural embeddedness	• Captures the impact of the structure of relations around actors on their tendency to cooperate with one another	
Positional embeddedness	• Captures the impact of the positions organizations occupy in the overall structure of the network on their decisions about new cooperative ties	
Structural embeddedness	Relational quality of interactor exchanges and the architecture of network ties	Uzzi (1996)
Categories of structural embeddedness		
Structural embeddedness	• Captures the extent to which an entity is entrenched in a network of relationships (enables the node to access more/less information)	Grewal et al. (2006)
Positional embeddedness	• Appraises the extent to which an entity is connected with other structurally embedded entities (makes reference to the information received or sent by the node)	
Junctional embeddedness	• Assess the extent to which an entity connects other entities (allows the node to access higher/lower quality information depending on its network position)	

2.2 Project Objectives

Joint R&D projects may have different objectives and hence, develop different activities, so the need for information and resources in each case may differ. As noted above, we analyze two broad categories of projects, exploration and exploitation projects. Thus, exploitation can be characterized as a process of routinization, which adds knowledge base and competence to the existing firms without changing the nature of activities. Exploration, in contrast, can be characterized as a break with an existing dominant design that shifts away from existing rules, norms and routines, in search of novel combinations. Gilsing et al. (2008) point out, from the firm's perspective, that exploration tasks involve the creation of technological knowledge which is new to the firm. Two characteristics, therefore, may possess the information in an exploration project: heterogeneity which allows partners to access novel information, and non-redundancy which avoids the overload of the information processing capacity and facilitates the ability to detect new alternatives. Since these two projects are quite different and require different resources and partners, we expect that structural attributes of networks will have an unequal impact on the formation and the performance of each network of projects.

2.3 Structural Embeddedness and Project Performance

When the structural embeddedness of a project is high, projects are connected to larger number of partners which allow to access to greater resources. Grewal et al. (2006), point out that the complex tasks associated with R&D development can be spread over more developers, resulting in better organization, and hence higher performance. On the other hand, structural embeddedness of a project implies that projects have access to a large amount of information. However, this information is not free of redundancies, which may hinder the ability to search for new alternatives (Koka and Prescott 2002). Hence, the influence of structural embeddedness should be negative as it ensures a greater quantity of information, and the value of project embeddedness should decline due to the homogeneous and redundant information.

In the case of a project manager, the structural embeddedness is higher when the project manager works on more projects. This large number of linkages implies that the project manager may be exposed to too much information, which may lead to cognitive overload and poorer work performance (Rosa et al. 1999; Gilsing and Nooteboom 2006). Grewal et al. 2006, point out that access to a greater quantity of information can be important in the case of mature or incremental R&D projects. Nevertheless, when the project manager is engaged in multiple projects, the different ways of managing these projects will result in the enrichment of its coordination role thereby increasing the project performance. Therefore, we propose:

Hypothesis 1a. The structural embeddedness of a project/project manager will positively influence the project performance of some projects and negatively influence the performance of others.

In the performance of the exploration and exploitation projects, we should note that in exploration projects, the influence of structural embeddedness will be less positive than in exploitation projects. While structural embeddedness ensures a greater quantity of information, this position may saturate the project/project manager's ability to create new alternatives as a result of the large amount of information received by participating in multiple projects. Therefore, the value of project/project manager structural embeddedness should decline in exploration projects due to the homogeneous and redundant information managed. Accordingly, we propose:

Hypothesis 1b. The likelihood that structural embeddedness positively influences project/project manager performance will be higher in exploitation than in exploration projects.

2.4 Junctional Embeddedness and Project Performance

When the junctional embeddedness of a project is high, this implies access to different flows of information through partners who have participated in mutually unconnected projects (Gulati 1995; Hargadon and Sutton 1997; Gilsing and Nooteboom 2006). This provides, on the one hand, access to heterogeneous information, and on the other, non-redundant information which as March (1991) points out, are the conditions needed to create new knowledge. Following Gilsing et al. (2008), the interconnection with different projects allows partners to access alternative ways of thinking and enables them to create new combinations for new uncertain alternatives. In sum, the influence of high junctional embeddedness should be positive as it tends to facilitate the fundamental process of searching for new alternatives, which may lead to improved project performance. Nevertheless, other studies highlight the positive effect on performance that past experiences and the trust level achieved between partners have in the development of projects (Poppo and Zenger 2002; Liu et al. 2009). Therefore, from this latter point of view, junctional embeddedness may affect negatively the project

It is the project manager who plays the key role of coordinating the overall project. Thus, the junctional embeddedness of project managers allows them to access different flows of information as they act as a node of connection between mutually unconnected projects (Hargadon and Sutton 1997), and hence, to have the possibility of receiving heterogeneous and non-redundant information. This will encourage the project manager to create or rethink alternatives through the information received and due to the central role that s/he plays, leading to improved project performance. Koka and Prescott (2002) and Gilsing et al. (2008) point out that higher quality information available to the manager results in greater technical solutions and will increase the likelihood of the success of the project. Nevertheless, the junctional embeddedness of the project manager will hinder his/her coordination tasks as a result of managing mutually unconnected projects. Therefore, we propose:

Hypothesis 2a. The junctional embeddedness of a project/project manager will positively influence the project performance of some projects and negatively influence the performance of others.

In exploitation projects, the influence of junctional embeddedness should be less positive than in the case of exploration projects, as a result of the low level of relationship between partners: on the one hand it increases the difficulty of creating a cooperative climate and the existence of trust between the partners, which may hinder the performance of the project; and, on the other hand, it hinders the interaction between partners, and affects the transmission and the appropriability of knowledge. Accordingly, we propose:

Hypothesis 2b. The likelihood that junctional embeddedness positively influences project/project manager performance will be higher in exploration than in exploitation projects.

3 Data and Measures

3.1 Data Collection

To test the above hypotheses, we collected our data from the Networks of Excellence database within the 6th EU Framework Program. The EU Framework Programs funded by the European Commission have played an important role in setting favorable conditions for the enlargement of R&D cooperation by the development of European joint projects (Mytelka 1991). In this sense, public institutions promote the development of technological networks as part of their technological policy, with the purpose of enhancing the country's competitiveness and technological ground (Vekstein 1999; Löfsten and Lindelöf 2005). Other countries have also followed these types of policies, such as Japan, in which technological projects developed in network are encouraged by the Japanese administration (Sakakibara 1997), and similarly the American administration, as is described in numerous researches (Miyata 1996; Doz et al. 2000). In Europe, technological policies are included in the R&D Framework Programs which promote the establishment of innovation networks for the development of technological projects (Mytelka 1991). R&D networks are considered as an organizational and economic reality, which makes this a promising field for scientific research (Gulati 1998; Hagedoorn et al. 2000; Branstetter and Sakakibara 2002).

Networks of excellence have played a prominent role in overcoming the fragmentation of the European research system and strengthening the European position in specific research areas. The creation of these networks was supported with European financing, but their activities have not become dependent as a result of this support. In fact, the European funding has complemented resources deployed by the participants, taking the form of fixed grants for integration. Networks have therefore carried out "clusters" of integrated projects in different research areas. These are long term and have a multidisciplinary nature, and the participants have been paid on the basis of the degree of integration achieved and the number of partners actively participating in the network, rather than as a result of R&D output.

Networks of excellence involve at least three legal entities from three different Member States or Associated States, though in practice the number of participants is substantially higher than three and generally no fewer than six. Larger networks might involve hundreds of researchers. Others might be of a much more limited size, on the basis of the goals pursued and the critical mass needed to achieve these goals. The dataset provides detailed information on 247 research projects and 2,770 participants. The 247 projects were distributed over 7 priority areas with a clear preponderance in the area of Information Society Technologies.

According to Rothaermel and Deeds (2004), exploration and exploitation projects are related to and built on each other: exploration develops into exploitation, and exploration emerges from exploitation. Thus, exploration concludes with the product development process and exploitation finalizes when the product is on the market. Using these criteria, we have selected as exploration projects those whose objectives were "research and technological development"—projects obtaining new knowledge intended to develop or improve products, processes or services—and as exploitation projects those whose objectives were "demonstration projects"—projects improving the viability of new technologies offering potential economic advantage in the market. The sample size was selected at random by stratified sampling, proportional to groups of type of project (exploration/exploitation). Data were collected through a mail survey. The questionnaire was distributed using the Dillman Total Design Method (Dillman 1978) and addressed to the company's CEOs and project managers. We obtained a sample of 741 usable responses (431 for project managers of exploration projects and 310 for project managers of exploitation projects). In order to test the potential for non-response bias, we compared key attributes of respondents to those of the targeted population sample by employing t-test and ANOVA analysis. Moreover, we also compared first, second and third-wave responses by using ANOVA analysis. The results of the t-test for attributes, revealed no significant differences between respondent and non-respondent groups, and also, that there was no significant difference between the three groups in terms of the two measures.

To evaluate the consequences of different location or form embeddedness in joint project performance, we rely on two-mode affiliation networks (Faust 1997; Grewal et al. 2006) based on partners and projects, and two types of projects, exploration and exploitation projects. Joint R&D projects are developed by partners who are related to one another because they work together on projects, and projects are related to one another because they share partners. We use the project as the unit of analysis (because R&D projects have different partners) and assess partner embeddedness by measuring the embeddedness of the project manager (Grewal et al. 2006).

Following the methodology proposed by Grewal et al. (2006), we constructed two affiliation matrixes. The first matrix for exploration projects with 431 rows (partners) and 67 columns (projects), and a second affiliation matrix for exploitation projects with 310 rows (partners) and 95 columns (projects), providing an appropriate sample of projects to represent the EU Networks of Excellence (Cordis 2004).

3.2 *Measures*

3.2.1 Measure of Independent Variables

To capture the network embeddedness of projects and project managers we consider a network to be a set of points, called nodes or vertices, with connections between them, called links, ties or edges. Two-mode Affiliation networks allow one to study the dual perspective of the project and the project manager.

Following Grewal et al. (2006), we consider an affiliation network A in which the rows represent the partners and the columns represent the projects with 1 when a partner belongs to a project and 0 otherwise. Thus, we can obtain the valued matrixes for partners (X^{Pt}) and projects (X^{Pj}) as:

$$X^{Pt} = AA'$$

$$X^{Pj} = A'A$$

where A' is the transpose matrix.

The network embeddedness, which is determined by the diverse measures of centrality, examines the different contributions/receptions of each partner/project to the network. These measures have been used by Gulati and Gargiulo (1999), Wasserman and Faust (1999) and Grewal et al. (2006) in the context of network analysis.

1. For structural embeddedness we use the degree centrality (Grewal et al. 2006). Degree centrality (Faust 1997) for partner i is defined as $CD({X^{Pt}}_i)$ and for project j is defined as $CD({X^{Pj}}_j)$ such as

$$CD({X^{Pt}}_i) = {X^{Pt}}_{ii}$$

$$CD({X^{Pj}}_j) = {X^{Pj}}_{jj}$$

   where the network has i partners and j projects; the degree centrality for a partner i is the ith diagonal element of X^{Pt}, and in a similar manner is calculated for projects.

2. For junctional embeddedness we use betweenness centrality (Grewal et al. 2006). For calculating betweenness centrality, that is, the shortest path between two partners or projects, Freeman (1979) proposes a two-step procedure. First, calculating "partial betweenness" (p_i) of partners, that is, the number of pairs of partners whose geodesic paths contain the partner *i*; and then using this partial betweenness to calculate partner betweenness. Betweenness centrality for partner *i* is then given as

$$CB(X^{Pt}{}_{i}) = \sum g_{jk}(p_i)/g_{jk}$$

where g_{jk} is the number of geodesic paths between partners *j* and *k*, and $g_{jk}(p_i)$ is the number of geodesic paths between *j* and *k* that contain *i*, with $j < k$.

The measures have been processed by UCINET software version 6.118 (Borgatti et al. 2002).

3.2.2 Measure of Dependent Variables

The dependent variable, *network performance*, measures the probability of successful performance of exploration and exploitation joint R&D projects. Traditionally, project efficiency has been assessed using the multiple measure of set of cost, time and performance (Swink et al. 2006). A project was considered efficient if it was completed within its budget estimated, within its initially scheduled time frame, and performed as it was designed to (Olk 2002). Much of the cooperation performance research has relied on managers' evaluations of partnership success (Saxton 1997). Doing so is appropriate when respondents represent top managers (Olk 2002). Because the respondents in our sample were project managers very well informed about the joint project in which they were involved, we were confident that it was proper to rely on managerial perceptions of joint project results. Moreover, Geringer and Herbert (1991) found strong correlations between subjective and objective measures of cooperation performance. Hence, to evaluate this variable, following the works of Meyer (1994), Pinto et al. (1993), and Warr et al. (1979), we used a perception measure of network performance that assessed team' satisfaction, which is defined as "the degree to which association with the team is considered a worthwhile, productive, and satisfying experience by team members" (Sarin and Mahajan 2001, p. 37). To form this measure we asked partners in each particular joint R&D project about their perceptions of how the project was performed. Network performance was operationalized using a selection of performance indicators marked by the European Commission (Cordis 2004) of what could be expected from a networking aiming at achieving a satisfactory level of integration at the end of the funding period. To evaluate this variable, the European Commission relied on project managers' evaluations using a five-point Likert scale (5 = completely achieved, 4 = mainly achieved, 3 = partially achieved, 2 = scarcely achieved, 1 = not achieved at all). The scale item reflected: (1) the communications inside the network; (2) the sharing and common management of equipment, installations, and infrastructures; (3) the common management of human resources; (4) the common knowledge management; (5) the network management; and (6) the continuity of the network after the conclusion of the Community's funding period.

4 Results

Table 2 presents the correlation coefficients for all variables used in this study and Table 3 presents the results of the estimation. We estimated five different model specifications taking into account both exploration and exploitation projects. Model (M1) contains the dependent variable as a function of the degree centrality for both the project and project manager in exploitation projects, and betweenness centrality for project and project manager in exploration and exploitation projects. Model 2 captures the dependent variable as a function of the degree centrality for project and project manager in exploration projects, and betweenness centrality for project and project manager in exploration and exploitation projects. Model 3 captures the dependent variable as a function of the degree centrality for project and project manager in exploration and exploitation projects, and betweenness centrality for project and project manager in exploitation projects Model 4 captures the dependent variable as a function of the degree centrality for project and project manager in exploration and exploitation projects, and betweenness centrality for project and project manager in exploration projects Model 5 captures the dependent variable as a function of the degree centrality for project and project manager in exploration and exploitation projects, and betweenness centrality for project and project manager in exploration and exploitation projects. Model fits are acceptable with significant chi-square values ($p < 0.01$) and R^2 values ranging from 0.372 to 0.501.

From the results of models 1–5, positive relationships were found between degree and betweenness centrality of the projects in exploration and exploitation projects and their performance. However, our results do not confirm these positive relationships between location, in the case of the project manager, and performance, which supports the Hypotheses 1a and 2a. In fact, in exploration projects, we observe a significant positive relation between the location of the project [degree centrality ($\beta = 0.156$; $p < 0.10$), betweenness centrality ($\beta = 0.417$; $p < 0.01$)] and its performance. Similarly, we observe a significant positive relation between the location of the project manager as betweenness centrality ($\beta = 0.301$; $p < 0.05$), and performance. On the other hand, in exploitation projects, we find a positive and significant relation between degree centrality in the case of project ($\beta = 0.212$; $p < 0.05$), and project manager ($\beta = 0.335$; $p < 0.05$), and performance. Similarly, our results show a significant positive relation between the location of the project as betweenness centrality ($\beta = 0.118$; $p < 0.10$), and performance. These findings are consistent with previous studies (Grewal et al. 2006) which suggest that the location of projects and project managers has a different impact on the performance of projects, as a function of the kind of projects

Table 2 Correlation matrix between variables

	1	2	3	4	5	6	7	8	9
Degree centrality exploration project	–	0.305	0.037	0.015	0.166	−0.091	0.079	0.002	0.201
Degree centrality exploration project manager		–	0.102	0.163	0.028	0.017	0.155	0.050	0.136
Degree centrality exploitation project			–	0.193	0.099	0.019	0.015	0.070	0.325
Degree centrality exploitation project manager				–	0.098	0.158	0.123	−0.073	0.187
Betweenness centrality exploration project					–	−0.101	0.124	0.173	0.390
Betweenness centrality exploration project manager						–	0.128	0.143	0.270
Betweenness centrality exploitation project							–	0.097	0.125
Betweenness centrality exploitation project manager								–	0.104
Performance									–

Table 3 Results of regression analyses predicting social impact in exploration/exploitation joint R&D projects

Measure	Variable type	Variable name	Performance				
			M1	M2	M3	M4	M5
Degree centrality	Exploration	Project	–	0.173*	0.110*	0.141*	0.156*
		Project manager	–	0.028	0.086	0.043	0.092
	Exploitation	Project	0.234**	–	0.199*	0.189*	0.212**
		Project manager	0.273**	–	0.299**	0.304**	0.335**
Betweenness centrality	Exploration	Project	0.341***	0.389***	–	0.376***	0.417***
		Project manager	0.291**	0.256**	–	0.250**	0.301**
	Exploitation	Project	0.133*	0.125*	0.110*	–	0.118*
		Project manager	0.011	0.023	0.097	–	0.036
R^2			0.428	0.372	0.341	0.454	0.501

*$p < 0.10$; **$p < 0.05$; ***$p < 0.01$

considered; it also corroborates the contingency character of the form of embeddedness. It is worthy to note the different impact of structural variables on performance in the case of project and project managers. The results show that centrality degree, in the case of exploration projects, and betweenness centrality in exploitation projects do not have a positive and significant impact on performance from the point of view of the project manager. These differences may be due to the way in which we have measured the network performance. Thus, being a measure of team satisfaction, it collects the role that project manager develops in the integration of the partners in the network (Sarin and Mahajan 2001). Accordingly, as suggested by the literature, the coordination of the overall project development activity and the transmission of high quality knowledge among partners (Williamson 2002; Gulati 1998) is a key role of project managers. Consequently, the meaning of structural and junctional variables on the performance is more powerful from projects than from project managers' view.

Following Liu et al. (2009), the critical test of the relationship is used to examine the differential effects that the project and project manager locations (centrality/ betweenness) have on performance (Hypotheses 1b and 2b). We examined these effects through the proportion of variance explained by these mechanisms. Firstly, we determined whether degree centrality influences project/project manager performance in exploration and exploitation projects (Hypothesis 1b). Thus, we obtain ΔR^2 as follows from the regression results of Model 5 and Model 1: $\Delta R^2_{\text{Model 5 - Model 1}} = R^2_{\text{Model 5}} - R^2_{\text{Model 1}} = 0.501 - 0.428 = 0.073$. Here $\Delta R^2_{\text{Model 5−Model 1}}$ represents the proportion of the variance of performance of exploration projects that can be explained by project and project manager degree centrality. Similarly, $\Delta R^2_{\text{Model 5−Model 2}}$ represents the proportion of the variance of performance of exploitation projects explained by project and project manager degree centrality. In order to determine which locations are perceived to have a greater impact on improving the performance of exploration and exploitation projects, we determined the balance of proportion of the variance explained by each location. Since $\Delta R^2_{\text{Model 5−Model 2}} > \Delta R^2_{\text{Model 5−Model 1}}$, we can conclude that degree centrality has greater impact on exploitation than exploration projects, which corroborates Hypothesis 1b. Thus, we can observe the positive and significative impact that the structural embeddedness has on exploitation projects, which confirms that the redundancy of the information has less impact on the performance of exploitation projects than on exploration projects.

Similarly, we have determined project/project manager performance both in exploitation and exploration projects (Hypothesis 2b). We obtain ΔR^2 as follows from the regression results of Model 5 and Model 3. Here $\Delta R^2_{\text{Model 5−Model 3}}$ represents the proportion of the variance of exploration projects performance that can be explained by the project and project manager betweenness degree. Similarly, $\Delta R^2_{\text{Model 5−Model 4}}$ represents the proportion of the variance of exploitation project performance explained by project and project manager degree centrality. Since $\Delta R^2_{\text{Model 5−Model 3}} > \Delta R^2_{\text{Model 5−Model 4}}$, we can conclude that betweenness centrality has greater impact on exploration than exploitation projects, which corroborates Hypothesis 2b. Thus, we can observe the positive and significative impact that the junctional embeddedness has on exploration projects, which confirms that the heterogeneity of partners has a positive impact on the performance of the exploration projects but not in the case of exploitation projects.

5 Discussion and Conclusion

This research analyzes the effects of the location of projects/project managers on the performance of R&D collaboration networks. We identify two R&D network types and show that for each the embeddedness of projects and project managers significantly affects network performance. We find that structural embeddedness positively influences in the joint R&D performance, being its impact higher in exploitation than in exploration projects. Otherwise, junctional embeddedness also positively influences in the joint R&D performance but its effect is higher in exploration than in exploitation projects. These findings support our hypotheses regarding the network contingency of the positioning-performance relationship. Our research contributes to a better our understanding of ways in which the form of embeddedness impacts on the performance of joint R&D projects. Studying two types of R&D network provides a more comprehensive knowledge of projects/ project managers positioning. Findings in previous research (Koka and Prescott 2002) suggest that network embeddedness provides access to different quality and quantity information, which is important for network performance (Gulati 1998); yet our research shows that the positioning-performance relationship is not always positive. Negative or positive performance outcomes may accrue depending upon the distinctiveness of the R&D project objectives (e.g., exploration vs. exploitation), coupled with the extent to which each project/project manager is structurally embedded.

As theoretical implications, our research extends social capital theory to analyse the impact of structural variables on the performance of joint R&D projects. Thus, social exchange theory emphasizes that interactions between partners provide social capital through the partnership, which has an impact on the performance of such projects. Moran (2005) suggested the complex nature of the relationship between social capital and the efficiency of joint projects, and, consistent with the social capital theory, our results confirm that project and project manager embeddedness have a differential impact on the performance of exploration and exploitation projects, which corroborates the complex of the network effect (Grewal et al. 2006). Therefore, our results are consistent with the logic of social exchange theory, and are complementary to previous research, adding new empirical evidence on the complex relation between network effect and performance of projects, highlighting the contingent nature existing between network effect and project performance and, allowing the consideration of the optimisation of interactions in the network.

Understanding the determinants of network performance is the most important managerial issue for project managers and institutions. Our findings provide insight as to how projects/projects manager might gain performance benefits by being in a network position. As we hypothesized, our results showed that the relationship between network position and project/project manager performance can be either positive or negative, depending upon the R&D project objectives in which the partnership is embedded. Our results suggest that a project/project manager pursuing an exploration objective needs to be involved in a highly junctional embedded network to achieve high performance, and that a project/project manager pursuing an exploitation objective needs to be involved in a structurally embedded network to achieve better performance. Our findings also encourage managers to think more broadly about the competitive objectives of their R&D projects. To successfully position a project/project manager in a competitive technological environment, we suggest that merely selecting a distinctive set of partners is not enough. Competitive positioning also involves not only a consideration of complex interactions between partners in a social context, but also the technological objectives of networks where partners are more or less embedded. Such a network effect within a partner' relationships provide critical information and knowledge for R&D projects. Understanding the network effect is an important task for managers who try to position their R&D projects strategically in a competitive environment.

However, this study is not devoid of limitations which should be addressed in future attempts. Furthermore, one should be cautious in generalizing the findings. First, we used only R&D projects developed within the framework of EU programs devoted to promoting partnerships, thus we cannot substantiate our claims and findings beyond these sponsored projects. It therefore seems useful to consider other types of projects in transversal and longitudinal studies as well, which could provide new evidence into the effects of the evolution of interorganizational relationships on R&D. Future research efforts directed at studying the network effect in joint projects in other contexts and locations would also be interesting. Moreover, future research should examine other dimensions of embeddedness to look for additional factors which might offer new insights from a social capital perspective.

References

Ahuja G (2000) The duality of collaborations: inducements and opportunities in the formation of interfirm linkages. Strateg Manage J 21:317–343

Borgatti SP, Everett MG, Freeman LC (2002) UCINET for Windows: software for social network analysis. Analytic Technologies, Harvard

Branstetter LG, Sakakibara M (2002) When do research consortia work well and why? Evidence from Japanese panel data. Am Econ Rev 92(1):143–159

Burt RS (1992) Structural holes: the social structure of competition. Harvard University Press, Cambridge

Coleman JS (1988) Social capital in the creation of human capital. Am J Sociol 94:95–120

Cordis (2004) Methodological guidance note on the performance indicators within Networks of Excellence. Cordis Data Base. http://europa.eu/comm/research/fp6/instruments_en.html

Dillman DA (1978) Mail and telephone surveys: the total design method. Wiley, New York

Doz YL, Olk PM, Ring PS (2000) Formation processes of R&D consortia: which path to take? Where does it lead? Strateg Manage J 21:239–266

Faust K (1997) Centrality in affiliation networks. Soc Netw 19:157–191

Freeman LC (1979) Centrality in social networks: conceptual clarification. Soc Netw 1:215–239

Geringer MJ, Herbert L (1991) Measuring performance of international joint ventures. J Int Bus Stud 22:249–263

Gilsing VA, Nooteboom B (2006) Exploration and exploitation in biotechnology. Res Policy 35:1–23

Gilsing V, Nooteboom B, Vanhaverbeke W, Duysters G, van den Oord A (2008) Network embeddedness and the exploration of Novel Technologies: technological distance, betweenness centrality and density. Res Policy 37:1717–1731

Granovetter M (1973) The strength of weak ties. Am J Sociol 78:1360–1380

Granovetter M (1985) Economic action and social structure: the problem of embeddedness. Am Sociol Rev 15:478–499

Grewal R, Lilien G, Mallapragada G (2006) Location, location, location: how network embeddedness affects project success in open source systems. Manage Sci 52:1043–1056

Gulati R (1995) Social structure and alliance formation pattern: a longitudinal analysis. Adm Sci Q 40:619–642

Gulati R (1998) Alliances and network. Strateg Manage J 19:293–317

Gulati R, Gargiulo M (1999) Where do interorganizacional networks come from? Am J Sociol 104:1439–1493

Hagedoorn J, Link A, Vonortas N (2000) Research partnerships. Res Policy 29:567–586

Hargadon A, Sutton R (1997) Technology brokering and innovation in a product development firm. Adm Sci Q 42:716–749

Koka BJ, Prescott JE (2002) Strategic alliances as social capital: a multidimensional view. Strateg Manage J 23:795–816

Krause DR, Handfield RB, Beverly BT (2007) The relationships between supplier development, commitment, social capital accumulation and performance improvement. J Oper Manage 25:528–545

Liu Y, Luo Y, Liu T (2009) Governing buyer–supplier relationships through transactional and relational mechanism: evidence from China. J Oper Manage 27:294–309

Löfsten H, Lindelöf P (2005) R&D networks and product innovation patterns—academic and non-academic new technology-based firms on Science Parks. Technovation 25:1025–1037

March JG (1991) Exploration and exploitation in organizational learning. Organ Sci 2:71–87

Meyer C (1994) How right measures help teams excel. Harv Bus Rev 72:95–103

Miyata Y (1996) An economic analysis of cooperative R&D in the United States. Technovation 16:123–131

Moran P (2005) Structural vs. relational embeddedness: social capital and managerial performance. Strateg Manage J 26:1129–1151

Mytelka LK (1991) States, strategic alliances and international oligopolies: the European ESPRIT Programme. In: Mytelka LK (ed) Strategic partnerships. Pinter Publishers, London, pp 182–210

Nahapiet J, Ghoshal S (1998) Social capital, intellectual capital, and the organizational advantage. Acad Manage Rev 23:242–266

Olk P (2002) Evaluating strategic alliance performance. In: Contractor FJ, Lorange P (eds) Strategies and alliances. Elsevier, Kidlington

Pek-Hooi S, Roberts EB (2005) Technology alliances and networks: an external link to research capability. IEEE Trans Eng Manage 52:419–428

Pinto MB, Pinto JK, Prescott JE (1993) Antecedents and consequences of project team cross-functional cooperation. Manage Sci 39:1281–1297

Poppo L, Zenger T (2002) Do formal contracts and relational governance function as substitutes or complements? Strateg Manage J 23:707–725

Portes A (1998) Social capital: its origins and application in modern sociology. Annu Rev Sociol 22:1–24

Rosa JA, Porac JF, Runser-Spanjol J, Saxon MS (1999) Sociocognitive dynamics in a product market. J Mark 63:64–77

Rothaermel F, Deeds D (2004) Exploration and exploitation alliances in biotechnology: a system of new product development. Strateg Manage J 25:201–221

Ruef M, Aldrich HE, Carter NM (2003) The structure of founding teams: homophily, strong ties, and isolation among U.S. entrepreneurs. Am Sociol Rev 68:195–222

Sakakibara M (1997) Evaluating government sponsored R&D consortia in Japan: who benefits and how? Res Policy 26:447–453

Sarin S, Mahajan V (2001) The effect of reward structures on the performance of cross-functional product developments teams. J Mark 65:35–53

Saxton T (1997) The effects of partner and relationship characteristics on alliance outcomes. Acad Manage J 40:443–461

Sctock GN, Tatikonda MV (2000) A typology of project-level technology transfer processes. J Oper Manage 18:719–737

Swink M, Talluri S, Pandejpong T (2006) Faster, better, cheaper: a study of NPD project efficiency and performance tradeoffs. J Oper Manage 24:542–562

Thamhain HJ (2003) Managing innovative R&D teams. R&D Manage 33:297–311

Tsai W, Ghosal S (1998) Social capital and value creation: the role of intrafirm networks. Acad Manage J 41:464–477

Uzzi B (1996) The sources and consequences of embeddedness for the economic performance of organizations: the network effect. Am Sociol Rev 61:674–698

Vekstein D (1999) Defense conversion, technology policy and R & D networks in the innovation system of Israel. Technovation 19:615–629

Warr P, Cook J, Wall T (1979) Scales for the measurement of some work attitudes and aspects of psychological well-being. J Occup Psychol 52:129–148

Wasserman S, Faust K (1999) Social network analysis: methods and applications. Cambridge University Press, Cambridge

Williamson OE (2002) The theory of the firm as governance structure: from choice to contract. Econ Perspect 16:171–196

Zukin S, DiMaggio PJ (1990) Structures of capital: the social organization of the economy. Cambridge University Press, New York

The Impact of Relational Norms on Key Relational Outcomes in Supplier–Buyer Relationships

Muhammad Zafar Yaqub

Abstract Previous research in the relational exchange theory has discussed efficacy of the relational norms as an important driver of the performance of exchange relationships. A number of studies, in different business contexts, have shown that the relational norms (altogether, the relational governance) positively affect the relational outcomes like satisfaction, trust and/or commitment (altogether, the relationship quality) that eventually lead to an enhancement in the value co-created by the exchange partners. However, most of these studies have theoretically and/or empirically treated relational norms and/or the intermediate relational outcomes as complex/abstract higher-order constructs. The author argues that such a treatment of these relational constructs may lead to the loss of useful information about the interaction of individual norms with the various facets/dimensions of these relational outcomes. While discussing its necessity and merits, the paper, by elaborating upon the interaction among these antecedent and outcome constructs complements and further extends the argument advanced in the relational exchange theory. The results from an empirical study conducted in the context of supplier-intermediate buyer dyadic relationships offer some interesting insights into the dynamics of relational exchanges in the downstream structural arrangements. The paper specifically contributes to the relational exchange literature by empirically demonstrating the phenomena like unilateral-relationalism and the relationality-paradox over and above presenting an unprecedented discussion on the association among the individual relational norms and the key relational outcomes.

Keywords Relational exchange • Relational norms • Relationality-paradox • Relationship quality • Unilateral-relationalism

---

M.Z. Yaqub (✉)
Center for Business Studies, University of Vienna, Bruenner Str. 72, A-1210 Vienna, Austria
e-mail: muhammad.zafar.yaqub@univie.ac.at

T. Ehrmann et al. (eds.), *Network Governance*, Contributions to Management Science,
DOI 10.1007/978-3-7908-2867-2_4,

1 Introduction

Granovetter (1992) argued that economic institutions are socially constructed i.e. they emerge from actions of socially situated individuals embedded in networks of personal relations with economic as well as non-economic goals. Similarly, from an in-depth case study, Larson (1992) concluded that economic transactions cannot be separated from the social context in which they take place. As such, it is highly inappropriate to view firms as atomistic entities competing (for profits) against each other in an impersonal marketplace (Gulati et al. 2000). In Granovetter's (1992) opinion, economic action and outcomes (like all social actions and outcomes) are affected by the (1) actors' history of dyadic relations (the relational-embeddedness argument), and (2) structure of the actors' overall network of relations (the structural-embeddedness argument). According to him, the central theme in economic sociology is the necessity of trust and trustworthy behavior (a function of the past interactions as well as the future expectations) for even the normal functioning (let alone the superior performance) of economic action and/or the institutions. Quite consonant with this economic-sociological account, Achrol (1991), in early 1990s, forecasted the rise of "true marketing-companies" within networks of functionally specialized organizations whose norm-driven interrelationships would be held together and coordinated by "market-driven focal organizations" by means of norms of sharing, and commitment based on trust. This conceptualization reveals two important facets along which research in the relational exchange theory (RET) progressed in the subsequent years. These two facets include;

1. **Relationalism (Relational Norms Perspective)** that refers to the degree of relational-orientation prevalent in the exchange environment and is measured on a (Discrete-Relational) continuum based on a mix of relationship-preserving norms.
2. **Commitment-Trust Theory (CTT)** that reveals the mediating role of trust and commitment between the antecedents and consequences of (successful) business relationships. A closer approximation (or extension) of the CTT being widely used in the marketing and strategic management research is relationship quality (RQ) model. Athanasopoulou (2009) in his critical review of the relationship quality literature has treated trust and commitment to constitute a bi-dimensional relationship quality construct.

According to Roehrich et al. (2002), the stability and success of an exchange relationship, to a substantial extent, is determined by conductivity of the overall atmosphere of that exchange. According to Mcneil (1978) and Yaqub and Vetschera (2011), the key to the development of such an atmosphere is to put in place (as governance mechanism) a relational contract based on an adaptive mix of relationship-preserving norms. Relational exchange theory (RET) reveals relational norms as a distinct form of governance (the relational governance) that prescribes commitment and proscribes opportunism in exchange relationships (Joshi and Stump 1999; Morgan and Hunt 1994). Bercovitz et al. (2006) argue that an adequate compliance to the relational norms leads to benefits like smoother coordination, increased adaptability within the exchange relationship, reduced opportunism and increased efforts from transacting parties. Yaqub and Vetschera (2011) argue that an adequate compliance to the relationship-preserving norms not only reduces transaction costs by substituting more elaborate governance but also contributes to the revenue/value by promoting a trust-inspired commitment.

Though most of the scholars and the practitioners would assume that effective relationship management (RM) efforts (like creating a highly relational environment) from the focal firms create strong relational bonds which eventually enhance their performance outcomes (Crosby et al. 1990; Morgan and Hunt 1994), still some business executives have embraced nothing more than sheer disappointment from their RM efforts (Colgate and Danaher 2000). Some researchers have gone even farther by suggesting that in certain situations, RM may even have a negative impact on the performance (De Wulf et al. 2001; Hibbard et al. 2001). Palmatier et al. (2006) argue that the effectiveness of RM efforts in enhancing relationship performance may vary depending upon the specific RM strategy and the exchange context.

An enhancement in the co-created value, perhaps, could be regarded as the most important (relational) outcome and/or indicator of the superior performance of cooperative exchange relationships. According to De Wulf et al. (2001) and Sirdeshmukh et al. (2002), efficient and effective RM efforts (like creating a highly relational environment through promoting relational norms) enhance the relationship value through the creation of strong relational bonds. A number of studies such as Artz and Brush (2000), Aulakh et al. (1996), Ivens (2004), Joshi and Stump (1999), Kaufman and Stern (1988), Vázquez et al. (2007) and Zhang et al. (2003) have, in a variety of business contexts, shown a positive association between adherence to relational norms and the (relationship) value enhancement while employing relationship quality or its individual determinants (satisfaction, trust and commitment) as the mediating constructs. As such satisfaction, trust and commitment (or together the relationship quality) has been the most important intermediate relational outcomes as revealed in much of the marketing and strategy literature. However, most of these studies have treated these intermediate relational outcomes at more abstract levels whereas these, in fact, are quite complex and/or multi-faceted constructs. Consequently, the association among the individual relational norms and various facets of the key (intermediate) relational outcomes is still unknown at large. This study bridges this research gap and augments the argument in the relational exchange theory about the nature of association among these relational constructs. The relevant research context has been the supplier-intermediate buyer dyadic relationships. The following paragraphs discuss the rationale and/or contribution of this research to the contemporary literature.

Frazier et al. (1988) pinpoint that the development and maintenance of relational norms requires substantial up-front investments of time, money and personnel from the focal firms. In Bercovitz et al. (2006) pinion, rather than following a "more-is-better" approach, it is advisable to follow the standard economic logic for achieving adequate levels of "relationality" as the benefits from relational behaviors accrue at diminishing rate while the cost of nurturing such behaviors accrues at increasing rates. Paulin et al. (1999) also argue that the context of exchange may influence instrumentality, relevance and the relative efficacy of the individual norms in ensuring the desired performance levels. Consequently, there exists a need for the research to focus on the development of a fine-grained understanding about which relational norms under what circumstances may or may not have a significant impact on specific relational outcomes. Such an understanding can facilitate the management practitioners in ensuring a precision in their RM effort so as to increase its efficiency through minimizing the wastage of time, efforts and energies required to garner valuable business relationships.

Morgan and Hunt (1994) describe the scope of focal firms' exchanges relationships to include supplier partnerships, lateral partnerships, internal partnerships and the buyer partnerships. Buyer partnerships are further sub-divided into partnerships with (1) the ultimate customers, and (2) the intermediate customers. Even though there is proliferation of research on the relational dynamics of the first type, the latter has received only scarce attention in literature so far (Yaqub and Vetschera 2011). The model introduced here makes up for this deficiency by discussing dynamics of the supplier-intermediate buyer dyadic relationships. Another contribution of this research stems from its use of South Asian data. The findings of this research bolster the case for broad-based generalizability of essentially ethic theories developed in the West to other cultures.

2 The Conceptual Model

Figure 1 shows the conceptual model. It is theorized that the (six) relational norms affect relationship value (the ultimate relational outcome) through influencing the intermediate (relational) outcomes like satisfaction, trust and commitment (together, the relationship quality). Though we would very briefly review/discuss the previous studies that link these norms to the value enhancement through the relational mediators of the higher-order, we would not test these links. It is for this reason that they are shown as the dotted links. Sutton and Staw (1995) and Dicky et al. (2007) hold that it quite acceptable for researchers to propose more extensive models than they actually test in a single study.

By elaborating upon the central tenets of relational norms perspective and the commitment-trust theory/RQ Model, relational exchange theory (RET) explains the essence of relational governance as being an impetus to successful exchange relationships. Relational governance envisages the creation of a relational environment by putting in place a social contract based on a multitude of relationship-preserving norms (Blios and Ivens 2006). The criterion for the successful culmination of such an environment is its ability to promote satisfaction, trust and commitment (together the relationship quality) among the exchanging parties (Ivens 2004). Therefore, RET suggests that firms should consider the development and promotion of relationship quality through promoting an adherence to relational norms as one of their key strategic objectives.

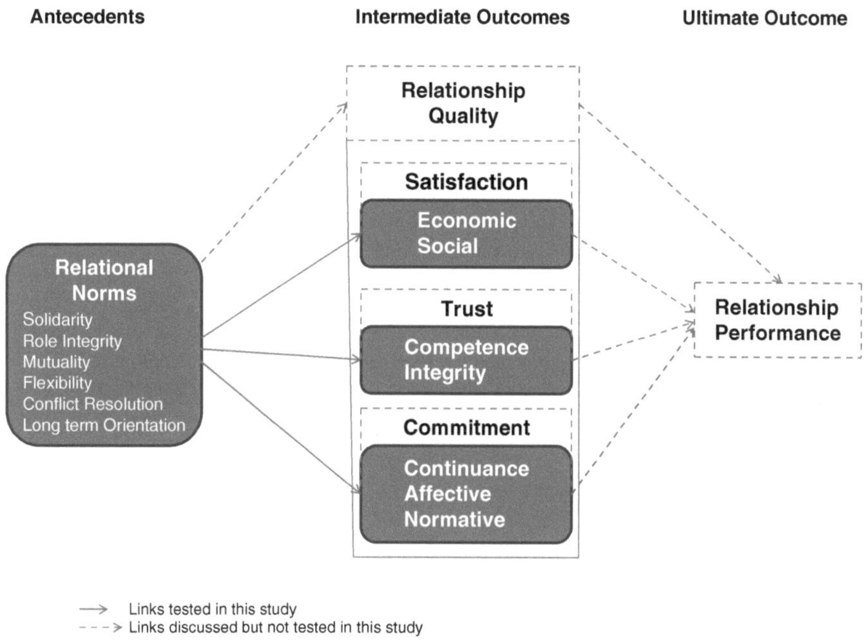

Fig. 1 The conceptual model

A number of studies, in a variety of business contexts, have attested to the efficacy of various relational norms in fostering satisfaction, trust and/or commitment. For example, Ivens (2004) found the relational behaviors like role integrity, flexibility and mutuality to be positively associated with the satisfaction in supplier–buyer relationships. Aulakh et al. (1996) revealed that trust mediates the relationship between relational norms such as continuity, flexibility and information exchange on one hand, and the performance of export partnerships on the other hand. Zhang et al. (2003) concluded that incorporating bilateral solidarity, maintaining flexibility and fostering information exchange with channel partners could have positive effects on trust in the context of international channels. Ivens (2004) reported a positive association between relational norms such as role integrity and mutuality, and the inter-firm trust. Ryu et al. (2007), in the context of manufacturer-supplier relationship, have revealed relational norms and satisfaction with supplier performance as antecedents of trust, which has further been described as an essential precursor of the manufacturer's long-term orientation (LTO). In consonance with these research findings about the nature of association among the higher-order levels of our subject constructs, we have hypothesized a positive association among the six relational norms and the various facets and/or dimensions of the three key relational outcomes (i.e. satisfaction, trust and commitment). The following section briefly discusses the constructs employed in our conceptual model.

Table 1 Aspects of focal suppliers' relational behavior

Norms	Description
Solidarity	Preservation of the relationship, particularly in the situations in which one partner is in predicament (Kaufman 1987)
Mutuality	The actor's attitude that the realization of one's own success passes through the partner's common success (Dant and Schul 1992)
Role integrity	Maintenance of complex multidimensional roles forming a network of relationships (Kaufman 1987)
Flexibility	The actor's readiness to adapt an existing implicit or explicit agreement on new environmental conditions (Noordewier et al. 1990)
Conflict resolution	Application of flexible, informal and personal mechanisms to the resolution of conflicts (Kaufman 1987)
Long-term orientation	An economic actor's desire and utility of having a long-term relationship with a specific exchange partner (Ganesan 1994)

Source: Ivens (2006)

2.1 *Relational Norms*

Norms are expectations about behavior that are partially shared by a group of decision makers and are directed towards collective goals (Jap and Ganesan 2000; Mcneil 1980). They constitute the expectations shared by exchange partners about what constitutes the "right" behavior(s) within the environment of their exchange relationship (Morgan and Hunt 1994). According to Kaufman and Stern (1988), norms that govern exchange behaviors in discrete transactions are different from those in the relational exchange. According to Blios and Ivens (2006), norms associated with discrete exchanges are more likely to create an environment where an exchange partner will give his own interests priority over those of the other party or even the cooperative gains. According to Bercovitz et al. (2006, p. 725); "……… *with discrete norms, partners adjust terms of trade through bargaining before entering short-term exchange arrangements* (Macneil 1978, 1980). *On the other hand, at the relational end of the spectrum norms support cooperative adaptation by stressing behaviors that will preserve and continue the relationship even when pure self-interest might suggest otherwise* (Macneil 1980)".

A number of relational norms have been discussed in the RET literature. Mcneil (1983) argues that various levels of "relationality" could be attained along a discrete-relational continuum, where each level characterizes a different mix of relational norms like role integrity, contractual solidarity, harmonization of relational conflict, supra-contractual relations, and proprietary of means. Kaufman and Stern (1988) reduced Mcneil's list to three norms which included solidarity, role integrity and mutuality. However, later studies added a number of relational norms such as information exchange, participation, fairness and flexibility (Blios and Ivens 2006; Heide and John 1992; Jap and Ganesan 2000) to this list and revealed them to be positively associated with superior performance of exchange relationships in a variety of business contexts (Kaufman and Stern 1988; Macneil 1980). Table 1 outlines six relational norms that have been selected for this study due to their (perceived) higher efficacy of affecting the relational outcomes as hypothesized in the context of this research. This is quite consistent with Cochet et al. (2008) who revealed that the relational norms become more useful when these are considered, by the exchange partners, to be increasingly relevant for their desired outcomes.

2.2 Relationship Quality and Its Individual Determinants

According to Henning-Thurau and Klee (1997), relationship quality refers to the appropriateness of an exchange relationship to fulfill needs of the actor(s) associated with that relationship. They describe relationship quality as the degree of appropriateness of a relationship to fulfill the (individual and collective) needs of the partners associated with that exchange relationship. According to Finn (2005), relationship quality model plays a critical role in the study of the maintenance of long-term relationships. According to Jap et al. (1999), Rajaobelina and Bergeron (2009) and Ural (2007), it captures the real essence of relationship management efforts. As already discussed, a number of studies such as Artz and Brush (2000), Aulakh et al. (1996), Ivens (2004), Joshi and Stump (1999), Kaufman and Stern (1988), Vázquez et al. (2007) and Zhang et al. (2003) have, in a variety of business contexts, shown a positive association between adherence to relational norms and the superior performance while employing relationship quality (or its individual determinants) as the mediating constructs.

Quite consistent with the pioneers Crosby et al. (1990), researchers such as Bejou, Wray and Ingram (1996), Huang and Chiu (2006), Lagace et al. (1991), Leuthesser (1997), Lin and Ding (2006), Rajaobelina and Bergeron (2009), Selnes (1998) and Sun (2010) have treated relationship quality as a two-dimensional higher order construct with satisfaction and trust being those two dimensions. Even though researchers like Henning-Thurau et al. (2002), Storbacka et al. (1994) and Wong and Sohal (2002) also used a bi-dimensional model of relationship quality but they paired commitment (instead of trust) with the satisfaction. However, some researchers like Baker et al. (1999), Garbarino and Johnson (1999), Ivens (2004), Ulaga and Eggert (2006) and Walter et al. (2003) have used a multidimensional model of relationship quality with satisfaction, trust and commitment being the three dimensions. The following section discusses the individual determinants/facets of the relationship quality.

2.2.1 Satisfaction

Satisfaction is generally referred to as a positive (affective) state which results from the appraisal of all aspects of the working relationship of an exchange partner with the other(s) (Geyskens et al. 1999). In a supplier–buyer dyadic relationship context, satisfaction can be viewed as the degree to which a supplier rises up to or exceeds expectations of the buyer in relation to its motives behind entering into an exchange relationship (Yaqub et al. 2010). Supplier–buyer relationships, like all other business relationships, are formed with the expectations of complementary benefits (Anderson and Jap 2005). According to Palmatier et al. (2006), buyers perceive value in such relationships only when they receive these (desired) benefits, which increase their willingness to continue, maintain and/or strengthen relational bonds with the focal supplier. Relational benefits have also been shown to positively affect the relational mediators by Morgan and Hunt (1994) and Reynolds and Betty (1999). As the scope of such benefits can be quite vast (including economic, social, informational, political and other dimensions), therefore, satisfaction has quite often been regarded as a multi-faceted construct in the marketing and the strategic management literature. In context of B2B relationships, Geyskens and Steenkamp (2000) reveal satisfaction as a two dimensional construct with the two sub-types being the economic satisfaction and the social satisfaction. Economic satisfaction refers to the evaluation of economic outcomes that flow from the relationship whereas social satisfaction refers to the psychological aspects of the relationship which consists of an exchange partners' evaluation of the personal contacts and interactions with the other partner (Geyskens and Steenkamp 2000).

2.2.2 Trust

From a relational perspective, inter-organizational trust has been defined as the expectation that an actor (1) can be relied on to fulfill obligations (Anderson and Weitz 1989), (2) will behave in a predictable manner, and (3) will act and negotiate fairly when the possibility for an opportunistic exploitation of the other exchange partner(s) arises (Anderson and Narus 1990). Whereas communication and fairness are crucial for the culmination of trust in the early phases of relationship development (Ferguson et al. 2005), relational trust, largely, stems from the quality of experience or interaction among the exchange partners (Ring and Van de Ven 1992). More specifically, relational trust results from mutually beneficent successive collaboration cycles among the exchanging parties. By transacting repeatedly, partners become familiar with one another and develop social attachments (Gulati 1995; Ring and Van De Van 1994) which foster stronger forms of trust (e.g. process-based trust and familiarity-based trust) as a consequence of opportunities to share information and learn about each partner's proclivities toward trustworthy behavior(s) (Gulati 1995, 1998).

Trust has always been regarded as a multifaceted construct that has been viewed differently from different theoretical perspectives. According to Dicky et al. (2007), trust has generally been defined in one of the two possible ways; (1) as a confident belief or expectation (i.e. a trusting belief), and/or (2) as a willingness or intention to depend on the trustee (i.e. a trusting intention). Trusting belief refers to the perception that the other party (trustee) will act in ways favorable to the trusting party (Boone and Holmes 1991), or that the trustee has ethical, efficacious or favorable characteristics (Hagen and Choe 1998). Some of these beliefs, as revealed in the literature, include: continuity of natural order, competence and fiduciary Barber (1983); dependability (Kumar 1996); ability, benevolence and integrity

(Mayer et al. 1995); competence, judgment and openness (Mishra 1996); reliability and predictability (Rempel et al. 1985). By contrast, trusting intention refers to a willingness to become vulnerable or dependent on the trustee (Baier 1986; Currall and Judge 1995) based on the expectation that it will not exploit this situation (Mayer et al. 1995). Keeping in view the nature and dynamics of the relationships investigated in this research, trust has been conceptualized as a bi-dimensional construct with its two facets being the competence-trust and the integrity trust. This view of conceptualizing and/or operationalizing the inter-firm trust is quite consistent with Barber (1983), Mayer et al. (1995) and Mishra (1996).

2.2.3 Commitment

Defined as an attitude that reflects the desire to continue a valued relationship (Moorman et al. 1992) and a willingness to make short-term sacrifices to maintain that relationship (Anderson and Weitz 1992), commitment has been examined quite extensively in consumer contexts (Verhoef et al. 2002), work-place contexts (Allen and Meyer 1990; Luthans 2006) and business-to business contexts (Gruen et al. 2000; Morgan and Hunt 1994). Extending Allen and Meyer's (1990) view of workplace commitment to the (business) exchange relationship context, we define commitment as a predisposition which comprises of an exchange partner's willingness to (1) stay long in the relationship, (2) accept the norms and values that govern the relationship, and (3) contribute maximally for the welfare of the exchange partners.

Whereas organizational researchers like Garbarino and Johnson (1999) and Morgan and Hunt (1994) view commitment as a unidimensional construct, a vast majority of researchers have treated it as a multidimensional construct in a variety of business contexts (Allen and Meyer 1990; Geyskens et al. 1996; Gundlach et al. 1995). If Geyskens et al. (1996) differentiated between affective commitment and calculative commitment, Allen and Meyer (1990), on the other hand, revealed three dimensions of commitment which included: continuance commitment (cost-based attachment), affective commitment (desire-based attachment) and normative commitment (obligation-based attachment). In this research, Allen and Meyer's (1990) multidimensional view of workplace commitment has been extended to the supplier-intermediate buyer dyadic relationship context.

3 Research Methodology

3.1 Data Collection

The sampled population designated for this study was all the private medical clinics operating in two cities (i.e. Bahawalpur and Rahimyar Khan) of Bahawalpur division of the Punjab province of Pakistan. A convenient and efficient access to the relevant resources (information, personnel etc.), as the researcher is a native, have been the major consideration behind choosing this area for the purpose of collecting the primary data for this research. In Pakistan, usually the medical clinics are identified, by the suppliers (the pharmaceutical companies) and the ultimate customers (the patients), with the owning/managing doctors of these clinics. Therefore, the same (i.e. managing doctors) were selected as the informants and/or the unit of analysis.

As the vast majority (71 %) of the total number of doctors were associated with the central hospitals (Bahawal Victoria Hospital Bahawalpur and Sheikh Zayed Hospital Rahimyar Khan) and many of them resided in the neighborhoods known as "medical colonies" of these two cities, therefore, area sampling was used as the technique to select respondents. The sampling frames were obtained from the Medical Superintendents (M.S.) of these hospitals. As the total number of elements in the sampling frames was small enough to allow an inclusion of all of them in the sample, therefore, the same was done which resulted in a sample size of 1098 doctors. However, only 803 of them could be reached at their wards and/or private clinics. 295 doctors could not be reached because they were on leave, their addresses were incorrect or they simply were not willing to cooperate.

As mail and/or telephone surveys are not customary data collection techniques in this area, therefore, students from the local business education institutions were engaged to conduct a personal investigation in both cities. Five teams each consisting of four under-graduate students coordinated by a course instructor from the local business education institution were set up. The questionnaires were personally delivered to 803 doctors at their wards and/or private clinics in the medical colonies of the two cities. In order to ensure efficiency and control against field workers' cheating, the questionnaires were collected back on daily basis after two weeks of the date of their disbursement. Follow-up calls were randomly made to the doctors to confirm authenticity of their response. The process lasted for about 3 weeks and resulted in the return of 362 filled-in questionnaires yielding a 45 % response rate. After performing necessary data-checks, 284 (usable) questionnaires were retained for further analysis.

3.2 *Measurement Scales*

A structured non-disguised questionnaire containing adapted versions of the measurement scales already established/used in the previous research was used for collecting the primary data. Guided by the previous research, the researcher explicitly (a priori) defined the measurement models i.e. it has been pre-specified which item(s) would load on which specific dimensions of which latent constructs. In order to ensure content/face validity, the (adapted) measurement scales were subjected to a review from a panel of five experts, two from pharmaceutical companies and three doctors (the intermediate buyers/informants).

In order to measure the six relational norms, the scales used by Fink et al. (2007) and Ivens (2006) were adapted according to the context of this study. The responses on all the individual items measuring each specific relational norm were recorded on a 5-point (Strongly disagree .... Strongly agree) Likert scale format where lower numbers represented varying levels of disagreement and vice versa. The disagreement was equated with low levels of the exhibition of that dimension of relationality in the exchange environment and vice versa. In order to measure the economic and the social satisfaction, the scales used by Ivens (2006) were adapted according to the context of study. The responses on the eight (8) items measuring these two types of satisfaction were recorded on a 7-point (Strongly disagree/ Strongly agree) Likert scale format where the lower numbers reflected varying levels of dissatisfaction and vice versa. In order to operationalize the two facets of trust, the scales used by Bansal et al. (2004), Dicky et al. (2007), Hess and Story (2005) and Voss et al. (2006) were adapted according to the context of this research. The competence-trust and integrity-trust were measured through five (5) and six (6) items respectively. The responses on all the eleven (11) items were recorded on a 5-point (Strongly disagree/Strongly agree) Likert scale format where lower numbers represented varying levels of disagreement and vice versa. A disagreement was equated with a lack of trust in the focal supplier and vice versa. Finally, each of the three sub-types of commitment were measured on six (6) items while adapting the measurement scales used by Allen and Meyer (1990), Bansal et al. (2004), Bagraim and Sader (2007), Blömer and Odekerken-Schröder (2006), Suliman and Iles (2000) and Voss et al. (2006). The responses on all the eighteen (18) items were recorded on a 5-point (Strongly disagree/strongly agree) Likert scale format where lower numbers represented varying levels of disagreement and vice versa. The disagreement was equated with a lack of intermediate-buyers' commitment with their respective focal supplier(s) and vice versa.

3.3 Data Analysis

Since we aggregated the items/indicators of various dimensions of latent constructs used/reported in some of the previous studies, a principal-component analysis (PCA) was carried out to purify the measurement scales and ascertain uni-dimensionality of the first-order factors/dimension of the higher-order latent constructs following the procedure suggested by Field (2009) and using SPSS as the software application. Later, a confirmatory factor analysis (CFA) was conducted in order to ascertain the measurement quality (validity and reliability) of the subject constructs. Finally, in order to gauge the nature of association among the subject constructs, a path-modeling was performed using SmartPLS Version 2.0M3 (Ringle et al. 2006).

4 Results

4.1 Sample Profile

Majority of the respondents were males (54 %), aged between 30 and 40 years (39 %), and had Medicinae Baccalaureus, Baccalaureus Chirurgiae (MBBS) i.e. Bachelor of Medicine (53 %) as their highest level of education. Majority of the clinics (64 %) had been established within last 10 years. The clinics on the average employed four workers. The average (weekly) number of visitors in these clinics was 179.

4.2 Analysis of the Measurement Model

The internal consistency reliability at the construct level was assessed on the basis of composite reliability (CR) measure developed by Werts et al. (1974), using the 0.8 threshold suggested by Nunnally and Bernstein (1994). At the indicator level, the reliability of (significant) individual items was judged on the basis of strength of the outer-loadings, meaning at least 0.6 and ideally 0.7 (Chin 1998). The indicators falling below the acceptable threshold were removed from their respective scales. All the constructs were eventually found to be reliable both at the construct level ($CR \geq 0.8$) and the indicators' level ($b \geq 0.6$, and $t \geq 1.96$). We used the average variance extracted (AVE) as a criterion to assess the convergent validity as suggested by Fornell and Larcker (1981). According to Götz et al. (2009) a value of at least 0.5 of AVE indicates sufficient convergent validity as it reveals that the latent variable explains (on the average) more than half of the variance of its indicators. All the constructs featured acceptable (in two cases, nearly acceptable) convergent validity ($AVE \geq 0.50$). They also exhibited sufficient discriminant validity with respect to the Fornel-Larker criterion (See Table 2), that holds that the square-root of AVE of the construct must be significantly greater than its correlation with other constructs, and through the examination of the cross-loading (i.e. the indicators' loadings were highest on the relevant construct viz-a-viz the cross loadings).

4.3 The Path Model Estimates (*βs*)

Table 3 shows the relevant statistics showing the nature of association among the subject constructs. As can be seen in Table 3, solidarity ($\beta = 0.15$, $p < 0.05$) and conflict resolution ($\beta = 0.16$, $p < 0.05$) were found to be significantly affecting the economic satisfaction whereas the social satisfaction has been found to be significantly affected by flexibility ($\beta = 0.19$, $p < 0.01$), role integrity ($\beta = 0.16$, $p < 0.05$) and conflict resolution ($\beta = 0.23$, $p < 0.001$). Conflict resolution has

Table 2 Discriminant validity

	The Fornell–Larker's Criterion (√AVE > R)												
	MU	**SL**	**FL**	**RI**	**CR**	**LT**	**ES**	**SS**	**CT**	**IT**	**CC**	**AC**	**NC**
MU	***0.73***												
SL	0.36	***0.81***											
FL	0.32	0.32	***0.69***										
RI	0.29	0.18	0.36	***0.73***									
CR	0.34	0.35	0.38	0.27	***0.79***								
LT	0.13	0.22	0.33	0.20	0.29	***0.80***							
ES	0.25	0.29	0.22	0.26	0.29	0.17	***0.77***						
SS	0.26	0.31	0.39	0.30	0.40	0.22	0.62	***0.75***					
CT	0.30	0.41	0.28	0.28	0.36	0.19	0.50	0.55	***0.72***				
IT	0.33	0.33	0.36	0.27	0.37	0.20	0.44	0.52	0.67	***0.71***			
CC	0.37	0.37	0.40	0.45	0.41	0.25	0.30	0.35	0.44	0.37	***0.70***		
AC	0.28	0.30	0.26	0.36	0.36	0.23	0.24	0.22	0.33	0.40	0.58	***0.69***	
NC	0.27	0.32	0.14	0.36	0.30	0.11	0.29	0.30	0.35	0.38	0.55	0.59	***0.72***

The diagonal values are √AVE, the rest are R

been the only norm that significantly affected both types of satisfaction with its effect being stronger than any other norm on both types of satisfaction. Mutuality and long term orientation showed no significant effect on any of the two facets of satisfaction. The overall extent of relationality in the exchange environment showed a superior influence on social satisfaction ($R^2 = 0.27$) compared to the economic satisfaction ($R^2 = 0.16$).

The influence of relational norms on the two facets/forms of trust has been almost the same, however, the set of the norms which affected both has been different. Whereas the competence-trust was found to be significantly affected by solidarity ($\beta = 0.25$, $p < 0.001$) and the conflict resolution ($\beta = 0.19$, $p < 0.01$), the integrity-trust, on the contrary, was found to be significantly affected by mutuality ($\beta = 0.15$, $p < 0.01$), solidarity ($\beta = 0.13$, $p < 0.05$), flexibility ($\beta = 0.15$, $p < 0.05$) and the collaborative conflict resolution ($\beta = 0.18$, $p < 0.01$). Whereas solidarity turned out to be the norm most strongly influencing competence-trust, it has been the collaborative conflict resolution that was found to be most influential on the integrity-trust.

Out of the three facets of commitment, continuance commitment was found to be more strongly influenced ($R^2 = 0.35$) by the perceived relationality. All relational norms (except long-term orientation) significantly affected i.e. mutuality ($\beta = 0.14$, $p < 0.05$), solidarity ($\beta = 0.11$, $p < 0.05$), flexibility ($\beta = 0.13$, $p < 0.05$), role integrity ($\beta = 0.27$, $p < 0.001$) and the conflict resolution ($\beta = 0.18$, $p < 0.01$) this form of commitment. Affective commitment was found to be significantly affected by role integrity ($\beta = 0.22$, $p < 0.001$) and conflict resolution ($\beta = 0.20$, $p < 0.01$) only. The normative commitment exhibited the least influence of the relational norms ($R^2 = 0.19$) with mutuality ($\beta = 0.14$, $p < 0.05$), role integrity ($\beta = 0.30$, $p < 0.001$) and conflict resolution ($\beta = 0.21$, $p < 0.01$) found to be significantly affecting this type of commitment. All three

Table 3 Path coefficients (β)

	Satisfaction		Trust		Commitment		
Relational norms	Econoimc	Social	Competence	Integrity	Continuance	Affective	Normative
Mutuality	n.s.	n.s.	n.s.	**0.151****	**0.140***	n.s.	**0.141***
Solidarity	**0.147***	n.s.	**0.254*****	**0.127***	**0.112***	n.s.	n.s.
Flexibility	n.s.	**0.193****	n.s.	**0.149***	**0.131***	n.s.	n.s.
Role integrity	n.s.	**0.159***	n.s.	n.s.	**0.269*****	**0.223*****	**0.299*****
Conflict resolution	**0.156***	**0.228*****	**0.186****	**0.184****	**0.181****	**0.199****	**0.205****
Long term orientation	n.s.	n.s.	n.s.	n.s.	n.s.	n.s.	n.s.
$\mathbf{R^2}$	0.16	0.27	0.25	0.24	0.35	0.23	0.19

*Significant at 5 % (two-tailed)
**Significant at 1 % (two-tailed)
***Significant at 0.01 % (two-tailed)
n.s. Non-significant

forms of commitment were more strongly influence by role integrity than any other relational norm. Finally, whereas the collaborative conflict resolution exhibited a significant positive influence on all outcome constructs, the long-term orientation (LTO) did not feature a significant relationship with any of them.

5 Discussion

Like most of the business relationships, perceptions about the realization of desired benefits affect an (intermediate) buyer's decision to enter, remain or make an exit from its relationships with a focal supplier. The existence of asymmetry in economic and/or relational power and/or the absence of distributive justice adversely affect the intermediate buyers' perceptions about the realization of (individual and/ or collective) benefits desired from the exchange relationship. When a focal supplier exhibits solidarity with an intermediate buyer (especially when it is feeling vulnerable) and accommodates it in the face of intensified (crushing) conflicts through a benevolent (collaborative) conflict resolution, it adds to the economic satisfaction of the intermediate buyer by creating an assurance about the protection of its (economic) interests. Similarly, an adequate compliance with the relational norms like flexibility, role integrity and collaborative conflict resolution positively affects the social satisfaction by signaling (to the intermediate buyer) the presence of a *sense of comradeship* in the focal supplier.

An adherence to the relational norms like mutuality, solidarity, flexibility and collaborative conflict resolution promote an integrity-based trust in the focal supplier. Mutuality breeds trust through increased fiduciary (alternatively, equity and distributive justice) as it prevents the focal supplier from optimizing its gains at the cost of the intermediate buyer. Similarly, a display of solidarity from a focal supplier, especially when an intermediate is highly vulnerable to an opportunistic exploitation, culminates trust by strengthening its integrity in the eyes of the intermediate buyer. Similarly, an exhibition of readiness to adapt the relationship to the changed circumstances even if it may lead to the re-writing of the charter of engagement together with the display of flexibility in accommodating an intermediate buyer in the face of conflicts that may have arisen from time to time signals an absence of the predisposition to an opportunistic exploitation and, therefore, strengthen the integrity-based trust in the focal supplier. On the contrary, an adherence to the relational norms like solidarity and collaborative conflict resolution foster a competence-based trust in the focal supplier.

When a focal supplier manages to create a highly relational environment by exhibiting relational behaviors like mutuality, solidarity, flexibility, role integrity and collaborative conflict resolution, it strengthens an intermediate buyer's belief that the relationship is efficacious enough to successfully achieve the desired goals, will prevail in the face of any crisis and would eventually result in the materialization of individual and collective benefits both in the short as well as the long-run. As such, these norms promote continuance commitment by proscribing (detrimental) unilateral behaviors like opportunism, free-riding, distributive injustice etc. which assures the intermediate buyers that they will equitably share the co-created value. In other words, these relational norms foster a higher continuance commitment by making it imperative to the intermediate buyer to continue to hang-in there as it is in its best interest to do so. The exhibition of role integrity, as it urges a supplier to adequately dispel its roles, rights and obligations in consonance with its charter of engagement with an intermediate buyer, and the collaborative conflict resolution, as it necessitates the focal actor to exercise flexibility and benevolence while addressing the (inevitable) conflicts, lead to the creation of a positive state-of-affect about continuing an exchange relationship (the affective commitment). Finally, a display of mutuality, role integrity and the collaborative conflict resolution from the focal supplier foster (in the intermediate buyers) the feelings, urge and/or need for a reciprocation to the benevolence of the focal supplier. As such, when (eventually) gotten internalized by the intermediate buyers with the passage of time, these relational norms strengthen the reciprocity-based (normative) commitment. It may be interesting to note that more than the affective and/or the normative commitment, the exhibition of high rationality from a focal supplier culminates the continuance type of commitment which means that more than an affect or reciprocity it is, in fact, the rationalization of being benefitted that leads to the longevity of relationships. In other words, the intermediate buyers consider those suppliers to be more beneficent who conduct their exchanges in a relationship preserving manner. Anyhow, irrespective of its type(s)/origin, the increase in satisfaction, trust and/or commitment (together, the relationship quality) perceived by the intermediate buyers eventually enhances the focal suppliers' performance.

It may also be interesting to note that it does not matter much (if not at all) to the intermediate buyers if or not if the focal firm itself exhibits a long-term orientation. They always want their focal suppliers to go an extra-mile while acting in a relationship-preserving manner for whatever time- no matter how short or long it is- they stay in that exchange relationship. The more relationship-preserving the focal suppliers seem, the more beneficial they are perceived and consequently the longer they are stayed with. Though owing to the norms of reciprocity, on can expect that relationality begets relationality, there might be some instances where it does not hold true. In other words, (though it may not be a wide spread phenomenon but) certain exchange contexts may characterize a *relationality-paradox*. In certain instances, actors get into exchange relationships without considerable ex-ante optimism and/or commitment (Yaqub 2009; Yaqub et al. 2009). Consequently, they perceive and treat the cooperative relationships as short-lived endeavors. They exhibit a short-term orientation because they either expect lower (individual or collective) gains over the life time of their association with the focal actor(s) or they expect the relationships to dissolve due to their own or other partners' cheating. For them, (certain and tangible) short-run benefits outweigh the (intangible and uncertain) long-run benefits. In this situation, they not only try to (ex-ante) negotiate for undue terms but also feel to have an incentive to (ex-post) exploit the focal actor(s) as quickly as possible and exit the relationship (Yaqub 2009). Nevertheless, it is always unpleasant for the focal actors to display a high relational-orientation to the actors who themselves are transaction-oriented. We would call it the *unilateral-relationalism*. Here, the focal actors' going of an extra-mile is only regarded as an incremental benefit (akin a bonus) and as such it fails to inspire any positive state-of-affect in the transaction-oriented exchange partner. In Das and Teng's (2000) opinion, this trade-off between short-term and long-term gains is a natural point of tension for many business relationships. Kronman (1985) noted that the focal actors are usually unwilling to make the short-term sacrifices necessary to preserve a relationship if they do not expect it to be profitable over the long-run.

6 Conclusion

The success of supplier–buyer relationships depends, to a substantial extent, upon quality of the ecosystem in which transactions take place. An ideal ecosystem provides higher levels of social, economic and political egalitarianism. In such an environment, the exchange partners equitably benefit from their efforts for arriving at win-win solutions for their economic and social problems, and ultimately end-up in attaining a state which leaves everyone at least as well-off (in social, economic, and political sense) as they were before becoming a part of that exchange relationship. The relational norms such as mutuality, solidarity, flexibility, role integrity and (collaborative) conflict resolution significantly affect the (perceived) quality of supplier–buyer relationships (manifested through satisfaction, trust and commitment) which eventually results in the superior performance of these relationships which implicates the focal actors to ensure an adequate adherence to the relationship-preserving norms. While management, in the focal firms, can put in place directives and incentives to develop and/or promote these relational norms, they mainly evolve over time as a consequence of the exchange partners' transacting experiences.

The maintenance of relational norms requires substantial up-front investments of time, money and personnel from the focal supplier. As relational norms become effective when these are perceived by the exchange partners to be increasingly relevant and instrumental for the attainment of their individual as well as collective goals, therefore it is better to focus only on the most relevant norms. Therefore, rather than following a "more-is-better" approach, it is advisable to follow the standard economic logic for achieving adequate levels of "relationality" as the benefits from the relational behaviors accrue at diminishing rate while the cost of ensuring such behaviors accrues at increasing rates. In the face of growing/escalated financial crunch, there is an ever-increasing pressure on the managers to make it sure that their RM efforts are precisely targeted. A fine-grained understanding of the (relational) dynamics of business exchanges could profoundly facilitate the managers in achieving the precision in their relational efforts so as to increase their efficiency and/or minimize the wastage of such efforts.

7 Limitations and Future Research

The association among relational norms and various facets of the three relational outcomes i.e. satisfaction, trust and commitment were studied only in the context of supplier-intermediate buyer dyadic relationships. The associations within the larger networks of relationships were not addressed. Therefore, the generalizeability of the results may be limited only to the downstream relationships. Similarly, some of the inferences drawn from the data seem to be highly context-specific when seen relative to the associations propounded by the relational view of networks.

Relational norms perspective suggests that the strength of relational norms prevalent in an exchange environment affects the level of cooperative behavior and relationship performance (Cannon et al. 2000). Research in RET has debated a lot on the efficacy of relational norms as antecedents and trust and/or commitment (or relationship quality) as the mediators for successful inter-firm relationships (Mcneil 1980; Kaufman and Stern 1988; Morgan and Hunt 1994). However, Palmatier et al. (2007) have espoused that relational norms may be a necessary but not a sufficient condition for superior performance of business exchanges. In other words, violating norms results in underperformance, but following norms does not necessarily guarantee a superior performance. Consequently, the relational norms need to be augmented with other drivers of relationship performance. Future research may endeavor to identify such drivers and study their joint effect on the intermediate and/or ultimate relational outcomes. Lastly, though our framework has, primarily, been intended at explaining relational dynamics of the supplier-intermediate buyer dyadic relationships, yet with little adaptation it can be generalized not only across upstream partnerships but also across other (more) elaborate forms of strategic structural arrangements like virtual organizations, strategic alliances etc. Future research may investigate the generalizeability and applications of this model in these structural contexts.

References

Achrol R (1991) Evolution of the marketing organization: new forms for turbulent environments. J Mark 55(4):77–93

Allen NJ, Meyer JP (1990) The measurement and antecedents of affective, continuance and normative commitment to the organization. J Occup Psychol 63(1):1–18

Anderson E, Jap SD (2005) The dark side of close relationships. MIT Sloan Manage Rev 46(3): 75–82

Anderson E, Weitz BA (1989) Determinants of continuity in conventional industrial channel dyads. Mark Sci 8(4):310–323

Anderson E, Weitz BA (1992) The use of pledges to build and sustain commitment in distribution channels. J Mark Res 29:18–34

Anderson JC, Narus JA (1990) A model of distributor firm and manufacturer firm working partnerships. J Mark 54(1):42–58

Artz KW, Brush TH (2000) Asset specificity, uncertainty and relational norms: an examination of coordination costs in collaborative strategic alliances. J Econ Behav Organ 41(4):337–362
Athanasopoulou P (2009) Relationship quality: a critical literature review and research agenda. Eur J Mark 43(5/6):583–610
Aulakh PS, Kotabe M, Sahay A (1996) Trust and performance in cross-border marketing partnerships: a behavioral approach. J Int Bus Stud 27(5):1005–1032
Baier A (1986) Trust and anti-trust. Ethics 96(2):231–260
Bagraim J, Sader R (2007) Family-friendly human resource practices and organizational commitment. Manage Dyn 16(4):2–11
Baker TL, Simpson PM, Siguaw JA (1999) The impact of suppliers' perceptions of reseller market orientation on key relationship constructs. J Acad Mark Sci 27(1):50–57
Bansal HS, Irving PG, Taylor SF (2004) A three-component model of customer commitment to service providers. J Acad Mark Sci 32(3):234–250
Barber B (1983) The logic and limits of trust. Rutgers University Press, New Brunswick
Bejou D, Wray B, Ingram TN (1996) Determinants of relationship quality: an artificial neural network analysis. J Bus Res 36(2):137–143
Bercovitz J, Jap SD, Nickerson JA (2006) Antecedents and implications of cooperative exchange norms. Organ Sci 17(6):724–740
Blios KJ, Ivens BS (2006) Measuring relational norms: some methodological issues. Eur J Mark 40(3/4):352–365
Bloemer J, Odekerken-Schröder G (2006) The role of employee relationship proneness in creating employee loyalty. Int J Bank Market 24(4):252–264
Boone SD, Holmes JG (1991) The dynamics of interpersonal trust: resolving uncertainty in the face of risk. In: Hinde RA, Groebel J (eds) Cooperation and prosocial behavior. Cambridge University Press, Cambridge, pp 190–211
Cannon JP, Achrol RA, Gundlach GT (2000) Contracts, norms, and plural form governance. J Acad Mark Sci 28(2):180–194
Chin WW (1998) The partial least squares approach for structural equation modeling. In: Marcoulides GA (ed) Modern methods for business research. Lawrence Erlbaum Associates, Mahwah, pp 295–336
Cochet O, Dormann J, Ehrmann T (2008) Capitalizing on franchisee autonomy: relational forms of governance as controls in idiosyncratic franchise dyads. J Small Bus Manage 46(1):50–72
Colgate M, Danaher P (2000) Implementing a customer relationship strategy: the asymmetric impact of poor versus excellent execution. J Acad Mark Sci 28(3):373–385
Crosby LA, Evans KR, Cowels D (1990) Relationship quality in service selling: an interpersonal influence perspective. J Mark 54(3):68–81
Curral SC, Judge TA (1995) Measuring trust between organizational boundary role persons. Organ Behav Hum Decis Process 64(2):151–170
Dant RP, Schul PL (1992) Conflict resolution processes in contractual channels of distribution. J Mark 56(1):38–54
Das TK, Teng BS (2000) Instabilities of strategic alliances: an internal tensions perspective. Organ Sci 11(1):77–101
De Wulf K, Odekerken-Schröder G, Iacobucci D (2001) Investments in consumer relationships: a cross-country and cross-industry exploration. J Mark 65(4):33–50
Dicky MH, McKnight DH, George JF (2007) The role of trust in franchise organizations. Int J Organ Anal 15(3):251–282
Ferguson RF, Paulin M, Möslein K, Müller C (2005) Relational governance, communication and the performance of biotechnology partnerships. J Small Bus Enterp Dev 12(3):395–409
Field AP (2009) Discovering statists using SPSS (and sex and drugs and rock 'n' roll), 3rd edn. Sage, London
Fink RC, Edelman LF, Hatten KF (2007) Supplier performance improvements in relational exchanges. J Bus Ind Mark 22(1):29–40
Finn A (2005) Reassessing the foundations of customer delights. J Serv Res 8(2):103–116

Fornell C, Larcker DF (1981) Structural equation models with unobservable variables and measurement error: algebra and statistics. J Mark Res 18(3):328–388
Frazier GL, Spekman RE, O'Neil CR (1988) Just-in-time exchange relationships in industrial markets. J Mark 52(4):52–67
Ganesan S (1994) Determinants of long-term orientation in buyer–seller relationships. J Mark 58 (2):1–19
Garbarino E, Johnson MS (1999) The different roles of satisfaction, trust, and commitment in customer relationships. J Mark 63(2):70–87
Geyskens I, Steenkamp JEM (2000) Economic and social satisfaction: measurement and relevance to marketing channel relationships. J Retailing 76(1):11–32
Geyskens I, Steenkamp JEM, Kumar N (1999) A meta analysis of satisfaction in marketing channel relationships. J Mark Res 36(2):223–238
Geyskens I, Steenkamp JEM, Scheer LK, Kumar N (1996) The effects of trust and interdependence on relationship commitment: a trans-Atlantic study. Int J Res Mark 13(4):303–315
Götz O, Liehr-Gobbers K, Krafft M (2009) Evaluation of structural equation models using the partial least squares (PLS) approach. In: Vinzi VE, Chin WW, Heneseler J, Wang H (eds) Handbook of partial least squares: concepts, methods and applications. Springer, Berlin
Granovetter M (1992) Problems of explanation in economic sociology. In: Nohria N, Eccles RG (eds) Networks and organizations: structure, forms, and action. Harvard University Press, Boston, pp 25–56
Gruen TW, Summers JO, Acito F (2000) Relationship marketing activities, commitment, and membership behaviors in professional associations. J Mark 64(3):34–49
Gulati R (1995) Does familiarity breed trust? The implications of repeated ties for contractual choice in alliances. Acad Manage J 38(1):85–112
Gulati R (1998) Alliances and networks. Strateg Manage J 19(4):293–317
Gulati R, Nohria N, Zaheer A (2000) Strategic networks. Strateg Manage J 21(3):203–215
Gundlach GT, Achrol RS, Mentzer JT (1995) The structure of commitment in exchange. J Mark 59 (1):78–92
Hagen JM, Choe S (1998) Trust in Japanese interfirm relations: institutional sanctions matter. Acad Manage Rev 23(3):589–600
Heide J, John G (1992) Does norms matter in marketing relationships? J Mark 56(2):32–44
Henning-Thurau T, Klee A (1997) The impact of customer satisfaction and relationship quality on customer retention: a critical reassessment and model development. Psychol Mark 14(8): 737–764
Henning-Thurau T, Gwinner KP, Gremler DD (2002) Understanding relationship marketing outcomes: an integration of relationship benefits and relationship quality. J Serv Res 4(3): 230–247
Hess J, Story J (2005) Trust-based commitment: multidimensional consumer-brand relationships. J Consum Mark 22(6):313–322
Hibbard JD, Brunel FF, Dant RP, Iacobucci D (2001) Does relationship marketing age well? Bus Strategy Rev 12(4):29–35
Huang HH, Chiu CK (2006) Exploring customer satisfaction, trust and destination loyalty in tourism. J Am Acad Bus 10(1):156–159
Ivens BS (2004) How relevant are different forms of relational behavior? An empirical test based on Mcneil's exchange framework. J Bus Ind Mark 19(5):300–309
Ivens BS (2006) Norm-based relational behaviors: is there an underlying dimensional structure. J Bus Ind Mark 21(2):9–105
Jap SD, Ganesan S (2000) Control mechanism and the relationship life cycle: implications for safeguarding specific investments and developing commitment. J Mark Res 37(2):227–245
Jap SD, Manolis C, Weitz BA (1999) Relationship quality and buyer-seller interactions in channels of distribution. J Bus Res 46(3):303–313
Joshi AW, Stump RL (1999) The contingent effect of specific assets investment on joint action in manufacturer-supplier relationships. J Acad Mark Sci 27(3):291–314

Kaufman PJ (1987) Commercial exchange relationships and the negotiator's dilemma. Negot J 3(1):73–80
Kaufman PJ, Stern LW (1988) Relational exchange norms, perceptions of unfairness, and retained hostility in commercial litigation. J Conflict Resolut 32(3):534–552
Kronman A (1985) Contract law and the state of nature. J Law Econ Organ 1(1):5–32
Kumar N (1996) The power of trust in manufacturer-retailer relationships. Harv Bus Rev 74(6): 92–106
Lagace RR, Dahlstrom R, Gassenheimer JB (1991) The relevance of ethical salesperson behavior on relationship quality: the pharmaceutical industry. J Pers Selling Sales Manage 11(4):39–47
Larson A (1992) Network dyads in entrepreneurial settings: a study of the governance of exchange relationships. Adm Sci Q 37(1):76–104
Leuthesser L (1997) Supplier relational behavior: an empirical assessment. Ind Mark Manage 26(3):245–254
Lin CP, Ding CG (2006) Evaluating group differences in gender during the formation of relationship quality and loyalty in ISP service. J Organ End User Comput 18(2):38–62
Luthans F (2006) Organizational behavior. McGraw-Hill, New York
Mayer R, Davis J, Schoorman F (1995) An integrative model of organizational trust. Acad Manage Rev 20(3):709–734
Mcneil IR (1978) Contracts: adjustment of long-term economic relations under classical, neo-classical, and relational contract law. Northwest Univ Law Rev 72(6):854–905
Mcneil IR (1980) The new social contract. Yale University Press, CT
Mcneil IR (1983) Values in contract. Northwest Univ Law Rev 75(6):1018–1063
Mishra AK (1996) Organizational responses to crisis: the centrality of trust. In: Kramer RM, Taylor TR (eds) Trust in organizations: frontiers of theory and research. Sage, Thousand Oaks, pp 261–287
Moorman C, Zaltman G, Deshpande R (1992) Relationships between providers and users of market research: dynamics of trust within and between organizations. J Mark Res 29(8): 314–328
Morgan RM, Hunt SD (1994) The commitment-trust theory of relationship marketing. J Mark 58 (3):20–38
Noordewier TG, John G, Nevin JR (1990) Performance outcomes of purchasing arrangements in industrial buyer-vendor relationships. J Mark 54(4):80–93
Nunnally JC, Bernstein IH (1994) Psychological theory, 3rd edn. McGraw-Hill, New York
Palmatier RW, Dant RP, Grewal D (2007) A comparative longitudinal analysis of theoretical perspectives of interorganizational relationship performance. J Mark 71(4):172–194
Palmatier RW, Dant RP, Grewal D, Evans KR (2006) Factors influencing the effectiveness of relationship marketing: a meta-analysis. J Mark 70(4):136–153
Paulin M, Ferguson RJ, Salazar AMA (1999) External effectiveness of service management: a study of business-to-business relationships in Mexico, Canada and the USA. Int J Serv Ind Manage 10(5):409–429
Rajaobelina L, Bergeron J (2009) Antecedents and consequences of buyer-seller relationship quality in the financial services industry. Int J Bank Mark 27(5):359–380
Rempel JK, Holmes JG, Zanna MP (1985) Trust in close relationships. J Pers Soc Psychol 49(1): 95–112
Reynolds KE, Betty SE (1999) Customer benefits and company consequences of customer-salesperson relationships in retailing. J Retailing 75(1):11–31
Ring PS, Van De Ven AH (1992) Structuring cooperative relationships between organizations. Strateg Manage J 13(7):483–498
Ring PS, Van De Van AH (1994) Developmental processes of cooperative interorganizational relationships. Acad Manage J 19(1):90–118
Ringle CM, Wende S, Will A (2006) SmartPLS 2.0M3. Available at http://www.smartpls.de

Roehrich G, Spencer R, Valette-Florence P (2002) The nature of relationship atmosphere and links with the value of relationships: the case of Europe and Asia. In: Proceedings from the 18th IMP conference in Perth, Australia
Ryu S, Park JE, Min S (2007) Factors of determining long-term orientation in interfirm relationships. J Bus Res 60(12):1225–1233
Selnes F (1998) Antecedents and consequences of trust and satisfaction in buyer-seller relationships. Eur J Mark 32(3/4):305–322
Sirdeshmukh D, Singh J, Sabol B (2002) Consumer trust, value, and loyalty in relational exchanges. J Mark 66(1):15–37
Storbacka K, Strandvik T, Gro¨nroos C (1994) Managing customer relationships for profit: the dynamics of relationship quality. Int J Serv Ind Manage 5(5):21–38
Suliman A, Iles P (2000) Is continuance commitment beneficial to organizations? Commitment-performance relationship: a new look. J Manag Psychol 15(5):407–426
Sun H (2010) Transferring attributes of e-commerce systems into business benefits: a relationship quality perspective. J Electron Commer Res 11(2):92–109
Sutton RI, Staw BM (1995) What theory is not. Adm Sci Q 40(3):371–384
Ulaga W, Eggert A (2006) Relationship value and relationship quality: broadening the nomological network of business-to-business relationships. Eur J Mark 40(3/4):311–327
Ural T (2007) The antecedents and consequences of relationship quality according to stages of the relationship between exporters and importers. Probl Perspect Manage 5(3):111–138
Vázquez R, Iglesias V, Rodríguez-del-Bosque I (2007) The efficacy of alternative mechanisms in safeguarding specific investments from opportunism. J Bus Ind Mark 22(7):498–507
Verhoef PC, Franses PH, Hoekstra JC (2002) The effect of relational constructs on customer referrals and number of services purchased from a multiservice provider: does age of relationship matter? J Acad Mark Sci 30(3):202–216
Voss KE, Johnson JL, Culln JB, Sakano T, Takenouchi H (2006) Relational exchange in US-Japanese marketing strategic alliances. Int Mark Rev 23(6):610–635
Walter A, Muller T, Helfert G, Ritter T (2003) Functions of industrial supplier relationships and their impact on relationship quality. Ind Mark Manage 32(2):159–169
Werts CE, Linn RL, Jöreskog KG (1974) Intra-class reliability estimates: testing structural assumptions. Educ Psychol Meas 34(1):25–33
Wong A, Sohal A (2002) An examination of the relationship between trust, commitment and relationship quality. Int J Retail Distrib Manage 30(1):34–50
Yaqub MZ, Hussain D, Yaqub RMS (2009) Making strategic networks pluralistic neighbourhoods. Int J Knowl Cult Change Manage 9(2):113–128
Yaqub MZ (2009) The antecedents, consequences, and control of opportunistic behavior in strategic networks. J Bus Econ Res 7(2):15–31
Yaqub MZ, Vetschera R (2011) The efficacy of relational governance and value-creating relational investments in revenue-enhancement in downstream networks. In: Hendrikse G, Tunnanen M, Windsperger J, Cliquet G (eds) New developments in the theory of networks: franchising, alliances and cooperatives. Springer, Berlin
Yaqub MZ, Malik MA, Shah H (2010) The roles of satisfaction, trust and commitment in economic value-creation in strategic networks. Eur J Econ Financ Adm Sci 18:133–145
Zhang C, Cavusgil ST, Roath A (2003) Manufacturer governance of foreign distributor relationships: do relational norms enhance competitiveness in the export market? J Int Bus Stud 34(6):550–566

The Impact of Trust on the Choice of Knowledge Transfer Mechanisms in Clusters

Marijana Srećković and Josef Windsperger

Abstract This study examines the impact of trust on the use of knowledge transfer mechanisms of cluster firms by deriving hypotheses from a relational governance perspective. Specifically, we analyse the influence of trust on the use of face-to-face knowledge transfer mechanisms in cluster relationships. Based on the relational view of governance, it is argued that trust may influence the choice of knowledge transfer mechanisms of the cluster companies in the following way: first, if trust reduces relational risk, an increase in trust will reduce the firms' use of face-to-face knowledge transfer mechanisms. Second, if trust increases knowledge sharing between the cluster partners, it will increase the firms' use of face-to-face knowledge transfer mechanisms. The hypotheses are tested by using data from 118 companies in the Italian textile and fashion sectors. Our data from the Italian textile and fashion sector supports the hypothesis that experience-based trust increases knowledge sharing between the cluster partners by increasing the use of face-to-face knowledge transfer mechanisms. It also supports the knowledge-based hypothesis that tacitness influences the choice of knowledge transfer mechanisms. The paper extends the knowledge-based view of the choice of knowledge transfer mechanisms by showing that trust is an additional determinant of the knowledge transfer strategy.

Keywords Cluster relationships • Knowledge transfer mechanisms • Relational view of governance • Trust

---

M. Srećković (✉)
Department of Real Estate Development and Project Management, Faculty of Architecture and Planning, Vienna University of Technology, Gusshausstrasse 30, A-1040 Vienna, Austria
e-mail: marijana.sreckovic@tuwien.ac.at

J. Windsperger
Center for Business Studies, University of Vienna, Brünner Str. 72, A-1210 Vienna, Austria
e-mail: josef.windsperger@univie.ac.at

T. Ehrmann et al. (eds.), *Network Governance*, Contributions to Management Science,
DOI 10.1007/978-3-7908-2867-2_5,

1 Introduction

Knowledge transfer between network partners, such as joint venture, license, franchise and cluster partners, is a key to gaining and sustaining a competitive advantage (e.g. Maskell and Malmberg 1999; Driffield and Munday 2000; Maskell 2001; Levin et al. 2002; Hult et al. 2004; Li 2004; Tallman et al. 2004; Bahlmann and Huysman 2008; Mu et al. 2008; Arikan 2009; Niu 2010; Windsperger and Gorovaia 2011; Lee et al. 2012). The success of cluster relationships depends on the effectiveness of the transfer of the know-how between cluster partners. Trust therefore plays a critical role for the performance of affiliated firms (Liao 2010). This study examines the impact of trust on the choice of knowledge transfer mechanisms of cluster firms by developing hypotheses based on the information richness theory and the relational view of governance.

Information richness theory offers information richness as a criterion to evaluate the transfer capacity of the communication media as knowledge transfer mechanisms (Daft and Lengel 1986; Büchel and Raub 2001; Sexton et al. 2003; Sheer and Chen 2004; Vickery et al. 2004). Information richness increases with the following attributes of a knowledge transfer mechanism: feedback capability, availability of multiple cues (voice, body, gestures, words), language variety, and personal focus (emotions, feelings). In cluster relationships, knowledge transfer mechanisms with a relatively higher degree of information richness (HIR) include seminars, workshops, committees, conference meetings, and visits. Knowledge transfer mechanisms with a relatively lower degree of information richness include written documents, fax, email, intra- and internet and other electronic media.

According to the relational view of governance (e.g. Gulati 1995; Dyer and Singh 1998; Zaheer et al. 1998; Poppo and Zenger 2002; Gulati and Nickerson 2008), trust reduces relational risk and increases information sharing, therefore influencing the use of knowledge transfer mechanisms. We hypothesize two trust effects: if trust reduces relational risk, it will reduce the cluster partners' need to use more HIR-knowledge transfer mechanisms. On the other hand, if trust increases knowledge sharing, it will increase the cluster partners' use of HIR-knowledge transfer mechanisms.

Although many researchers have examined the problem of knowledge transfer in network relationships in the last two decades (e.g. Nonaka 1994; Simonin 1999a, b; Albino et al. 1999; Bresman et al. 1999; Argote and Ingram 2000; Altinay and Wang 2006; Jensen and Szulanski 2007; Szulanski and Jensen 2006; Haas and Hansen 2007; Becerra et al. 2008; van Wijk et al. 2008; Minguela-Rata et al. 2010; Winter et al. 2011), this literature does not investigate the determinants of the choice of knowledge transfer mechanisms in inter-organizational networks. To the best of our knowledge, the works of Inkpen and Dinur (1998), Murray and Peyrefitte 2007; Windsperger and Gorovaia (2011), as well as Srećković and Windsperger (2011) are exemptions. They develop and test a knowledge-based view by analyzing the relationship between knowledge characteristics and knowledge transfer mechanisms used in joint ventures, franchising and cluster relationships. According to the knowledge-based theory, the tacitness of the partner knowledge determines the degree of information richness of the knowledge transfer mechanisms. In this study, we extend the knowledge-based view of the choice of knowledge transfer mechanisms of Srećković and Windsperger (2011) by considering trust as an additional explanatory variable of the cluster firm's knowledge transfer strategy. Our empirical study tests the hypotheses by utilizing primary data from the Italian textile and fashion cluster that enables us to estimate the influence of trust on the knowledge transfer strategy of the cluster firms.

The paper proceeds as follows: Section 2 gives an overview of the relevant literature and derives the hypotheses from the relational view of governance. Section 3 tests the hypotheses with the data from the Italian cluster. Section 4 discusses the results and derives some conclusions.

2 Trust and Choice of Knowledge Transfer Mechanism in Clusters

If the know-how of the cluster partners is codifiable, trust has either a weak influence on the impact of knowledge attributes on the use of knowledge transfer mechanisms or none at all. This is shown because exchange hazards are very low and the cluster firms can explicitly specify the relevant knowledge in the contract (Levin and Cross 2004; Gulati and Nickerson 2008). If the know-how of the cluster partners is tacit, the contracts between the cluster partners are very incomplete and the cluster firms have difficulties in successfully applying partner-specific knowledge. Consequently, under highly tacit knowledge, trust has a major impact on knowledge transfer (Levin et al. 2002). Based on the relational view of governance (e.g. Zaheer and Venkatraman 1995; Dyer and Singh 1998; Lazzarini et al. 2008), we can differentiate two perspectives regarding the impact of trust on the use of knowledge transfer mechanisms:

Reduction of relational risk: Trust reduces the knowledge transfer hazards by decreasing relational risk (Gulati 1995; Yu et al. 2006). When the cluster partners trust each other, their tolerance level of perceived risk will be higher, and the cluster firms will more likely select knowledge transfer mechanisms with a lower degree of information richness (Lo and Lie 2008). Hence, under high trust, the cluster firms are likely to use less HIR-knowledge transfer mechanisms, as in this low relational risk situation low information-rich knowledge transfer mechanisms facilitate sufficient knowledge sharing. Conversely, when distrust exists between the cluster partners, their tolerance level of perceived risk will be lower, and the cluster firms will be more likely to select knowledge transfer mechanisms with a higher degree of information richness that transfer more knowledge in order to reduce the degree of relational uncertainty. We derive the following hypothesis:

Hypothesis 1 (H1). The higher trust, the less likely the use of HIR-knowledge transfer mechanisms becomes.

Increase in knowledge sharing: Trust overcomes communication barriers and facilitates knowledge sharing and increases therefore the use of all modes of knowledge transfer (Blomqvist et al. 2005; Yeh et al. 2006; Seppänen et al. 2007; Bohnet and Baytelman 2007; Lazzarini et al. 2008). In addition, greater communication due to the use of more HIR-knowledge sharing mechanisms may lead to more trust between the cluster partners (Anderson and Narus 1990; Dyer and Chu 2000; Blomqvist et al. 2005; Fink and Kraus 2007; Ben-Ner and Putterman 2009). Consequently, under high trust, the cluster firms use more HIR-knowledge transfer mechanisms, because trust creates an incentive for intense and open communication. As a result, we can derive the following hypotheses:

Hypothesis 2 (H2). The more trust exists, the more likely the use of HIR-knowledge transfer mechanisms becomes.

2.1 Inter-organizational Experience as Moderator

According to Reagans and McEvily (2003), the frequency of communication through inter-organizational experience influences the knowledge transfer process. We hypothesize that inter-organizational experience moderates the relationship between trust and the use of knowledge transfer mechanisms. We distinguish two effects: (A) If trust reduces relational risk, more experience with the network partner will result in a stronger decrease in the use of HIR-knowledge transfer mechanisms when inter-organizational experience increases. (B) If trust overcomes communication barriers and facilitates knowledge sharing, inter-organizational experience will increase the positive impact of trust on the use of HIR-knowledge transfer mechanisms. Therefore, depending on the role of trust as a relational risk reduction or a knowledge sharing mechanism, we can derive the following hypotheses:

Hypothesis 1A (H1A). The negative impact of trust on the use of HIR-knowledge transfer mechanisms increases with inter-organizational experience.

Hypothesis 2A (H2A). The positive impact of trust on the use of HIR-knowledge transfer mechanisms increases with inter-organizational experience.

3 Empirical Analysis

3.1 Sample and Data Collection

The empirical study uses data from the Italian textile and fashion industry. Italian industrial districts are a very important contributor to the Italian Economy, and considering the fashion and textile industry, Italy is one of the leading exporting countries in this field.[1] In 2011, textile and fashion districts have accounted for 28.8 % of the working population in Italy.[2]

---

[1] See http://mefite.ice.it/settori/Tessile.aspx?idSettore=02000000 [retrieved 20.11.2011].

[2] See http://www.istat.it/en/ [retrieved 20.11.2011].

The empirical setting for testing these hypotheses is the Italian textile and fashion cluster situated in the Province of Prato in Tuscany. In 2009, the textile and clothing sector in the Prato district had an estimated workforce of 30.200 people and 7.582 business firms, which accounted for a turnover of 3,872 million Euros in that sector. "Prato is one of the areas in Central and Northeast Italy (the so-called "Third Italy") where centuries-old craft skills have successfully merged with modern industrial growth. Originating between the nineteenth and twentieth centuries, the industrialization process underwent a rapid acceleration after World War II and was fully established by the 1970s. During this period of development, Prato grew to become Europe's most important textiles and fashion centre, and the most advanced example—or prototype—of that particular form of organization of production that is the industrial district. One feature of industrial districts, and of the Prato district as well, is the specialization and distribution of work among small business firms; this segmentation finds its recomposition in a "culturally and socially constituted" local market whose competitiveness is based more on the economical aspects of the area itself than on those of the single undertakings."[3]

We started our empirical work by analyzing textile and fashion companies working in Italian industrial districts. First, we contacted exclusively companies from the fashion cluster situated in the Prato district. The identification of cluster firms was based on two sources: (1) the online data bases (e.g., "Unione Industriale Pratese")[4] and (2) the Italian Chamber of Commerce. In total, 426 residential cluster firms were contacted by mail. 144 companies accessed the online questionnaire, but only 34 firms responded to most of the questions. Despite several attempts, ranging from multiple reminders to non-respondents and personal contacts via telephone, the response rate remained low. In order to increase the response rate and enlarge the sample, it was necessary to contact firms from other clusters as well. For this purpose, the so-called "snowball technique" (Churchill and Iacobucci 2005) was used. A leading multinational fashion corporate group which is in cooperation with retailers and producers in the Italian industrial districts was contacted. General managers of the single affiliates were asked to contact exclusively with executive directors of target cluster firms, and to spread the questionnaire among cluster partners who might be interested in cooperating. General managers and executive directors were judged to be the most suitable respondents, or key informants, as they are the top decision makers in the company regarding the organization of the knowledge transfer between the partner firms. Key informants should occupy roles that make them knowledgeable regarding the issues being researched (John and Reve 1982). This procedure led to an additional 131 questionnaires, i.e., questionnaires in which the majority of questions apart from the general company description have been answered. Unfortunately, the online questionnaire tool allowed skipping single questions or question batteries, thus the problem occurred

[3] See http://www.ui.prato.it/unionedigitale/v2/english/presentazione%20distretto%20inglese.pdf [retrieved 20.11.2011].

[4] See http://www.ui.prato.it/unionedigitale/v2/default.asp [retrieved 20.11.2011].

that some respondents answered the questionnaire only in parts. However, the extension of the sample led to a satisfying sample size for all analyses. The questionnaire took approximately 10 min to complete on the average. We received 118 completed responses—a response rate of 27.70 %. We examined the non-response bias by investigating whether the results obtained from the analysis were driven by differences between the group of respondents and the group of non-respondents. Non-response bias was estimated by comparing early versus late respondents (Armstrong and Overton 1977), where late respondents serve as proxies for non-respondents. No significant differences emerged between the two groups of respondents. In addition, based on Podsakoff et al. (2003), we used Harman's single-factor test to examine whether a significant amount of common method variance exists in the data. After we conducted factor analysis on all items and extracted more than one factor with eigenvalues greater than one, we felt confident that common method variance is not a serious problem in our study.

3.2 Measurement

To test the hypotheses, the following variables are important: knowledge transfer mechanisms, trust, and control variables (see Appendix).

Knowledge Transfer Mechanisms: Our study conceptualizes information richness of knowledge transfer mechanisms in accordance with the Daft and Lengel's approach (Daft and Lengel 1984). We measure high information richness (HIR) by the extent to which the partner firms use face-to-face knowledge transfer mechanisms, such as committees and other formal meetings. The general managers were asked to rate the use of these knowledge transfer mechanisms on a five-point scale. The higher the score, the higher is the company's use of these HIR-knowledge transfer mechanisms (HIR) (see Appendix).

Trust (TRUST): According to the relational view of governance, trust may influence the use of knowledge transfer mechanisms in two ways: under the substitutability view, trust is a substitute for the use of formal knowledge transfer mechanisms (Gulati 1995; Yu et al. 2006). Therefore, it mitigates the knowledge transfer hazards and reduces the extent of formal knowledge transfer mechanisms (Lo and Lie 2008). Consequently, cluster companies are likely to use less HIR-knowledge transfer mechanisms when trust exists between the cluster partners, and use more HIR-knowledge transfer mechanisms when mistrust exists. Under the complementarity view, trust facilitates knowledge sharing and increases the use of all knowledge transfer modes (Seppänen et al. 2007; Liao 2010). Therefore, under a high level of trust, cluster partners use more HIR-knowledge transfer mechanisms, because trust creates an incentive for intense communication. Adapted from the relevant literature (e.g. Seppänen et al. 2007), TRUST was measured with a five-items scale (see Appendix) (Cronbach alpha = 0.89).

Table 1 Descriptive statistics

	Mean	Std. deviation	N
COMPLEX	2.79	0.8158	116
TRUST	3.05	0.9074	118
AGE	32.10	35.3157	118
NUM_EMPLOYEES	80.28	19.5147	115

3.2.1 Control Variables

Complexity (COMPLEX): Kogut and Zander (1993, p. 633) define complexity "as the number of critical and interacting elements embraced by an entity or activity". Similarly, Sorenson et al. (2006) define complexity in terms of the level of interdependence inherent in the subcomponents of a piece of knowledge (see Simonin 1999a, b). When the system knowledge is more complex, it is considered more tacit. Applied to the cluster relationships, complexity is high when the application of the partner knowledge requires a large number of heterogeneous, complicated and interdependent tasks. Likewise, it is also high when cluster partners have to master diverse techniques in order to successfully apply the partner knowledge. To summarize, when the knowledge of the cluster firms is more complex, it is considered more tacit. Adapted from Zander and Kogut (1995), we use a battery of four items to measure complexity of system-specific knowledge. Reliability passes the threshold of 0.7 (see Appendix).

Age of the Cluster Company (*AGE*): Age is a proxy for inter-organizational learning and experience (Gulati and Sytch 2008). Inter-organizational experience moderates the impact of trust on the choice of knowledge transfer mechanisms.

Size (SIZE): The number of employees is a proxy for the size of the firm. The larger the firm size, the more face-to-face knowledge transfer mechanisms are used.

3.3 Results

Table 1 presents the descriptive statistics for the sample in the Italian textile and fashion cluster. To test the hypotheses we carry out a regression analysis. We conduct an ordinary least squares regression analysis (OLS) with HIR as the dependent variable. HIR refers to the use of committee meetings and other formal meetings of the cluster members (top-managers, cluster managers). We conduct an OLS regression analysis (a) with the control variables and (b) with the complete model. The explanatory variables refer to TRUST and TRUST*AGE. Control variables refer to the age of the cluster companies (AGE), the size of the company (SIZE) and complexity of knowledge (COMPLEX). Table 2 presents the correlations of the variables we use in the regression analysis. In addition, the variance inflation factors are well below the rule-of-thumb cut-off of 10 (Netter et al. 1985). In summary, we do not find any collinearity indication.

Table 2 Correlations

	COMPLEX	TRUST	AGE	NUM_EMPLOYEES
COMPLEX	1			
TRUST	0.445**	1		
AGE	0.003	−0.136	1	
NUM_EMPLOYEES	−0.047	0.054	0.308**	1

***p < 0.01; **p < 0.05; *p < 0.1

Table 3 Regression results for HIR

HIR	Model 1
Intercept	−3.075*** (0.118)
AGE	0.293*** (0.003)
SIZE	0.206** (0.000)
COMPLEX	0.199** (0.088)
	F = 8.854
	R Square = 0.202
	Adj.R Square = 0.179
	N = 113

***p < 0.01; **p < 0.05; *p < 0.1; values in parentheses are standard errors

We estimate the following regression equation:

$$\mathrm{HIR} = \alpha + \beta_1\,\mathbf{AGE} + \beta_2\,\mathbf{SIZE} + \beta_3\,\mathbf{COMPLEX} + \beta_4\,\mathbf{TRUST} + \beta_5\,\mathbf{TRUST} * \mathbf{AGE}$$

In the first step, we conduct the regression analysis with the control variables (see Table 3). HIR varies positively with age (AGE), size (SIZE) and with complexity (COMPLEX). The positive and significant coefficient of size (SIZE) confirms that larger firms use more HIR-knowledge transfer mechanisms. The highly significant and positive coefficient of age (AGE) confirms that inter-organizational learning and experience (Gulati and Sytch 2008) have a strong influence on the use of HIR. Complexity (COMPLEX) varies positively and significantly with HIR. This is consistent with the knowledge-based hypothesis that an increase in tacitness of knowledge results in the use of more HIR-knowledge transfer mechanisms (Srećković and Windsperger 2011).

In the second step, we include TRUST and TRUST*AGE and all control variables (see Model 2 in Table 4). Specifically, the interaction effect TRUST*AGE considers the impact of inter-organizational experience on the relationship between trust and the use of HIR-knowledege transfer mechanisms (Lazzarini et al. 2008; Gulati and Sytch 2008). The coefficient of TRUST*AGE is positive and significant. This is consistent with our hypothesis H2A. If trust overcomes communication barriers and facilitates knowledge sharing, inter-organizational experiences increases the positive impact of trust on the cluster partners' use of HIR-knowledge transfer mechanisms (Seppänen et al. 2007). This means that experience-based trust plays an important role in the knowledge transfer process by strengthening face-to-face communication. In addition, consistent with the knowledge-based view, the use of HIR-knowledge transfer mechanisms varies positively with complexity (COMPLEX).

Table 4 Regression results for HIR

HIR	Model 2
Intercept	0.570*** (0.280)
AGE	−0.310 (0.009)
SIZE	0.159* (0.000)
COMPLEX	0.199** (0.099)
TRUST	−0.208 (0.158)
TRUST*AGE	0.649** (0.003)
	F = 6.228
	R Square = 0.232
	Adj.R Square = 0.195
	N = 113

***$p < 0.01$; **$p < 0.05$; *$p < 0.1$; values in parentheses are standard errors

4 Discussion and Implications

This study examines the impact of trust on the use of HIR-knowledge transfer mechanisms in cluster relationships. Based on the relational view of governance, trust influences the choice of knowledge transfer mechanisms of the cluster firms. Our data from the Italian textile and fashion sector supports the hypothesis that experience-based trust increases knowledge sharing between the cluster partners by increasing the use of face-to-face knowledge transfer mechanisms. Consistent with Srećković and Windsperger (2011), our data also supports the knowledge-based view of the choice of knowledge transfer mechanisms. Overall, we can conclude that trust and knowledge attributes (tacitness) are important determinants of the choice of knowledge transfer mechanism in cluster relationships.

What is the contribution of this study to the relevant literature? Although many researchers in the field of the knowledge-based view of the firm have examined the problem of internal and inter-organizational knowledge transfer (Nonaka 1994; Albino et al. 1999; Ancori et al. 2000; Argote et al. 2003; Bresnen et al. 2003; Jensen and Szulanski 2007; Szulanski and Jensen 2006; Haas and Hansen et al. 2007; van Wijk et al. 2008; Paswan and Wittmann 2009), most of these studies do not investigate the determinants of the choice of knowledge transfer mechanisms. In the context of cluster relationships, Srećković and Windsperger (2011) developed a knowledge-based perspective by analyzing the impact of tacitness on the use of knowledge transfer mechanisms. This study investigates the impact of trust on the cluster firm's choice of knowledge transfer mechanisms from a relational governance perspective. We extend the results of Srećković and Windsperger (2011) by considering trust as an additional determinant of the cluster firm's knowledge transfer strategy.

This study has also managerial implications: first, for successful knowledge transfer, cluster firms have to consider both *tacitness* of knowledge and *trust* as important determinants of the choice of the knowledge transfer mechanisms. If the partner-specific knowledge is characterized by a high degree of tacitness, more HIR-knowledge transfer mechanisms should be used to successfully transfer the partner-specific knowledge to the other network partner. Furthermore, the cluster firms' choice of knowledge transfer mechanisms depends on the degree of *trust* between the cluster partners. Therefore, interorganizational experience strengthens the role of trust as a facilitator of face-to-face knowledge sharing between the cluster firms. Consequently, cluster relationships characterized by experience-based trust are better able to use face-to-face knowledge transfer mechanisms to increase the knowledge exchange between the network partners.

Appendix: Measures of Variables

Higher-IR-knowledge transfer mechanisms (HIR)	To what extent does the cluster company use the following knowledge transfer mechanisms: committee meetings, other formal meetings of cluster members (top-managers, district managers) (1, no use at all; ... 5, very frequent use)
Trust (TRUST) Coefficient alpha: 0.893	Please specify in which extent the following statements correspond to the relationships between your company and the cluster partners: (1, strongly disagree; ... 5, strongly agree) Trust 1: There is a distinct relationship of trust between your company and your cluster partners. Trust 2: There prevails an atmosphere of openness and honesty between your company and your cluster partners. Trust 3: The exchange of information inside the cluster goes beyond the stipulated extent. Trust 4: The collaboration between your company and cluster partners relies on a cooperative basis. Trust 5: We comply with verbal agreements, even if these could be at our disadvantage. Trust 6: The recommendations of your cluster partners with the goal to enhance collaboration are usually heard and discussed Trust 7: The recommendations of your partners in terms of alteration/innovation are heard and discussed inside the cluster.
Complexity (COMPLEX) Coefficient alpha: 0.710	Please specify to which extent you agree with the following statements (1, strongly disagree; ... 5, strongly agree) Complex 1: Cluster partners must learn a vast amount of activities, in order to be able to adopt successfully the transmitted know-how. Complex 2: The techniques and methods used to adopt transmitted know-how are heterogeneous. Complex 3: The techniques and methods used to adopt transmitted know-how are very difficult. Complex 4: The techniques and methods used to adopt transmitted know-how are highly interdependent.
Age (AGE)	Age of the cluster firm
Size (SIZE)	Number of employees

References

Albino V, Garavelli AC, Schiuma G (1999) Knowledge transfer and inter-firm relationships in industrial districts: the role of the leader firm. Technovation 19(1):53–63

Altinay L, Wang CL (2006) The role of prior knowledge in international franchise partner recruitment. Int J Serv Ind Manage 17(5):430–443

Ancori B, Bureth A, Cohendet P (2000) The economics of knowledge: the debate about codification and tacit knowledge. Ind Corp Change 9(2):255–287

Anderson E, Narus J (1990) A model of the distributor firm and manufacturer firm working relationships. J Mark 54(1):42–58

Argote L, Ingram P (2000) Knowledge transfer: a basis for competitive advantage in firms. Organ Behav Hum Decis Process 82(1):150–169

Argote L, McEvily B, Reagans R (2003) Managing knowledge in organizations: an integrative framework and review of emerging themes. Manage Sci 49(4):571–582

Arikan AT (2009) Inter-firm knowledge exchanges and the knowledge creation capability of clusters. Acad Manage Rev 34:658–677

Armstrong JS, Overton TS (1977) Estimating non-response bias in mail surveys. J Mark Res 14(3): 396–402

Bahlmann MD, Huysman MH (2008) The emergence of a knowledge-based view of clusters and its implications for cluster governance. Inf Soc 24:304–318

Becerra M, Lunnan R, Huemer L (2008) Trustworthiness, risk, and the transfer of tacit and explicit knowledge between alliance partners. J Manage Stud 45:691–713

Ben-Ner A, Putterman L (2009) Trust, communication and contracts: an experiment. J Econ Behav Organ 70:106–121

Blomqvist K, Hurmelinna P, Seppänen R (2005) Playing the collaboration game right – balancing trust and contracting. Technovation 25:497–504

Bohnet I, Baytelman Y (2007) Institutions and trust: implications for preferences, beliefs and behaviour. Rationality Soc 19:99–135

Bresman H, Birkinshaw J, Nobel R (1999) Knowledge transfer in international acquisitions. J Int Bus Stud 30(3):439–462

Bresnen M, Edelman L, Newell S, Scarbrough H, Swan J (2003) Social practices and the management of knowledge in project environments. Int J Proj Manage 21:157–166

Büchel B, Raub S (2001) Media choice and organizational learning. In: Antal AB, Dierkes M, Child J, Nonaka I (eds) Handbook of organizational learning and knowledge. Oxford University Press, Oxford

Churchill G, Iacobucci D (2005) Marketing research. South-Western, Thomson

Daft RL, Lengel RH (1984) Information richness: a new approach to managerial behavior and organizational design. In: Cummings LL, Staw BM (eds) Research in organizational behavior. JAI Press, Homewood, pp 191–233

Daft RL, Lengel RH (1986) Organizational information requirements, media richness and structural design. Manage Sci 32(5):554–571

Driffield N, Munday M (2000) Industrial performance, agglomeration, and foreign manufacturing investment in the UK. J Int Bus Stud 31(1):21–37

Dyer J, Chu W (2000) The Determinants of trust in supplier-automaker relationships in the U.S., Japan, and Korea. J Int Bus Stud 31(2):259–285

Dyer JH, Singh H (1998) The relational view: cooperative strategy and sources of interorganizational competitive advantage. Acad Manage Rev 23(4):660–679

Fink M, Kraus S (2007) Mutual trust as a key to internationalization of SMEs. Manage Res News 30(9):674–688

Gulati R (1995) Does familarity breed trust? The implications of repeated ties for contractual choice in alliances. Acad Manage J 38(1):85–112

Gulati R, Nickerson JA (2008) Interorganizational trust, governance choice, and exchange performance. Organ Sci 19:688–708

Gulati R, Sytch M (2008) Does familiarity breed trust? Revisiting the antecedents of trust. Manag Decis Econ 29:165–190

Haas MR, Hansen MT (2007) Different knowledge, different benefits: toward a productivity perspective on knowledge sharing in organizations. Strateg Manage J 28(11):1133–1153

Hult GTM, Ketchen DJ, Cavusgil ST, Calatone RJ (2004) Knowledge as a strategic resource in supply chains. J Oper Manage 24(4):458–475

Inkpen AC, Dinur A (1998) Knowledge management processes and international joint ventures. Organ Sci 9(4):454–468

Jensen RJ, Szulanski G (2007) Template use and the effectiveness of knowledge transfer. Manage Sci 53(11):1716–1730

John G, Reve T (1982) The reliability and validity of key informant data from dyadic relationships in marketing channels. J Mark Res 19(4):517–524

Kogut B, Zander U (1993) Knowledge of the firm and the evolutionary theory of the multinational corporation. J Int Bus Stud 24(4):625–645

Lazzarini SG, Miller GJ, Zenger TR (2008) Dealing with the paradox of embeddedness: the role of contracts and trust in facilitating movement of committed relationships. Organ Sci 19(5): 709–728

Lee S, Kim BG, Kim H (2012) An integrated view of knowledge management for performance. J Knowl Manage 16:183–203

Levin DZ, Cross R, Abrams LC, Lesser EL (2002) Trust and knowledge sharing: a critical combination. Retrieved 09/10/2011 from http://www-935.ibm.com/services/in/igs/pdf/g510-1693-00-cpov-trust-and-knowledgesharing.pdf

Levin DZ, Cross R (2004) The strength of weak ties you can trust: the mediating role of trust in effective knowledge transfer. Manage Sci 50:1477–1490

Li S (2004) Location and performance of foreign firm in China. Manage Int Rev 44(2):151–169

Liao TJ (2010) Cluster and performance in foreign firms: the role of resources, knowledge, and trust. Ind Mark Manage 39:161–169

Lo S, Lie T (2008) Selection of communication technologies – a perspective based on information richness theory and trust. Technovation 28:146–153

Maskell P, Malmberg A (1999) Localised learning and industrial competitiveness. Cambridge J Econ 23(2):167–185

Maskell P (2001) Towards a knowledge-based theory of the geographical cluster. Ind Corp Change 10:921–943

Minguela-Rata B, Lopez-Sanchez JI, Rodriguez-Benavides MC (2010) Knowledge transfer mechanisms and the performance of franchise systems: an empirical study. Afr J Bus Manage 4:396–405

Mu J, Peng G, Love E (2008) Interfirm networks, social capital, and knowledge flow. J Knowl Manage 12:86–100

Murray SR, Peyrefitte J (2007) Knowledge type and communication media choice in the knowledge transfer process. J Manag Issues 19(1):111–133

Netter J, Wasserman W, Kutner MH (1985) Applied linear statistical models. Irwin, Homewood

Niu KH (2010) Organizational trust and knowledge obtaining in industrial clusters. J Knowl Manage 14:141–155

Nonaka I (1994) A dynamic theory of organizational knowledge creation. Organ Sci 5(1):14–37

Paswan AK, Wittmann MC (2009) Knowledge management and franchise system. Ind Mark Manage 38:173–180

Podsakoff PM, MacKenzie SB, Lee JY, Podsakoff NP (2003) Common method biases in behavioral research: a critical review of the literature and recommended remedies. J Appl Psychol 88:879–903

Poppo L, Zenger TR (2002) Do formal contracts and relational governance function as substitutes or complements? Strateg Manage J 23:90–118

Reagans R, McEvily B (2003) Network structure and knowledge transfer: the effects of cohesion and range. Adm Sci Q 48:240–267

Seppänen R, Blomqvist K, Sundqvist S (2007) Measuring inter-organizational trust – a critical review of the empirical research in 1990–2003. Ind Mark Manage 36:249–265
Sexton M, Ingirige B, Betts M (2003) Information technology-enabled knowledge sharing in multinational strategic alliances: media richness – task relevance fit. Working Paper, Research Institute of the Built and Human Environment. University of Salford, England
Sheer VC, Chen L (2004) Improving media richness theory: a study of interaction goals, message valence, and task complexity in manager-subordinate communication. Manage Commun Q 11(1):76–93
Simonin BL (1999a) Ambiguity and the process of knowledge transfer in strategic alliances. Strateg Manage J 20(7):595–623
Simonin BL (1999b) Transfer of marketing know-how in international strategic alliances: an empirical investigation of the role and antecedents of knowledge ambiguity. J Int Bus Stud 30(3):463–490
Sorenson O, Rivkin JW, Fleming L (2006) Complexity, networks and knowledge flow. Res Policy 35:994–1017
Srećković M, Windsperger J (2011) Organization of knowledge transfer in clusters: a knowledge-based view. In: Tuunanen M, Windsperger J, Cliquet G, Hendrikse G (eds) New developments in the theory of networks. Franchising, alliances and cooperatives. Springer, Berlin, pp 318–334
Szulanski G, Jensen RJ (2006) Presumptive adaptation and the effectiveness of knowledge transfer. Strateg Manage J 27(10):937–957
Tallman W, Jenkins M, Henry N, Pinch S (2004) Knowledge clusters and competitive advantage. Acad Manage Rev 29:258–271
Van Wijk R, Jansen JP, Lyles MA (2008) Inter- and intra-organizational knowledge transfer: a meta-analytic review and assessment of its antecedents and consequences. J Manage Stud 45:830–853
Vickery SK, Droge C, Stank TP, Goldsby TJ, Markland RE (2004) The performance implications of media richness in a business-to-business service environment: direct versus indirect effects. Manage Sci 50(8):1106–1119
Windsperger J, Gorovaia N (2011) Knowledge attributes and the choice of knowledge transfer mechanisms in networks: the case of franchising. J Manage Gov 15:617–640
Winter SG, Szulanski G, Ringov D, Jensen RJ (2011) Reproducing knowledge: inaccurate replication and failure in franchise organizations. Organ Sci 23(3):672–685
Yeh Y-J, Lai S-Q, Ho C-T (2006) Knowledge management enablers: a Case study. Ind Manage Data Syst 106:793–810
Yu C-MJ, Liao TJ, Lin ZD (2006) Formal governance mechanisms, relational governance mechanisms, and transaction-specfic investments in supplier-manufacturer relationships. Ind Mark Manage 35:128–139
Zaheer A, McEvily B, Perrone V (1998) Does trust matter? Exploring the effects of inter-organizational and interpersonal trust on performance. Organ Sci 19(2):141–159
Zaheer A, Venkatraman N (1995) Relational governance as an inter-organizational strategy: an empirical test of the role of trust in economic exchange. Strateg Manage J 16:373–392
Zander U, Kogut B (1995) Knowledge and the speed of transfer and imitation of organizational capabilities: an empirical test. Organ Sci 6(1):76–92

International Market Expansion of Retail Networks: Determinants of Market Entry Failures

Odile Streed and Gérard Cliquet

Abstract What causes an international retailer to divest from a specific country or region? The answer to that question is not consistent and retailers are usually left in the dark due to lack of definite frameworks to help them make appropriate foreign market entry decisions. Failure is a common and costly occurrence in international retailing and in a time of increased economic uncertainty this may have a dramatic impact on the overall organization. The purpose of this article is to investigate the determinants of market entry failures in emerging markets. Based on a sample of 112 cases, exploratory results show that time of entry, brand penetration, entropy level and local store density are strongly correlated to failure or success of an international retailer in emerging countries. Additionally, preliminary results in term of market entry mode choice tend to show that governance modes with low level of control such as franchising, licensing or minority joint ventures may be the best market entry choices for international retailers expanding into emerging countries.

Keywords Emerging countries • Entropy • Failure • Governance mode • International retailing • Market entry

1 Introduction

In February 2011, two major U.S. retailers Best Buy Inc. and Home Depot announced divesture from the Chinese market. While certain chains, such as Kentucky Fried Chicken or Tesco are for the most part thriving in emerging markets such as China,

O. Streed (✉)
Offutt School of Business, Concordia College, Moorhead, MN, USA
e-mail: streed@cord.edu

G. Cliquet
CREM UMR CNRS 6211 University of Rennes 1, Rennes Cedex, France

T. Ehrmann et al. (eds.), *Network Governance*, Contributions to Management Science,
DOI 10.1007/978-3-7908-2867-2_6, © Springer-Verlag Berlin Heidelberg 2013

Thailand or South Korea, others are struggling to achieve profitability. In spite of substantial investments, extensive marketing research, prestigious locations and lengthy employee trainings, many large foreign retailers are still facing intense difficulties in emerging markets. Nevertheless, these markets keep attracting international chains ranging from hypermarkets to luxury specialty stores or quick service restaurants.

Even though international retailing has inspired a growing number of articles (Dawson 1994; Robinson and Clarke-Hill 1995; Reijnders and Verhallen 1996; Dawson 2001; Burt et al. 2002; Palmer and Owens 2006; Picot-Coupey 2006, 2009; Gielens and Dekimpe 2007; Suh and Howard 2009), specific retail internationalization theories are still in their infancy (Burt et al. 2002). Therefore researchers have heavily borrowed from the manufacturing literature whose validity may be questionable in the field of retailing. Additionally, although Hollander (1970) already stressed the importance of the topic, a few but growing numbers of authors have published articles on retail failures (Vida and Fairhurst 1998; Alexander and Quinn 2002; Burt et al. 2002; Burt et al. 2003; Burt et al. 2004; Costil 2003; Wrigley and Currah 2003; Dupuis and Fournioux 2006; Etgar and Rachman-Moore 2007; Hang 2009). Moreover Hang (2009) argues that a consensus over the definition and the conceptualization of retail failure has not been reached (Gielens and Dekimpe 2001; Alexander and Quinn 2002; Burt et al. 2002, 2003). This reflects the need for specific frameworks dedicated to this facet of international retailing.

Moreover most current research on international retail failure has focused on developed countries (Burt et al. 2003; Palmer 2004) in spite of the substantial attraction of emerging markets on foreign retailers. This article aims to contribute to the under-researched problem of international retail failures in emerging markets by focusing on the following factors that have been more or less investigated in the literature as potential antecedents of failure or success: market entry mode choice, chain size, relative entropy, store density, financial soundness, cultural and psychic distance and market attractiveness. The purpose of this article is to start developing a set of exploratory guidelines specific to international retailers in emerging markets.

This paper will first review the theoretical background on the topic and articulate a set of hypotheses that will be tested in a subsequent section. Archival data and store failure narratives obtained through journal articles helped in developing the empirical research. These findings will be followed by a section dedicated to managerial and strategic implications.

2 Theoretical Backgrounds

2.1 Definition of International Retail Failure

Burt et al. (2003) argue that the knowledge obtained from the analysis of international retail failures was instrumental in the success of current market leaders such as Carrefour or Tesco. But for obvious reasons retailers do not emphasize nor publicize much of these failed attempts and focus instead on their success stories. It is therefore cumbersome to collect unbiased material on the topic and that may explain the scarcity of the research. Additionally as stated by Burt et al. (2002) the definition of the concept of retail failure is rather poor in the literature. A few international retailing researchers (Burt et al. 2003; Alexander et al. 2005) have therefore attempted to identify the various facets of divestment activities.

In essence, divestment does not necessarily entail country exit but could take multiple other forms such as store closure, sales of the chain in that region, termination of business agreements designing governance modes such as alliances, franchise contracts, licensing or joint ventures (Baroncelli and Manaresi 1997), and organizational restructuring. However, Burt et al. (2003) warn that the metrics used to measure failure could be misleading: for example store closures in a certain region may contribute to the success of the overall chain by allowing more profitable investments in other geographies with better strategic prospects for the organization; similarly, the takeover of certain foreign operations by a rival may also be highly profitable for a firm and deemed a financial success. But what is driving failure or success in a foreign market entry?

2.2 *Antecedents of Failure and Success*

Burt et al. (2003) summarizing the work of Benito (1997) list four potential causes to divestment: (1) market failures or issues with the economic, social and political environment, such as country openness and country risks; (2) competitive failure due to stronger competitors, and lack of firm differentiation; (3) operational failure where under-performance may result from the lack of skill to transfer domestic experience into foreign markets; and (4) business failure where a retailer may face difficulties in its home market and may lack governance competencies and financial means to maintain international operations.

2.2.1 Market Entry Decisions and Governance

Research has often linked exit decisions in international retailing to market entry modes (Li 1995; Chang and Singh 1999; Gielens and Dekimpe 2001). According to the literature (Li 1995; Burt et al. 2003) international retail joint ventures (IRJV) are particularly risky, principally when they take place between culturally distant partners. Palmer (2004) emphasizes the fact that they are multiple forms of IRJVs. He urges researchers to distinguish between equity joint ventures (EJV) and contractual joint ventures (CJV). Among EJVs, acquisition of an existing structure versus the creation of a greenfield entity may also have different outcomes. Alliances such as licensing or franchising are deemed safer than IRJVs (Burt et al. 2003). However the literature (Li 1995; Chang and Singh 1999; Gielens and Dekimpe 2001) recommends full ownership, through acquisition or greenfield investment as a preferred mode of entry. Additionally, Benito (1997) stresses that greenfield wholly owned subsidiaries (WOS) have a lower propensity to fail than acquired structures.

2.2.2 Other Antecedents of Failure and Success

Several authors have chosen to concentrate their efforts on other antecedents of failure or success: for example Couturier and Sola (2010) have investigated the impact of regulations and concluded that a country with few legal and environmental constraints that allows investors to freely choose their mode of entry is very favorable to success. The impact of cultural and psychic distance between the domestic and the host country is also actively researched (Sousa and Bradley 2005; Chen et al. 2009). Chen et al. (2009) believe that large cultural distance discourages firms to select joint-venture or fully owned acquisitions as their mode of entry, encouraging instead more flexible arrangements that allow for faster withdrawal in case of failure.

Domestic issues can also be potential antecedents for failure or success. According to Alexander and Quinn (2002) the reasons for divesture may reside in negative circumstances in the retailer's domestic market that may call for a portfolio adjustment. Additionally, it is interesting to consult the survival literature. One of its key theories is that market entry timing plays a significant role in the longevity of a firm in a foreign market (Golder and Tellis 1993; Pan and Chi 1999; Cui and Lui 2005; Johnson and Tellis 2008).

Retailers' characteristics in terms of industry, age and size may also influence the outcome. Burt et al. (2003) believe that the older and larger a corporation is the lower its risk to fail in a foreign market entry. Last, Bradach (1997) outlines the first challenge for a chain: its ability to add new units and Cliquet (1998) explains spatial strategies "as the need for growth by addition of new stores". Hence as the notion of territory coverage appears to be a valid measurement of retailer's success domestically spatial strategies can be a potential antecedent of failure or success in international retailing. Relative entropy will assess territory coverage (Cliquet 1998, p. 219) in this research in order to measure its impact on retailers' propensity to fail.

2.3 *Emerging Markets Specificities: The Case of China*

According to Beamish (1985) joint-ventures in emerging countries are problematic endeavors characterized by higher instability rate and managerial dissatisfaction than developed countries' ventures. Beamish stresses the need to consider these differences for governance purposes. According to Chen et al. (2009) firms entering transitional economies such as China should focus on reducing institutional risks instead of maximizing efficiency. Therefore the authors believe that cultural and institutional perspectives such as rules and norms developed by the government but also by society are more relevant in assessing success and failure than the traditional transaction cost theory.

Due to its size and the structure of its central government that empowers local governmental entities to make economic decisions (Lubman 1999), Chinese institutional laws along with cultural norms vary greatly from region to region. Liu (2008) partly attributes the success of Kentucky Fried Chicken (KFC) in China to the ability of their joint-venture partner in Shanghai to understand Chinese politics and to establish a strong relationship with the Chinese government including provincial, city and local government units. In 1992, China authorized foreign retailers to enter the Chinese market through IRJVs. Foreign equity was initially limited to 50 % in 1992 and increased to 65 % in 2002. This restriction was completely removed when China joined the world trade organization (WTO) in 2004 which benefited to Wal-Mart and Carrefour (Wang 2009). According to the National Statistics Bureau of China, newly appointed foreign retailers shifted their market entry strategy from the mandatory IRJV model to wholly owned foreign enterprise (WFOE) when the government lifted the investment restrictions in 2004.

3 Methodology

3.1 Measurements and Development of Hypotheses

Considering the current literature (see Table 1) and the availability of data the following factors were selected for this research: company size, financial performance (global level), market entry mode choice such as franchising, licensing, IRJV(minority or majority), greenfield investment or acquisition, time (order of entry, operation duration in the country), territorial coverage (entropy, store density), brand penetration in the host country, cultural and psychic distance with the host country, and country attractiveness (regulatory, political, economic environment in the host country).

3.1.1 Outcome of Market Entry

The dependent variable used in the regression and as a grouping factor in the independent sample t-test refers to the outcome of market entry, failure or success. The outcome variable was defined as follow: 0-failure and 1-success. For the purpose of this article, failure was equated to country exit for a specific chain and success with its continuing presence in the country at year seven. Consistent with the results obtained in previous studies (Gandolfi and Strach 2009), the original statistical analysis of the 77 cases of failure in this research indicated that the

Table 1 Review of the variables selected for the empirical assessment

Selected variables	Key dimensions	Authors
Company size	Transaction cost analysis (annual revenue)	Parsons (1996), Luo (2002) Burt et al. (2003) Johnson and Tellis (2008).
Financial performance	Transaction cost Analysis (profitability, credit rating)	Alexander and Quinn (2002) Cairns et al. (2008)
Entry mode	Level of control Local experience Existing customer base (Franchising, licensing, IRJV acquisition, greenfield, whole ownership)	Benito (1997), Quinn (1998) Chang and Singh (1999) Quinn and Doherty (2000) Quinn and Alexander (2002) Burt et al. (2002; 2003), Palmer (2004), Barkema et al. (1996), Hennart et al. (1998) Palmer and Owens (2006)
Time	Timing Order of entry Operation duration, Network effect	Golder and Tellis (1993) Pan and Chi (1999) Cui and Lui (2005) Johnson and Tellis (2008)
Territorial coverage	Geographical dimension (Store density, entropy)	Evans (1996), Cliquet (1998) Alexander and Quinn (2002), Wood (2002)
Penetration	Brand penetration in the host country	Dupuis and Fournioux (2006)
Cultural and psychic distance	Institutional theory (Normative dimension) Hofstede scores	Wilcox and O'Callaghan (1999) Kogut and Singh (1988) Sousa and Bradley (2005) Huang and Sternquist(2007)
Country attractiveness assessment	Institutional theory (Regulative dimension)	Wood (2002) Huang and Sternquist (2007) Myers and Alexander (2007) Couturier and Sola (2010)

average length of operation between market entry and country exit was 7 years and justifies therefore the following method of data collection: (1) at divestment time for the 77 cases of failure; and (2) at year seven of operation counting from the market entry year for the 35 cases of success. This procedure was particularly essential to follow in order to obtain meaningful results in terms of territorial coverage, and brand penetration.

3.1.2 Company Size

Firm's size may influence international expansion success for several reasons: (1) access to more resources, (2) wealth of knowledge and experience, and (3) ability to sustain losses (Parsons 1996; Luo 2002; Johnson and Tellis 2008). However the literature debates vigorously about this topic. Johnson and Tellis

(2008) for example mentioned the failure of Wal-Mart in South Korea and conclude that a large company size may facilitate success but is not a guarantee of success. Burt et al. (2003) argue that older and larger retailers have less risk to fail when expanding internationally. The following hypothesis was therefore established:

Hypothesis 1 (H1). Larger retailers have lower risks to fail.

Using the method described by Johnson and Tellis (2008), the company size measurements were obtained from retailers' annual reports. Annual revenue at time of divesture or on the seventh year of activity in the host country was converted into U.S. dollars and transformed into logarithms.

3.1.3 Financial Performance

According to Alexander and Quinn (2002) worsening financial performance is a key trigger for divestment activities in international retailing. This is particularly true for public companies that are facing market pressure for short-term profitability. Hypothesis two below reflects this observation:

Hypothesis 2 (H2). Retailers with strong financial performance have lower risks to fail.

Financial strength was assessed using the Moody's rating developed for large corporations. The 18 levels of ratings were coded the following way: Aaa, the highest rating for a firm was computed as 18, while the lowest rating of Caa-C was computed as 1 and withdrawn rating computed as 0.

3.1.4 Territorial Coverage: Entropy and Store Density

Territorial coverage has received minimal consideration in international market entry studies. However one may expect that a retailer is more likely to maintain operations in a given country if it reaches a level of territorial coverage that is relatively consistent with its other foreign locations. According to Cliquet (1998) growth through addition of new units is essential to the success of a retailer. One of the key measurements of spatial coverage is relative entropy. The following hypothesis was therefore formulated:

Hypothesis 3 (H3). It is proposed that higher relative entropy in a host country lowers the risk of failure in that specific country

Relative entropy is typically calculated the following way (Cliquet 1998):

$$E = -\sum_{i=1}^{k} f_i \log f_i$$

Where

E = entropy
k = number of levels (1 level in our case. The host country)
f_i = frequency of stores (for this specific chain) in area I ($f_i = n_i/N$)
n_i = number of stores in area I (the host country in our case)
N = total number of stores for the chain (worldwide in our case)

Store density, (per type of retailer) in the host country is a good complement to the entropy measurement since it allows the assessment of the degree of retail development and the level of competitive pressure in the host country. According to the "organizational ecologists" defined by Mellahi and Wilkinson (2004) as scholars who explain retail failure through external factors such as density, the higher the store density in a given retail category the higher the competition and the risk of failure (Pal et al. 2007). Hence hypothesis four was derived.

Hypothesis 4 (H4). Higher store density in the host country is positively correlated with failure.

Store density is measured this way: a ratio between the total number of hypermarkets, the only format considered in this research due to dataset limitations, at time of divestment (or 7 years after market entry for successful retailers) and the total population (in 100,000) of the host country that specific year and expressed as a density index. For example, an index of 2 means that there are two hypermarkets for 1 million inhabitants.

3.1.5 Brand Penetration

According to Arkolakis (2010) there is a strong negative relationship between overall brand penetration in a specific territory and relative marketing costs. This has a significant impact on the bottom line. This observation is expressed in hypothesis 5:

Hypothesis 5 (H5). Higher brand penetration in the host country is negatively correlated with failure.

The following formula calculates brand penetration rate in the host countries for each hypermarket chain:

$$f_i = n_i/N$$

where:
f_i = frequency of stores (for a specific chain) in area I (host country in our case)
n_i = number of stores (for a specific chain) in area I (hypermarkets in this research)
N = total number of similar stores in area I (Hypermarkets in this paper)

3.1.6 Cultural and Psychic Distance

Authors frequently use the measurement of cultural and psychic distance in their research on foreign market entry (Kogut and Singh 1988; Wilcox and O'Callaghan 1999; Mitra and Gloder 2002; Huang and Sternquist 2007; Johnson and Tellis 2008; Chen et al. 2009). Hypothesis 6 reflects the general consensus of these authors:

Hypothesis 6 (H6). The stronger the cultural distance between the domestic and the host country the higher the risk of failure.

This variable utilizes the four dimensions of the Hofstede scores for both domestic and host countries by following the method developed by Kogut and Singh (1988):

$$CD_{smt} = \sum_{j=1}^{4} \sqrt{(D_{jst} - D_{jmt})^2}$$

Where "CD_{smt} represents the country distance score between host country s and home country m in year t, D_{jst} is the score on dimension j for host country s, and D_{jmt} is the score on dimension j for home country m both measured in year t" (Johnson and Tellis 2008).

3.1.7 Country Attractiveness Assessment

Restrictive regulation, political risks and economic issues may lead to failure (Huang and Sternquist 2007; Myers and Alexander 2007; Chen et al. 2009). Multiple indexes are available. The market potential index (MPI) developed by Michigan State University has been thoroughly tested through academic research (Cavusgil 1997, 2004). The MPI was specifically designed for emerging countries and covers most of the time period that pertains to the dataset. The index uses the list of emerging countries annually published by the Economist as a frame for its analysis and measures market attractiveness according to seven dimensions: (1) market size, (2) market growth rate, (3) market intensity such as consumption expenditures, (4) market consumption capacity, (5) commercial infrastructure, (6) economic freedom such as trade policy, regulatory climate, taxation policy, (7) market receptivity to imports, transforming raw results into standardized data in order to compare to each other. The MPI values rank from 1 to 100 (100 being the perfect score).

The impact of the market attractiveness will be tested in hypothesis 7:

Hypothesis 7 (H7). The host country overall attractiveness is negatively correlated with failure.

3.1.8 Market Entry Mode Choice

The literature has often researched market entry mode as a potential antecedent of failure or success in market expansion. Palmer and Owens (2006) have focused on the IRJVs while other authors have chosen to investigate international store acquisition or franchising (Quinn 1998; Quinn and Doherty 2000; Quinn and Alexander 2002).

This research investigates the influence of the following modes of entry on the outcome of failure or success: (1) franchising or licensing; (2) IRJV without a controlling position; (3) IRJV with controlling position; (4) acquisition of a wholly owned subsidiary; and (5) greenfield organic growth.

There is no consensus in the literature in regard to the risk of one entry mode choice versus the other. While Li (1995) and Burt et al. (2003) consider that IRJVs present the highest risk, Benito (1997), along with Barkema et al. (1996) and Hennart et al. (1998) claim that acquisitions have the highest propensity to fail. Therefore hypothesis 8 will test the nature of this relationship.

Hypothesis 8 (H8). Market entry mode choice has a significant impact on the market entry outcome of success or failure.

3.1.9 Entry Timing and Order of Entry

The debate is raging between proponents of early market entry and late market entry: Pan and Chi (1999): argue that early entrants receive multiple benefits such as government incentives that are not available to late entrants. On the opposite Golder and Tellis (1993) believe that early entrants may face a long learning curve in emerging countries and Cui and Lui (2005) obtained mixed results in their research on China. Hypothesis 9 was therefore derived:

Hypothesis 9 (H9). The order of entry into the host country influences the outcome of success or failure.

Due to the limited size of the present sample, it was not possible to test this variable in a quantitative manner, hence the choice of a qualitative approach consisting in selecting and analyzing the cases of a few retailers. Results are presented in Table 6.

3.2 Data

This study uses a set of data compiled from multiple sources such as annual reports, statistical offices, and other reliable governmental offices such as the USDA as well as ratings and ranking tools such as the Hofstede's (1991) four cultural dimensions,

Table 2 Descriptive statistics

Variable	Number	Percentage
1. Outcome		
Failure	77	69
Success	35	31
2. Mode of entry		
Licensing or franchising	13	12
Joint venture (No majority)	49	44
Joint venture (Majority)	18	16
Wholly owned subsidiary (Acquisition)	5	4
Wholly owned subsidiary (Greenfield)	27	24
3. Country of origin (per region)		
USA	31	28
Europe	71	63
Other	10	9
4. Host country (per region)		
Latin America and Mexico	30	27
Asia	57	51
Central and Eastern Europe	10	9
Middle East and Africa	15	13
5. Type of retailer		
Food retailers (Hypermarkets)	55	49
Food retailers (Supermarkets)	27	24
Department stores	9	8
Specialty stores	21	19

Moody's credit scores and the Market Potential Index (MPI) developed by Michigan State University.

The 112 cases collected for the empirical research pertained to emerging countries and represented various sectors of retailing with an over-representation of food retailers (see descriptive properties of the data in Table 2). The sample comprised 77 cases of failure and 35 cases of success, spanning 37 years from 1974 to 2011. It is however important to keep in mind that this database includes missing or incomplete data. For example the Moody's rating was only available for 68 out of the 112 cases. Most retailers in this sample originated in Europe, and Asia was one of their main destinations for market expansion. A majority used joint-venture as their original mode of entry and this is certainly partly due to the restrictive regulation on full ownership in certain countries such as China. One may also note that most retailers acquired their joint-venture partner or obtained a majority control position after several years of operation in the host country.

4 Findings

Using independent sample t-tests for each selected variable the authors investigated differences between failed and successful market entries (see Table 3). Significant results were obtained in terms of relative entropy score, density, brand penetration

Table 3 t-Test results differences between failures and successes

Variables	Failure or success	N	Mean	Standard deviation	Sig.
H1 company size (in billion USD)	Failure	66	30.900	0.6411	0.405
	Success	32	38.904	0.4140	
H2 financial performance	Failure	51	7.76	4.402	0.205
	Success	16	9.04	5.046	
H3 territorial coverage (relative entropy)	Failure	61	0.03143	0.0317	0.004***
	Success	32	0.05414	0.403	
H4 territorial coverage (store density-hypermarkets)	Failure	37	2.297	1.869	0.080*
	Success	13	1.249	1.640	
H5 brand penetration (hypermarkets)	Failure	22	0.1159	0.1450	0.001***
	Success	12	0.3772	0.2666	
H6 cultural and psychic Distance	Failure	77	127.09	39.88	0.044**
	Success	35	109.91	44.68	
H7 country attractiveness	Failure	59	52.53	26.06	0.156
	Success	30	44.36	24.09	
Operation duration (in the country)	Failure	77	7.0	4.57	0.000***
	Success	35	14.7	5.47	

*$P < 0.10$; **$P < 0.05$; ***$P < 0.01$

and cultural distance. Calculated for descriptive purposes, the t-test on the operation duration variable revealed significant differences that are linked with the research procedure adopted in this paper. As an average "failed" retailers spent 7 years in the host country before divesting. The other retailers in the sample have been operating in their host country for an average of 14.7 years.

Additionally, as indicated in Table 4, Pearsons correlation tests yielded significant results: as expected there are significant relationships between cultural distance and failure or success; high cultural distance is negatively correlated with success. There is also a positive correlation between relative entropy and success. In addition, as anticipated a higher level of relative entropy occurs in successful chains and store density appears to be negatively correlated with success. Last, country attractiveness seems to be positively correlated with success.

The correlation between company size and the financial performance scores is indirectly relevant for this research. It links the company size with a higher Moody's score that could also connect to market entry success by helping the company to obtain additional funding. It is also interesting to note the negative correlation between relative entropy and company size: the larger the company, the lower the relative entropy level.

Last, a binary logistic regression help to understand the antecedents of market expansion failures.

$$Y = \alpha + \beta_1 X_1 + \beta_2 X_2 + \beta_3 X_3 + \beta_4 X_4 + \beta_5 X_5$$

Table 4 Correlation results

Variables	1	2	3	4	5	6	7
1. Outcome (failure)	1.000						
2. Company size	0.085	1.000					
3. Financial perf.	0.157	0.505***	1.000				
4. Entropy	0.298***	−0.269**	0.053	1.000			
5. Store density	−0.250*	0.053	0.345*	−0.088	1.000		
6. Cultural distance	−0.190**	−0.001	0.262**	−0.146	−0.088	1.000	
7. Country attractiveness	0.201*	0.300***	0.411***	−0.032	0.083	−0.092	1.000

*P < 0.1; **P < 0.05; ***P < 0.01

Table 5 Regression results

	Full-model estimates	
Variables	Logit parameter estimates	Sig.
Company size	−0.467	0.672
Financial performance	0.110	0.399
Market entry mode	−0.546	0.087
Territorial coverage (Relative entropy)	20.187	0.020
Cultural and psychic distance	−0.004	0.913

Where:

Y = Outcome (failure or success) where failure = 0, success =1.
α = Constant of the equation.
β = Coefficient of the predictor variables.
X_1 = Company size
X_2 = Financial performance
X_3 = Market entry mode choice where licensing/franchising = 1, IRJV (no majority) = 2, IRJV (majority) = 3, acquisition (full ownership) = 4, and greenfield (full ownership) = 5.
X_4 = Territorial coverage (relative entropy)
X_5 = Cultural and psychic distance (with the host country)

Results are presented in Table 5.

The full model displays a reasonable explanatory power of 71.7 and a pseudo R-square of 0.220. Although not significant, the results show a positive relationship between financial performance and success. Significant results include a positive relationship between entropy level and success, and a negative one with high-control market entry modes such as acquisitions.

Due to the number of missing cases the following variables were not considered in the full regression model: store density (hypermarkets) and country attractiveness. Separate logistic regressions were conducted for both variables and yielded significant result ($p < 0.1$) for store density whose estimated coefficient of −0.400 shows a negative relationship with the market entry outcome.

However there is no evidence of potential impact of the host country attractiveness (MPI score) on propensity to fail or succeed.

Considering the limited size of the sample, one may be cautious in drawing conclusions despite nevertheless significant statistical results.

H1 was not confirmed, there were no statistically significant differences between larger and smaller companies in terms of failure or success. However one may still note that successful retailers seem slightly larger than their counterparts.

H2 was not confirmed. Both the t-test and the regression results were not statistically significant. H3 was confirmed through the t-test, the Pearsons correlation test and the regression. It appears that the chains with the highest rate of relative entropy for one specific country are the most successful: for example when Tesco divested from Taiwan its entropy level was very low compared to the ones in Thailand and in South Korea after 7 years of operation. Hypothesis 4 was also supported. Store density (H4) varies dramatically between the two groups of retailers ("failed" or "successful"): as depicted through the t-test result, in average the store density was two times higher in the host countries where retailers divested. Similarly the Pearsons correlation test identified a negative correlation between store density and success. That may be due to the fact that successful retailers were often early entrants and therefore even after 7 years of activity, store density, and therefore competition were still fairly low in the host country. The results obtained in this research, in term of brand penetration (H5) are exploratory, due to the limited number of cases that were considered ($n = 36$) and the fact that only hypermarkets were included in the calculation. It appears however, as indicated in the t-test results that successful chains have a much higher penetration rate that their counterparts. Considering this observation in light of the previous comment on store density one may hypothesize that successful retailers may have benefited of a low store density that allowed them to establish themselves as market leaders and gain substantial spatial coverage and market-share. H6 was also supported, through the t-test, the Pearsons correlation but not through the logistic regression and indicated a negative correlation between high cultural distance and success. Although the t-test, measuring the differences according to the MPI score (H7) was not statistically significant it is counter-intuitive to realize that the MPI score was actually lower for the successful market entries than for the failures. However the correlation results presented in Table 4 indicated an opposite result that seems to indicate a potential impact of market attractiveness on the outcome. The cross-tabulation between failure and success and the market entry mode choices (H8) did not yield significant results but the logistic regression indicated a negative relationship between high-control market entry modes such as full ownership and success. It is important to note however that this research took into consideration the original mode of entry and not the governance mode in effect at divestment time. Most organizations seem to move from an IRJV model to full ownership as soon as the country's regulations soften. This may have impacted the results.

Analyzing a few select cases allowed the authors to assess the impact of order entry on failure or success (H9). It is of course preposterous to reach definite conclusions with this limited number of observations but it appeared that in seven cases out of eight, early entrants were rewarded with success. It is however interesting to note that a few chains such as Casino in Brazil and Thailand and Auchan in China seem to have benefited from their late entrance. Detailed results are available in Table 6.

5 Discussion, Managerial Implications and Research Limitations

This article contributes both to the international retailing literature and the research on emerging markets. More specifically studies on the topic of market entry failures are very scarce in the international retailing literature. Moreover most of the existing articles are based on qualitative data and not on a quantitative dataset. In spite of a small sample, this research allows us to identify potential new avenues to investigate in further research: time of entry, store density in the host country, brand penetration and relative entropy seem to play a significant role in terms of success or failure of international retailers in emerging countries. Market entry mode, cultural distance, global financial soundness, and host country attractiveness should also been taken into consideration to a lesser extent.

The debate is raging in regard to the time of entry. Exploratory results in the present article tend to demonstrate that early entrants are more successful. Entry timing seems to be important in this matter since more mature markets with higher store density that may in fact reflect stronger local competition, seem to negatively impact the outcome of the new market entry. Early entrants get the opportunity to increase their brand penetration and reach a higher entropy level due to a lower level of competition. Moreover it appears that retailers that have not reached a certain level of entropy and brand penetration in a given country by year seven are more likely to divest than others. It is a race against time. That would of course be particularly true if the retailer encounters at the same time financial tensions in its domestic market. The relative entropy score may in fact be reflective of the profitability level and overall relative contribution of the local venture as original market entry costs such as national advertising, public relation or other awareness building expenses would not be followed by a sufficient number of openings in the host country and may therefore compare negatively with other international locations across the chain. In that extent this may be why the size of the firm was an irrelevant antecedent to success or failure since the entropy rate is a relative measurement. In other words a smaller chain will reach a high relative entropy score in a given country much faster than a larger retailer. The large retailer would need to make very substantial investments in the host country in order to attain their "acceptable" entropy threshold in the first 7 years. Additionally large firms have a tendency to prefer governance modes that give them high control in their local subsidiaries. The downside of this strategy is that they have to fund their market expansion themselves and the costs may be prohibitive before they reach their

Table 6 Order entry as driver of failure or success

	Host country	Year of entry	Country exit	Still present
Carrefour	Argentina	1982		1
Walmart	Argentina	1995		2
Auchan	Argentina	1997	3	
Ahold	Argentina	1998	4	
Carrefour	Brazil	1975		1
Walmart	Brazil	1995		2
Ahold Brazil	Brazil	1996	3	
Jeronimo Martins	Brazil	1997	4	
Sonae Brazil	Brazil	1997	4	
Casino	Brazil	1999		5
Walmart	Mexico	1991		1
Priceclub	Mexico	1991	1	
K-Mart	Mexico	1993	2	
Carrefour	Mexico	1994	3	
Auchan	Mexico	1995	4	
Pricesmart	Mexico	2003	5	
Auchan	Russia	2002		1
Metro	Russia	2002		1
Marktkauf	Russia	2003	2	
Rewe	Russia	2004		3
Carrefour	Russia	2009	4	
Yaohan	China	1995	1	
Carrefour	China	1995		
Daiei	China	1995	1	
Ahold(Tops)	China	1996	2	
Metro	China	1996		
Walmart	China	1996		
Pricesmart	China	1997	3	
Makro	China	1997	3	
Auchan	China	1999		
Makro	South Korea	1996	1	
Carrefour	South Korea	1996	1	
Walmart	South Korea	1998	2	
Tesco	South Korea	1999		
Makro	Taiwan	1988	1	
Carrefour	Taiwan	1989		
Casino Geant	Taiwan	1998	3	
Tesco	Taiwan	2000	4	
Carrefour	Thailand	1994	1	
Tesco	Thailand	1995		
Auchan	Thailand	1997	3	
Ahold	Thailand	1997	3	
Delhaize (Food Lion)	Thailand	1997	3	
Casino	Thailand	1999		

critical mass threshold. It would be of great interest to develop research in international franchising trying for instance to show to which extent plural form organizations can help franchisors in managing chains (Dant et al. 2008) as it seems to be the case in domestic markets (Bradach 1997).

One may also argue that the mode of entry choice becomes therefore a significant component of success or failure. Franchising or licensing and to a lesser extent, minority IRJVs may help speed up spatial coverage in the host country and therefore increase both brand penetration and entropy score. Moreover a positive correlation between low control mode of governance and success was identified in this research. This exploratory finding that should be tested more extensively in further research slightly contradicts certain authors in the current literature who consider IRJVs as the riskiest mode of entry (Li 1995; Burt et al. 2003). Furthermore most retailers tend to consider IRJVs for example, as a necessary but temporary step when entering a new market. As soon as China entered the World Trade Organization and loosened the regulations on market entry, many international retailers chose to become majority shareholders or sole owners of their foreign business ventures.

Data availability is one of the major limitations for this type of research. Although limited, our dataset seemed nevertheless sufficient to assess our hypotheses by using quantitative methods. The key contribution of this paper has also been to investigate a new potential antecedent of failure or success of international retailing: the entropy threshold as a measure of territory coverage. This is a promising and fascinating field of research that may yield substantial managerial implications. Relative entropy may become a recognized metrics for evaluating foreign subsidiaries for a specific chain and to help assess progresses in terms of store openings in a given country.

6 Conclusion

At time of economic turbulence it is more essential than ever for organizations to accurately forecast and assess their international expansion progresses. It is also crucial to make the right choices when prioritizing their investment. The "me" too approach that may lead a company to enter apparently promising markets such as Brazil, China or Russia may just not work for all retailers. Strategies may require adaptations for each situation by taking into consideration criteria such as retailer's characteristics, host market specificities but also key learning from past experiences. Franchising, licensing and potentially minority IRJVs may become a primary choice for market entry as a way to limit risks in high uncertainty markets and an opportunity to quickly reach a suitable relative entropy level and brand penetration in a host country.

References

Alexander N, Quinn B (2002) International retail divestment. Int J Retail Distrib Manage 30 (2):112–125

Alexander N, Quinn B, Cairns P (2005) International retail divestment activity. Int J Retail Distrib Manage 33(1):5–22

Arkolakis C (2010) Market penetration costs and the new consumers margin in international trade. J Polit Econ 118(6):1151–1199

Barkema H, Bell J, Pennings J (1996) Foreign entry, cultural barriers and learning. Strateg Manage J 17:151–166

Baroncelli A, Manaresi A (1997) Franchising as a form of divestment: an Italian study. Ind Mark Manage 26:223–235

Beamish PW (1985) The characteristics of joint-ventures in developed and developing countries. Columbia J World Bus 20(3):13–19

Benito GRG (1997) Why are foreign subsidiaries divested? A conceptual framework. In: Bjorkman I, Forsgen M (eds) The nature of the international firm. Handelshojskolens Forlag, Copenhagen, pp 309–344

Bradach JL (1997) Using the plural form in the management of restaurant chains. Adm Sci Q 42:276–303

Burt SL, Mellahi K, Jackson TP, Sparks L (2002) Retail internationalization and retail failure: issues from the case of Marks and Spencer. Int Rev Retail Distrib Consum Res 12(2):191–219

Burt S, Dawson J, Sparks L (2003) Failure in international retailing: research propositions. Int Rev Retail Distrib Consum Res 13(4):355–373

Burt S, Dawson J, Sparks L (2004) The international divestment activities of European grocery retailers. Eur Manage J 22(5):483–492

Cairns P, Doherty AM, Alexander N, Quinn B (2008) Understanding the international retail divestment process. J Strateg Mark 16(2):111–128

Cavusgil ST (1997) Measuring the potential of emerging markets: an indexing approach. Bus Horiz 40(1):87–91

Cavusgil ST (2004) Complementary approaches to preliminary foreign market opportunity assessment: country clustering and country ranking. Ind Mark Manage 33:607–617

Chang SJ, Singh H (1999) The impact of modes of entry and resource fit on modes of exit by multibusiness firms. Strateg Manage J 20:1019–1035

Chen YR, Yang C, Hsu SM, Wang YD (2009) Entry mode choice in China's regional distribution markets: institution vs. transaction costs perspectives. Ind Mark Manage 38:702–713

Cliquet G (1998) Integration and territory coverage of the hypermarket industry in France: a relative entropy measure. Int Rev Retail Distrib Consum Res 8(2):205–224

Costil O (2003) L'heure des Choix à l'International pour les Distributeurs. LSA, 1797:14–16

Couturier J, Sola D (2010) International market entry decisions: the role of local market factors. J Gen Manage 35(4):45–63

Cui G, Lui HK (2005) Order of entry and performance of multinational corporations in an emerging market: a contingent resource perspective. J Int Mark 13(4):28–56

Dant RP, Perrigot R, Cliquet G (2008) A cross-cultural comparison of the plural forms in Franchise networks: USA, France, and Brazil. J Small Bus Manage 46(2):286–311

Dawson J (1994) Internationalisation of retailing operations. J Mark Manage 10(4):267–282

Dawson J (2001) Strategy and opportunism in European retail internationalization. Br J Manage 12(4):253–266

Dupuis M, Fournioux J (2006) Building an international retail strategy. In: Dawson JA, Larke R, Mukoyama M (eds) Strategic issues in international retailing. Routlege, New York

Etgar M, Rachman-Moore D (2007) Determinant factors of failures of international retailers in foreign markets. Int Rev Retail Distrib Consum Res 17(1):79–100

Evans W (1996) Retail Border Wars II: Canadian cases in international retailing, CSCA, Centre for the study of commercial activity. Ryerson Polytechnic University, pp 1–22

Gandolfi F, Strach P (2009) Retail internationalization: gaining insights from the Wal-Mart experience in South Korea. Rev Int Comp Manage 10(1):187–199
Gielens K, Dekimpe MG (2001) Do international entry decisions of retail chains matter in the long run? Int J Res Mark 18:235–259
Gielens K, Dekimpe MG (2007) The entry strategy of retail firms into transition economies. J Mark 71(2):196–212
Golder PN, Tellis G (1993) Pioneer advantage: marketing logic or marketing legend? J Mark Res 30(2):158–170
Hang H (2009) The emergence of international retailing in China: pre-conditions and contexts. VDM Verlag Dr. Muller Aktiengesellschaft & Co. KG, Saarbrucken
Hennart JF, Kim DJ, Zeng M (1998) The impact of joint venture status on the longevity of Japanese stakes in US manufacturing affiliates. Organ Sci 9:382–395
Hollander SC (1970) Multinational retailing. Michigan State University, East Lansing
Hofstede G (1991) Culture and organizations: software of the mind. McGraw-Hill, New York
Huang Y, Sternquist B (2007) Retailers' foreign market entry decisions: an institutional perspective. Int Bus Rev 16:613–629
Johnson J, Tellis GJ (2008) Drivers of success for market entry into China and India. J Mark 72 (2):1–13
Kogut B, Singh H (1988) The effect of national culture on the choice of entry mode. J Int Bus Stud 19(3):411–432
Li J (1995) Foreign entry and survival: effects of strategic choices on performance in international markets. Strateg Manage J 16:333–351
Liu W (2008) KFC in China: secret recipe for success. John Wiley & Sons (Asia) Pte, Singapore
Lubman S (1999) Bird in a cage: legal reform in China after Mao. Stanford University Press, Stanford
Luo Y (2002) Partnering with foreign businesses: perspectives from Chinese firms. J Bus Res 55:481–493
Mellahi K, Wilkinson A (2004) Organizational failure: a critique of recent research and a proposed integrative framework. Int J Manage Rev 5/6(1):21–41
Mitra D, Gloder P (2002) Whose culture matters? Near-market knowledge and its impact on foreign market entry timing. J Mark Res 39:350–365
Myers H, Alexander N (2007) The role of retail internationalization in the establishment of a European retail structure. Int J Retail Distrib Manage 35(1):6–19
Pal J, Medway D, Byrom J (2007) Analyzing retail failure from an historical perspective: a case study of A. Goldberg & and sons, Plc. Serv Ind J 26(5):513–535
Palmer M (2004) International retail restructuring and divestment: the experience of Tesco. J Mark Manage 20:1075–1105
Palmer M, Owens M (2006) New directions for international retail joint venture research. Int Rev Retail Distrib Consum Res 16(2):159–179
Pan Y, Chi PSK (1999) Financial performance and survival of multinational corporations in China. Strateg Manage J 20(4):359–374
Parsons AJ (1996) Nestle: the visions of local managers. The McKinsey Quarterly 2(2):5–29
Picot-Coupey K (2006) Determinants of international retail operation mode choice: towards a conceptual framework based on evidence from French specialized retail Chains. Int Rev Retail Distrib Consum Res 16(2):215–237
Picot-Coupey K (2009) Determinants of a retailer's choice of international expansion mode: conceptual model and empirical validation. Recherche et Applications en Marketing 24 (4):23–54
Quinn B (1998) Towards a framework for the study of franchising as an operating mode for international retail companies. Int Rev Retail Distrib Consum Res 8(4):445–467
Quinn B, Doherty AM (2000) Power and control in international retail franchising: evidence from theory and practice. Int Mark Rev 17(4/5):354–372

Quinn B, Alexander N (2002) International retail franchising: a conceptual framework. Int J Retail Distrib Manage 30(5):264–276
Reijnders WJM, Verhallen TMM (1996) Strategic alliances among small retailing firms: empirical evidence for the Netherlands. J Small Bus Manage 34(1):36–45
Robinson T, Clarke-Hill CM (1995) International alliances in European retailing. Int Rev Retail Distrib Consum Res 5(2):167–184
Sousa CMP, Bradley F (2005) Cultural distance and psychic distance: two peas in a pod? J Int Mark 14(1):49–70
Suh YG, Howard E (2009) Restructuring retailing in Korea: the case of Samsung-Tesco. Asia Pac Bus Rev 15(1):29–40
USDA. Retail Food Reports
Vida I, Fairhurst A (1998) International expansion of retail firms: a theoretical approach for future investigations. J Retailing Consum Serv 5(3):143–151
Wang S (2009) Foreign retailers in post-WTO China stories of success and setbacks. Asia Pac Bus Rev 15(1):59–77
Wilcox M, O'Callaghan E (1999) AER Rianta International (ARI). In: Dupuis M, Dawson JA (eds) European cases in retailing. Blackwell, Oxford
Wood S (2002) The limits of portfolio restructuring: lessons from regional consolidation in the US department story industry. Reg Stud 36(5):515–529
Wrigley N, Currah A (2003) The "stresses" of retail internationalization: lessons from Royal Ahold's experience in Latin America. Int Rev Retail Distrib Consum Res 13:221–243

Networks and Their Goals: Implications for Strategic Chain Management

Taras Gagalyuk and Jon H. Hanf

Abstract As suggested by the proponents of the relational view of strategic management, the advantages of an individual firm are often linked to the advantages of the network of relationships in which the firm is embedded (Dyer and Singh, Acad Manage Rev 23:660–679, 1998). The aim of this study is to develop and test a model of goal achievement in chain networks. The model includes the relationships between goal achievement at firm level and network level, the network management's goal achievement and its determinants. We test our model in the context of supply chain networks in the food industry. In particular, our study examines the relationships (1) between a food manufacturer and its independent (upstream) suppliers and (2) between the food manufacturer and its independent (downstream) customers.

Keywords Alignment of actions • Alignment of interests • Network goals • Supply chain networks

1 Introduction

As suggested by the proponents of the relational view of strategic management, the advantages of an individual firm are often linked to the advantages of the network of relationships in which the firm is embedded (Dyer and Singh 1998). Accordingly, there is an ongoing discussion on how to manage a firm's network of relationships

T. Gagalyuk (✉)
Ukrainian Agribusiness Club, Provulok Tbiliskiy 4/10, office 506, 03055 Kiev, Ukraine
e-mail: gagalyuk@agribusiness.kiev.ua

J.H. Hanf
Geisenheim University, Campus Geisenheim, Von-Lade-Str. 1, D-65366 Geisenheim, Germany
e-mail: jon.hanf@hs-rm.de

T. Ehrmann et al. (eds.), *Network Governance*, Contributions to Management Science,
DOI 10.1007/978-3-7908-2867-2_7,

successfully, i.e. such that the firm's competitive advantage is sustained (Gulati et al. 2000; Kale et al. 2002; Dyer and Hatch 2006).

Brinkhoff and Thonemann (2007) show in an empirical study that the unclear definition of common goals can be regarded as one of the major sources for failure of networks. In this context it seems, that the discussion on network management has not exhaustively addressed the "network management–network success–firm success" cause-and-effect relationship. Given that success generally means the achievement of goals, we argue that the link between networks and their success has been understudied, in particular, because of incomplete understanding of network goals. In fact, most empirical studies that focus on the network success or performance address the achievement of goals by an individual firm participating in a network and analyse the role of network-related "collective constructs" such as inter-firm trust, commitment and relational norms (Medlin 2006, p. 860) in achieving those goals. Yet, goals that are set at the network level, i.e. collectively pursued outcomes, are mainly neglected although their presence and relevance in inter-organisational relationships has been widely emphasised (e.g. van de Ven 1976; Pitsis et al. 2004; Provan and Kenis 2007; Winkler 2006).[1]

As shown by Medlin (2006) studying collective constructs needs to be undertaken with regard to both collective and self-interest outcomes. Focussing on solely the goals of an individual firm in a network will provide biased results with respect to management styles that are actually based on self and collective interests. Thus, without simultaneous consideration of goals at the firm and network levels and without an understanding of how the network should be managed in this respect, the whole network's success will remain under-defined and the validity of the derived implications can be questioned.

Therefore, the aim of this study is to develop and test a model of network success that includes the relationships between goal achievement at firm level and network level. Furthermore, it includes the network management's goal achievement and its determinants. We test our model of the whole network's success in the context of supply chain networks in the food industry. In particular, our study examines the relationships (1) between a food manufacturer and its independent (upstream) suppliers and (2) between the food manufacturer and its independent (downstream) customers.

The paper proceeds as follows: First, we delineate the theoretical foundations of supply chain network management building on prior research on management of

[1] Empirical investigations of the achievement of collective goals have been undertaken in the context of dyads, e.g. in strategic alliances (Ariño 2003; Schreiner et al. 2009), strategic partnerships (Mohr and Spekman 1994) and dyadic supply chain relationships (Paulraj and Chen 2005). Provan and Milward (1995) have considered network-level goals in their study of network effectiveness in public sector. Gellynck et al. (2008) measured supply chain performance in the traditional food sector as the level of achievement of goals common to all the supply chain parties. However, to our best knowledge, empirical analysis of collective goal achievement at the network level has not been yet carried out in the strategic management context. Neither has it combined the achievement of network-level goals with the achievement of firm-level goals.

procurement relationships to generate hypotheses that constitute our conceptual model. Next, we test the model and discuss the results. Finally, we derive some implications.

2 Theory and Hypotheses

Among networks in which firms are embedded, there has been a growing interest in networks of procurement relationships (Choi and Kim 2008). Accordingly, the discussion on network management has taken place in the context of so-called supply chain networks[2] (Wathne and Heide 2004; Hanf and Dautzenberg 2006). In this paper, we use the term "supply chain network" as defined by Hanf et al. (2009, p. 46): "A supply chain network involves long-term and recurrent, formal and informal relationships of material, resource, financial and information exchange among more than two participants of the supply chain that are strictly coordinated by the focal firm and aim at fulfilment of certain strategic tasks." This definition generally follows the logic of strategic network theorists (Gulati et al. 2000; Jarillo 1988) who state that a focal firm is in control of a network of other firms and operates as a hub firm, channel, or network captain, and is concerned with the management of the network (Ritter et al. 2004, p. 178).

In this context, the main challenge for the focal actor in managing the supply chain network is adaptation to uncertainty which depends on how the connected relationships are organised (Jap and Ganesan 2000; Wathne and Heide 2004). For example, a manufacturer's ability to adapt in a flexible manner to uncertainty in the downstream relationship can be contingent upon its effectiveness in structuring the relationship with its upstream supplier and vice versa. In this regard, Gulati et al. (2005) have posited that adaptation in the procurement relationship involves fulfilment of the coordination and cooperation tasks. The coordination task is the alignment of actions, i.e. enabling a joint action, whereas the cooperation task is the alignment of interests, i.e. motivation of the exchanging parties. Furthermore, Hanf and Dautzenberg (2006) have shown that individual and collective interests as well as individual and collective actions are interrelated in supply chain networks and, therefore, interests and actions must be aligned at the firm, dyadic and network levels simultaneously (Duysters et al. 2004). Both, partnering and supply chain management strategies have to be derived from the overall collective strategy (Hanf and Dautzenberg 2006, p. 80).

The strategic management literature has mainly addressed collective strategies in the context of their orientation towards reduction of variation in inter-organisational environment (Bresser and Harl 1986). However, in the strategic network context, collective strategies aim not only to shape the network processes

[2] The synonymous terms "supply network" (Harland et al. 2001) and "netchain" (Theuvsen 2004; Omta et al. 2001) have been also increasingly used in the supply chain management literature.

and relationships but also to achieve certain network goals (Sydow and Windeler 1998, p. 268). In a strategic network in which a focal firm is responsible for the correctness of attributes of the final product (Hanf and Dautzenberg 2006), a collective strategy will be most often goal-oriented. To exemplify, Tesco, the largest food retailer in the UK, has formed its beef supply chain network setting up effective guidelines for managing relationships with suppliers and customers. Following these guidelines, Tesco has enhanced long-term vertical and horizontal cooperation among the network members and effectively informed the public about mad cow disease and food issues in general. As a result, the retailer has been successful in selling British meat and in establishing equivalent standards for meat produced on its behalf overseas and imported into the UK (Lindgreen and Hingley 2002, p. 166). In this example, the collective strategy has aimed to achieve the food safety and animal welfare goals as well as the economic goals of Tesco and its suppliers who have benefited from meeting the retailer's standards.

The notion of goal-orientation by collective strategies can be supported by the findings of organisation studies. For instance, Locke (2004) has shown that the focus on goals generates insight into the design of organisational structures and incentive systems, whereas Ethiraj and Levinthal (2009) have emphasised that goals are necessary to direct and coordinate behaviour. Consequently, we posit that network goals may be used to define what the supply chain network's success is. Beyond that, we suggest that a collective strategy may be perceived as a framework of activities to sustain a network's success because it aims at the achievement of network goals. We further describe the dimensions of network success in detail and develop hypotheses on the antecedents of network goal achievements.

2.1 Network Goals

To consider network goals, one needs a multiple-constituencies approach (Provan and Kenis 2007) because there are multiple parties to a network. This includes each participating firm as an independent organisation, the network's management, and the community, i.e. consumers, non-governmental organisations, and the government (Ariño 2003, p. 68). Similarly to Ariño (2003), in this paper we solely focus on the goals of network members and network management by assuming that they are constrained by the goals of other constituencies and, therefore, reflect them insofar as they are constrained by them. Accordingly, we work out a model in the following paragraphs. The model is presented in Fig. 1.

2.1.1 Network Members' Goals

The interrelationship of individual and collective interests implies that the success of individual network members is critical to the success of the whole network and, conversely, positive outcomes for the whole network contribute to the single firm's

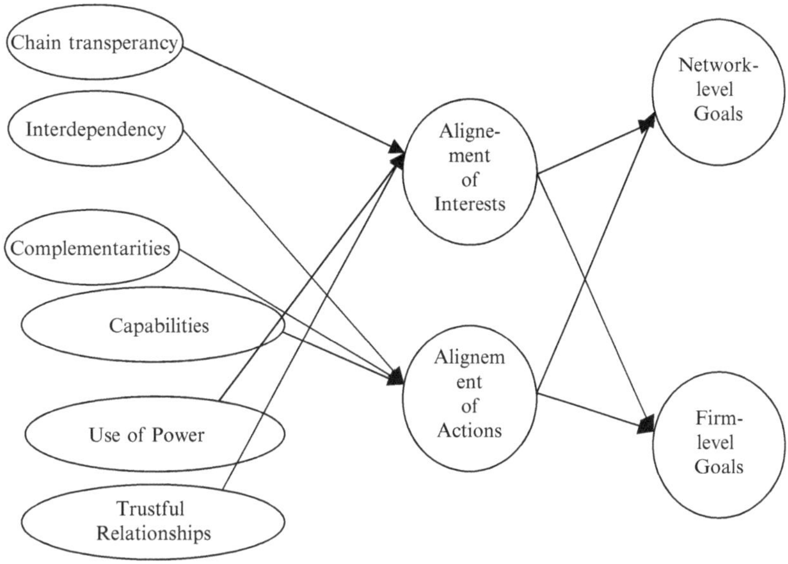

Fig. 1 Model of supply chain success

success. Thus, success of a supply chain network will involve the achievement of the network members' goals which can be addressed in relation to the abovementioned levels of a network. At the firm level, firms are setting their individual goals whereas they are setting the collective goals at the network level.

We understand **network-level goals** to be the predefined set of outcomes that are collectively pursued by all network members. Further they can be achieved only if all network members work together. Although such shared goals have rarely been addressed in empirical analyses (Sydow and Windeler 1998), their examples can be found in the food industry. For example, various aspects of food safety and quality addressing primarily the increasing consumers' demands and the risk of food scandals Despite the fact that we define network-level goals as the collective outcomes, in strategic networks they have to be seen as viable and acceptable primarily by the powerful stakeholders (Sydow and Windeler 1998, p. 274). As a type of strategic network, a supply chain network is most often deliberately established by a powerful chain captain, either distributor- or manufacturer-brand owner, who selects appropriate supply chain partners to develop products under its brand (Belaya and Hanf 2009). We therefore suggest that the network-level goals are at first hand defined by a powerful focal firm.

Arguing in this manner, we do not aim to contest the importance of **firm-level goals**, i.e. goals which single firms want to achieve for themselves by participating in a network. Instead, we emphasise that goals of the whole supply chain network involve network-level and firm-level goals. The network-level goals are set by the focal actor and are jointly pursued by all the network members. The firm-level goals are set by individual network participants that exert their individual efforts in pursuit of these goals within a given network.

2.1.2 Goals of the Network Management

The achievement of goals of network members requires members to synchronise their actions as well as to consent on goals and procedures to achieve goals. In this context, the above conceptualised alignment of actions and alignment interests can be seen as goals of the network's management.

The **alignment of actions** is necessary to implement concerted, joint actions needed to capitalise on the specialised but interdependent activities of partners (Schreiner et al. 2009). In the context of strategic networks, the firms need to combine and integrate their resources and knowledge across organisational boundaries to create competitive advantage (Gulati et al. 2000). Consequently, there exists a high interdependence of tasks between partners that involves managing a complex and overlapping division of labour, linking their specific activities with each other, and making regular mutual adjustments. In such a situation, the greater the joint efforts taken by the partners to manage their activities, and/or the more a partner becomes involved in activities that are traditionally considered the other's responsibility and vice versa, the greater their ability to compete successfully with the marketplace (Schreiner et al. 2009, p. 1402). The aligned actions will consequently imply that partners provide timely and reliable responses to each other's work-related needs, being responsive to concerns arising at the firm level of individual partners as well as at the network level. Accordingly, we hypothesise:

Hypothesis 1 (H1). The alignment of actions has a directly positive effect on the achievement of network-level goals.

Hypothesis 2 (H2). The alignment of actions has a directly positive effect on the achievement of firm-level goals.

The **alignment of interests** is meant to be the establishment of good working relationships among the parties. It addresses factors such as the degree of compatibility of firms' cultures and decision-making styles, a convergence of business views, and other organisational characteristics (Ariño et al. 2001). The alignment of interests of the network members facilitates higher levels of trustful relationships, commitment and low levels of conflict among members so that confidence in the reliability and integrity of the partners is gained. Furthermore, the alignment of interests enables organisations to gather high-quality information about the others and creates strong disincentives for opportunistic behaviour (Ariño et al. 2001). Finally, interest alignment can be defined as the degree to which the members of the organisation, e.g. a strategic network, are motivated to behave in line with organisational goals (Gottschalg and Zollo 2007). We therefore hypothesise that:

Hypothesis 3 (H3). The alignment of interests has a directly positive effect on the achievement of network-level goals.

Hypothesis 4 (H4). The alignment of interests has a directly positive effect on the achievement of firm-level goals.

2.2 *Determinants of Goal Achievement*

In order to evaluate strategic networks, Gulati et al. (2000) have proposed to consider three types of relational characteristics: network structure, network membership, and tie modality. Network structural characteristics describe the overall pattern of relationships in the network. Network member characteristics include the identities, resources, access, and other features of the network actors. Tie modality is the set of institutionalised rules and norms that govern appropriate behaviour in the network (Gulati et al. 2000, p. 205). Based on the ideas of Gulati et al. (2000), we analyse respective constructs that reveal how the network structure, network membership, and tie modalities affect the achievement of goals of the network management.

2.2.1 Network Structural Characteristics

Supply chain networks consist of a multitude of participating firms. Therefore, the embedded upstream and downstream flows of resources and information have to cross various stages of the chain while the involved firms differ widely in size. As a result, supply chain networks are highly complex systems and they bear the high risk of failure. Hence, reducing complexity is one of the most important tasks. In particular, the supply chain network's management has to consider comprehensively the levels of transparency and interdependence.

Transparency refers to the extent of coverage from upstream industries to downstream industries within the supply chain and how apparent information is to downstream industries (Theuvsen 2004, p. 125). Dyer and Singh (1998) have emphasised the role of transparency in transferring knowledge among partners. Because of the complex nature of supply chain networks, their structure is often not made public to all network members, and a feeling of anonymity may appear. Such missing transparency of the network structure increases the probability of free-riding. Transparency is associated with the establishment of strong ties and open communication. Therefore, it will be primarily conducive to enabling the partners' knowledge of each other's decision-making styles, and certainty in intentions of each other. We accordingly hypothesise that:

Hypothesis 5 (H5). Higher levels of transparency have a directly positive effect on the alignment of interests.

Interdependence is acknowledged by firms when they join forces to achieve mutually beneficial outcomes (Mohr and Spekman 1994). Supply chain network's structure is characterised by high numbers of interdependencies among members.

In this respect, a firm's performance depends on how it environs itself with other companies (i.e., its suppliers and customers). Because the magnitude of interdependencies is mainly disproportional at the different stages of the supply chain, the establishment of a joint action is an extremely difficult task. Furthermore, beyond the firm's set of first-level contacts, there is a limited amount of intentionality possible on the part of the focal firm in terms of coordinating the wider network (Gulati et al. 2000). Based on these arguments, we hypothesise:

Hypothesis 6 (H6). Higher levels of interdependence have a directly negative effect on the alignment of actions.

2.2.2 Network Membership Characteristics

Research on networks focuses primarily on the interrelationships of firms but single enterprises can be regarded as initial elements of networks because collaborations do not exist without them. Each partner in a network dedicates its unique resources and capabilities which, when combined with partners' resources and capabilities, can create inimitable and non-substitutable value (Dyer and Singh 1998). We therefore express the network membership characteristics by the constructs of firms' complementarities and coordination capabilities.

Network members' complementarities create incentives for firms to collaborate (Park and Ungson 2001). Noteworthy, collaborations do not inevitably create advantages for the involved firms. Instead, especially during their establishment, they absorb resources. Consequently, without the firms' willingness to cooperate, collaboration will not prevail. Thus, firms have to recognise collaboration not as a constraint but as a means to access complementary resources. Furthermore, since supply chain networks are formed to last over a long period, complementarities are not only essential at the beginning of collaboration but throughout the whole period. Thus, complementarities in culture and strategies (Park and Ungson 2001) combined with resource complementarities (Dyer and Singh 1998) will be conducive to action alignment among the network members.

Hypothesis 7 (H7). Network members' complementarities have a directly positive effect on the alignment of actions.

Coordination capabilities of firms include necessary skills and abilities to establish learning routines, build up unique and network-specific knowledge, use modern information technologies, etc. Despite collaboration is determined by the complementary abilities of the involved firms, only a part of the firm's strategic resources is sensitive to synergy (Dyer and Singh 1998). Therefore, higher coordination capabilities of the network members have the potential to enhance their concerted action (Schreiner et al. 2009). As a result, we hypothesise:

Hypothesis 8 (H8). Higher levels of coordination capabilities have a directly positive effect on the alignment of actions.

2.2.3 Tie Modalities

We acknowledge that the ultimate tie modalities will be reflected by the extent of interest alignment, it is important to clarify how inherent distinctions among actors are smoothed to preclude the negative consequences of relationships. To overcome problems of opportunistic behaviour by the network members, some scholars pose that it is feasible to exert power (Payan and McFarland 2005), the others recommend to employ trust-based enforcement mechanisms (Dyer and Singh 1998). Furthermore, several studies emphasise that the use of non-coercive power (e.g., rewards, recommendations, etc.) has positive impact on the relationships while the use of coercive power (e.g., punishment, threats, etc.) negatively affects the relationships (Leonidou et al. 2008). We verify these suggestions by analysing the effects of **trustful relationships** and **non-coercive power** on the alignment of interests.

Hypothesis 9 (H9). Trustful relationships have a directly positive effect on the alignment of interests.

Hypothesis 10 (H10). Use of non-coercive power has a directly positive effect on the alignment of interests.

3 Methodology

This section explains the survey design, the operationalisation of variables, and the statistical procedure used to analyse the data.

3.1 Survey Design

To test the model, data was collected from branded food manufacturers in Ukraine from September 2009 to November 2009. We assume a branded food manufacturer to be a focal company in a network of firms that work together to bring the branded product to the market. The branded food manufacturer is responsible for the attributes of the branded product and, therefore, is knowledgeable about the network to a large extent. The database of the firms was obtained from a locally based market research company. Totally, the database comprised 359 firms.

A questionnaire was designed based on a review of literature on such variables as strategic partnership, supply chain and strategic alliance performance.[3] Then, the questionnaire was pretested with five food chain specialists. Those specialists were buying and quality managers of international food retailers, one CEO of an

[3] The questionnaire can be provided by authors upon request.

international standardisation body and a CEO of a non-governmental organisation being active in the food business. The respondents were asked to make their comments on the order of questions, wording and format of the questionnaire. Their feedback was considered to modify the questionnaire.

Overall 101 telephone interviews were conducted. This resulted in a response rate of 28 %. Each interview lasted about 20 min on average.

3.2 *Measures*

We now turn to operationalise the variables used in the model. Corresponding measures were obtained from the literature on performance of supply chains, strategic alliances, strategic partnerships and inter-organisational relationships.

Network members' goal achievement. These measures assess the degree of fulfilment of goals at the network and firm levels from the perspective of the focal firm. In each case, a four point-scale measuring the informants' assessment from "very dissatisfied" to "very satisfied" was employed. At the network level, the focal company's overall satisfaction with cooperation (Ariño 2003) as well as its assessment of overall satisfaction by its suppliers and customers were used as the measures of network-level goal achievement. We also employed the focal actor's satisfaction with the total chain quality and sales of the branded product as measures at the network level. At the firm level, we operationalised goal achievement by satisfaction of the focal firm's suppliers and customers with knowledge gained within a network, reputation from cooperation as well as with profit generated within a network (Schreiner et al. 2009).

Network management's goal achievement. We operationalised the alignment of interests by the following measures: focal company's satisfaction with communication within a network (Mohr and Spekman 1994), confidence in reliability of the partners (Schreiner et al. 2009) and the extent of suppliers' and customers' relation-specific investments (Dyer and Singh 1998). The alignment of actions was measured by the responsiveness of suppliers and customers and their willingness to perform necessary tasks (Provan and Kenis 2007; Schreiner et al. 2009). In each case, a four point-scale measuring the focal firms' assessment from "very dissatisfied" to "very satisfied" was employed.

Network Structural Characteristics. Transparency was measured by the focal company's degree of awareness of suppliers' and customers' decision-making styles and by the degree of openness of the focal firm's decision-making styles to suppliers and customers. The measures of interdependence were drawn form Mohr and Spekman (1994) and include the extent to which the focal firm is able to easily substitute its suppliers and buyers and vice versa (reverse coded). For both, transparency and interdependence, a four point-scale from "totally disagree" to "totally agree" was employed.

Network Membership Characteristics. The cultural and the strategic fit (Park and Ungson 2001) of suppliers and customers measured the network members' complementarities. Coordination capabilities were operationalised by the suppliers' and customers' agreement on task distribution and by their firm size (Schreiner et al. 2009). For both, complementarities and coordination capabilites, we used a four point-scale from "totally disagree" to "totally agree."

Tie Modalities. We measured trustful relationships by the focal firm's willingness to always inform its suppliers and customers about future steps and by the suppliers' and buyers' perception of favourability of participation in a network (Mohr and Spekman 1994). The use of non-coercive power was measured by frequency of placing bonuses and providing recommendations to suppliers and customers (Payan and McFarland 2005). For trustful relationships, a four point-scale measuring the informants' assessment from "totally disagree" to "totally agree" was employed; the use of non-coercive power was measured by a four point scale from "very rarely" to "very frequently."

3.3 Path Analysis

To test the model, we used the Partial Least Squares (PLS) technique for Structural Equation Modeling using the SmartPLS software 2.0.1 (Henseler et al. 2009). Our decision to use PLS was based on its advantages compared to other techniques, i.e., the possibility to analyse small size samples in the absence of distribution assumptions. PLS involves analysis of two forms of variables, i.e., the latent and the manifest variables. Manifest variables that make no significant contributions to the respective latent variables are progressively removed and the analysis is repeated until all the manifest variables are significant (Gyau and Spiller 2009).

4 Results

In this section, we test the model and represent the estimated results.

4.1 Testing the Measurement Model

The fit of the measurement model in PLS is evaluated with regard to the inner and the outer models. Individual item reliabilities and convergent validity of the model provide information about the fit of the outer model. The individual item reliabilities are evaluated via the factor loadings of the items on their constructs. According to Hair et al. (1998), an item is considered insignificant and removed from the model if its factor loading is less than 0.4.[4] We also calculated the

[4] Due to space limitations, we do not provide tables with the results of the measurement model testing. The tables can be provided by authors upon request.

composite reliability of the measurements to evaluate internal consistency of the measurements. All the composite reliability indices for the constructs exceed the recommended 0.7 homogeneity criterion. The convergent validity was estimated by calculating the Average Variance Extracted (AVE). The recommended threshold of 0.5 (Bagozzi and Yi 1988) was exceeded for all the constructs indicating that the chosen indicators are explained by their respective constructs.

The fit of the inner model was evaluated by the discriminant validity criterion which means that every construct is significantly different from the others. The first way to analyse discriminant validity is a comparison of item loadings and cross loadings. If all loadings are higher than cross loadings, then the construct significantly differs from the others. The second way is to compare the square root of the AVE with the correlation between the construct and the other constructs. The square root of the AVE should be higher than the correlation between the constructs (Gyau and Spiller 2009). In both cases our results support the fit of the inner model.

4.2 The Structural Model

The structural model was evaluated based on the R2 and the significance of the path coefficients. The variances explained (R2) for each of the endogenous variables were as follows: achievement of network-level goals 0.520, achievement of firm-level goals 0.213, alignment of interests 0.341, and alignment of actions 0.216. Considering the complexity of the research model, the results for the achievement of network-level goals which we brought to the forefront of our argumentation are indicating good fit. In addition, rather moderate R2 values for the achievement of firm-level goals and the alignment of actions reflect the complex nature and manifold determinants of these constructs. To determine the significance of the path coefficient we used the bootstrap method with 200 re-samplings. Path coefficients and their significance are shown in Table 1. The standardised path coefficients can be used to analyse the degree of accomplishment of the hypotheses. One may multiply the path coefficients by the correlation coefficient between the latent variables to obtain an approximate measure of the variance of the construct explained by the latent predictive variable (Gyau and Spiller 2009). We show the result in Table 1. Using this approach, one might consider values of less than 1.5 % as not making significant contribution to their respective latent variables (Gyau and Spiller 2009, p. 30). Based on this criterion, we accepted nine out of the ten hypotheses that were formulated.

5 Discussion

This study contributes to the research on food chain and network management by investigating goals that are pursued in supply chain networks. Prior empirical research has been characterised by numerous efforts to analyse supply chain and

Table 1 Results of the structural model

Hypotheses	Constructs	Expected sign	b	Correlation coefficient (r)	b*r
H1	Alignment of actions → Network-level goals	+	0.644**	0.667	0.430
H2	Alignment of actions → Firm-level goals	+	0.384**	0.403	0.155
H3	Alignment of interests → Network-level goals	+	0.275*	0.328	0.090
H4	Alignment of interests → Firm-level goals	+	0.226**	0.258	0.058
H5	Transparency → Alignment of interests	+	0.271**	0.358	0.097
H6	Interdependence → Alignment of actions	−	−0.338**	−0.408	0.138
H7	Complementarities → Alignment of actions	+	0.087	0.020	0.002
H8	Coordination capabilities → Alignment of actions	+	0.185*	0.335	0.062
H9	Trustful relationships → Alignment of interests	+	0.379**	0.375	0.142
H10	Non-coercive power → Alignment of interests	+	0.326*	0.312	0.102

*Significance of the path coefficients is based on t-values at the $p < 0.05$ level
**Significance of the path coefficients is based on t-values at the $p < 0.0001$ level

network performance. However, these efforts have rarely addressed network-level goals, i.e. goals that are jointly pursued by all network members. This is in contrast to theoretical studies that have conceptualised the achievement of shared goals as the measure of network effectiveness (Provan and Kenis 2007). Building on those theoretical studies as well as on few empirical contributions, we provide a theoretical account of goals that are set in supply chain networks. We conceptualise goals of a whole supply chain network as those set by members at the firm and network levels. Moreover, we relate the achievement of network-level and firm-level goals of network members to the achievement of goals of the network management. The latter include the alignment of interests and the alignment of actions which, if fulfilled simultaneously, pave the way for the achievement of both network-level and firm-level goals of the network participants. Furthermore, we hypothesise that the interest and action alignment are contingent upon a number of network characteristics.

The achievement of network-level goals is to a large extent explained by how properly cooperation and coordination problems are solved by the network management. Our results support hypotheses H1 and H3 which state that the alignment of actions and the alignment of interests, respectively, have a positive effect on the achievement of members' goals at the network level. This finding supports the ideas by Gulati et al. (2005) and Hanf and Dautzenberg (2006) who highlight the strategic value of viewing chain management as a multifaceted construct that consists of cooperation and coordination elements at the different levels. In particular, the alignment of actions has a strong and significant effect on the achievement of network-level goals emphasising the role of a joint action and strong coordination skills by the focal actor in achieving shared outcomes.

Hypotheses H2 and H4 are also supported, i.e. both the alignment of actions and the alignment of interests have a significant positive effect on the achievement of firm-level goals. Thus, as supposed by Medlin (2006), the joint action and collective interest constructs are closely linked to individual constructs in business relationships. In our case, this implies that successful chain management has beneficial outcomes also at the firm level of suppliers and customers. Although the respective effects (path coefficients) are weaker at the firm level than at the network level, the effect of interest alignment (t-value) on firm-level goals is even more significant than on network-level goals. As a result, we suggest that the aligned interests, i.e. good working relationships are important to improve perceptions by single suppliers and customers with regard to the achievement of their individual goals.

Interest alignment is, in turn, subject to a significant positive effect by higher levels of transparency, trustful relationships, and non-coercive power as proposed in hypotheses H5, H9, and H10, respectively. These results are consistent with the findings of earlier research. For example, Deimel et al. (2008) have revealed that high levels of transparency are associated with explicitness and clearness of information as well as with partner commitment. Handfield and Bechtel (2002) have shown that trustful relationships have a significant effect on partner responsiveness. Leonidou et al. (2008) have found that the exercise of non-coercive power is negatively related to conflict in inter-firm working relationships.

The alignment of actions is negatively affected by higher levels of interdependence. This result supports our hypothesis H6 and is consistent with the results of Mohr and Spekman (1994) who have modelled positive relation between higher level of interdependence and partnership success but found no significant correlation. Although interdependence is usually addressed as enabler of collaboration, we analyse this construct in the context of a whole supply chain network. Accordingly, higher interdependence in relationship with e.g. a supplier implies higher contingency upon volatilities in supplier's relationships with its suppliers. As a result, action alignment appears to be complicated. In this regard, we have also taken into account specifics of our research setting. Despite the wide scope of vertical coordination practices and the growing use of chain management concepts, the business environment in Ukraine is highly volatile with persisting infrastructural problems (Gagalyuk and Hanf 2009).

The remaining hypotheses (H7 and H8) state that network members' complementarities and higher levels of coordination capabilities have a directly positive effect on coordination. Only the latter of these constructs has a significant influence on the alignment of actions. This indicates that the suppliers' and customers' abilities to identify and build consensus about task requirements in a network contribute to successful resolution of coordination problems and establishment of a joint action. This result coincides with the findings of Schreiner et al.

(2009) who have confirmed the positive link between alliance management capability and joint action. At the same time, higher levels of coordination capabilities exhibit a rather moderate effect on the alignment of actions (path coefficient = 0.185). This supports the proposition by Provan and Kenis (2007) who have suggested that networks with lead organisations will demonstrate a moderate need for network-level competencies of members as lead organisation is better suited to address network-level demands and needs.

In general, our results are with regard to the effects of network structure, membership and tie modalities on the outcomes of supply chain network management not surprising as they correspond to the findings of the other authors. The results, however, must be evaluated with some caution as we surveyed only focal firms. Accordingly, future research should take a direction similar to what Gellynck et al. (2008) did in the traditional food sector in the EU, i.e. it should encompass all network participants. Yet, it should also take account of both, shared and individual goals of network members. Another issue for future research would be the comparison of the goal achievement among different supply chain networks. Hereby, the approach we used, i.e. focus on perceived rather than objective measures seems to be suitable because different supply chain networks have distinctive features and, thus, objective measures will mean little without a benchmark for comparison.

Overall, we think that the most important finding of our study is that the network-level goals really exist in strategic networks and must be considered along with firm-level goals. As such, the topic of network goals has to be of particular interest for firms which are responsible for introduction and implementation of the network's strategy (often referred to as a collective strategy). Those focal firms or chain captains have to be particularly concerned that the network participants agree upon network-level goals and work together to achieve them. Thus, shared goals have to be clearly formulated and explicitly addressed, and a certain degree of compatibility between network-level and firm-level goals has to be reached. Therefore, the interests and the actions of the involved parties have to be aligned with consideration of the parties' social and organisational characteristics. Special attention should be paid to the development of network management capabilities that would enable capturing of goals of the whole network.

Acknowledgement The authors would like to thank the foundation Alfred Toepfer Stiftung F.V.S. for financial support of this research.

References

Ariño A, de la Torre J, Smith Ring P (2001) Relational quality: managing trust in corporate alliances. Calif Manage Rev 44:109–131

Ariño A (2003) Measures of strategic alliance performance: an analysis of construct validity. J Int Bus Stud 34:66–79

Bagozzi R, Yi Y (1988) On the evaluation of structural equation models. J Acad Mark Sci 13:989–1006
Belaya V, Hanf JH (2009) The two sides of power in business-to-business relationships: implications for supply chain management. Mark Rev 9:361–381
Bresser RKF, Harl JE (1986) Collective strategy: vice or virtue? Acad Manage Rev 11:408–427
Brinkhoff A, Thonemann UW (2007) Perfekte Projekte in der Lieferkette. Harv Bus Manag 7:6–9
Choi TY, Kim Y (2008) Structural embeddedness and supplier management: a network perspective. J Supply Chain Manage 44:5–13
Deimel M, Frentrup M, Theuvsen L (2008) Transparency in food supply chains: empirical results from German pig and dairy production. J Chain Netw Sci 8:21–32
Duysters G, Heimeriks KH, Jurriens JA (2004) An integrated perspective on alliance management. J Chain Netw Sci 4:83–94
Dyer JH, Singh H (1998) The relational view: cooperative strategy and sources of interorganizational competitive advantage. Acad Manage Rev 23:660–679
Dyer JH, Hatch NW (2006) Relation-specific capabilities and barriers to knowledge transfers: creating advantages through network relationships. Strateg Manage J 27:701–719
Ethiraj SK, Levinthal D (2009) Hoping for A to Z while rewarding only A: complex organizations and multiple goals. Organ Sci 20:4–21
Gagalyuk T, Hanf JH (2009) Impact of retail internationalization on agribusiness: the case of Ukraine. J East–west Bus 15:96–118
Gellynck X, Molnár A, Aramyan L (2008) Supply chain performance measurement: the case of the traditional food sector in the EU. J Chain Netw Sci 8:47–58
Gottschalg O, Zollo M (2007) Interest alignment and competitive advantage. Acad Manage Rev 32:418–433
Gyau A, Spiller A (2009) An integrated model of buyer-seller relationship performance in agribusiness: the partial least squares approach. J Chain Netw Sci 9:25–41
Gulati R, Lawrence PR, Puranam P (2005) Adaptation in vertical relationships: beyond incentive conflicts. Strateg Manage J 26:415–440
Gulati R, Nohria N, Zaheer A (2000) Strategic networks. Strateg Manage J 21:203–216
Hair J, Anderson R, Black W (1998) Multivariate data analysis. Prentice Hall, Upper Saddle River
Handfield RB, Bechtel C (2002) The role of trust and relationship structure in improving supply chain responsiveness. Ind Mark Manage 31:367–382
Hanf JH, Dautzenberg K (2006) A theoretical framework of chain management. J Chain Netw Sci 6:79–94
Hanf JH, Dautzenberg K, Gagalyuk T, Belaya V (2009) Network approach to supply chain management: terms, scope of issues and lines of development. Russ Manage J 7:43–68
Harland CM, Lamming RC, Zheng J, Johnsen TE (2001) A taxonomy of supply networks. J Supply Chain Manage 37:21–27
Henseler J, Ringle CM, Sinkovics RR (2009) The use of partial least sqaures path modeling in international marketing. Adv Int Mark 20:277–319
Jap SD, Ganesan S (2000) Control mechanisms and the relationship life cycle: implications for safeguarding specific investments and developing commitment. J Mark Res 37:227–245
Jarillo JC (1988) On strategic networks. Strateg Manage J 9:31–41
Kale P, Dyer JH, Singh H (2002) Alliance capability, stock market response and long-term alliance success: the role of the alliance function. Strateg Manage J 23:747–767
Leonidou LC, Talias MA, Leonidou CN (2008) Exercised power as a driver of trust and commitment in cross-border industrial buyer–seller relationships. Ind Mark Manage 37:92–103
Lindgreen A, Hingley M (2002) Monitoring buyer-seller relationships: the Tesco scorecard. In: Paradoxes in food chains and networks: Proceedings of the Fifth International Conference on Management in AgriFood Chains and Networks. Wageningen Academic Publishers, Wageningen, The Netherlands, pp 165–175
Locke EA (2004) Linking goals to monetary incentives. Acad Manage Exec 18:130–133
Medlin CJ (2006) Self and collective interest in business relationships. J Bus Res 59:858–865

Mohr J, Spekman R (1994) Characteristics of partnership success: partnership attributes, communication behavior, and conflict resolution techniques. Strateg Manage J 15:135–152

Omta AWF, Trienekens JH, Beers G (2001) Chain and network science: a research framework. J Chain Netw Sci 1:1–6

Paulraj A, Chen IJ (2005) Strategic supply management and dyadic quality performance: a path analytical model. J Supply Chain Manage 41:4–18

Park SH, Ungson GR (2001) Interfirm rivalry and managerial complexity: a conceptual framework of alliance failure. Organ Sci 12:37–53

Payan JM, McFarland RG (2005) Decomposing influence strategies: argument structure and dependence as determinants of the effectiveness of influence strategies in gaining channel member compliance. J Mark 69:66–79

Pitsis TS, Kornberger M, Clegg S (2004) The art of managing relationships in interorganizational collaboration. M@n@gement 7:47–67

Provan KG, Kenis P (2007) Modes of network governance: structure, management, and effectiveness. J Public Adm Res Theory 18:229–252

Provan KG, Milward HB (1995) A preliminary theory of network effectiveness: a comparative study of four community mental health systems. Adm Sci Q 40:1–33

Ritter T, Wilkinson IF, Johnston WJ (2004) Managing in complex business networks. Ind Mark Manage 33:175–183

Schreiner M, Kale P, Corsten D (2009) What really is alliance management capability and how does it impact alliance outcomes and success? Strateg Manage J 30:1395–1419

Sydow J, Windeler A (1998) Organizing and evaluating interfirm networks: a structurationist perspective on network processes and effectiveness. Organ Sci 9:265–284

Theuvsen L (2004) Transparency in netchains as an organizational phenomenon: exploring the role of interdependencies. J Chain Netw Sci 4:125–138

Van de Ven AH (1976) On the nature, formation and maintenance of relations among organizations. Acad Manage Rev 1:24–36

Wathne KH, Heide JB (2004) Relationship governance in a supply chain network. J Mark 68:73–89

Winkler I (2006) Network governance between individual and collective goals: qualitative evidence from six networks. J Leadersh Organ Stud 12:119–134

Part B
Cooperatives

The Resilience of the Cooperative Form: Cooperative Beehiving by Swedish Cooperatives

Karin Hakelius, Kostas Karantininis, and Li Feng

Abstract The paper identifies the phenomenon of cooperative beehiving. Members de-associate themselves from large cooperatives and form smaller entities, just as bees swarm out of the old crowded beehive in search for a new one. We show in the framework of transaction cost theory that the exiting farmers are those who have experience and advantages in organizing cooperatives and are willing to take risks as entrepreneurs. The new beehives, organized also as cooperatives, rely heavily on outsourcing and start-up assistance plans. Two cases from the Swedish agrifood industry illustrate our claims.

Keywords Cooperative • Cooperative beehiving • Swedish agrifood industry • Transaction costs

> "When the population of a hive rises, one portion of the bees leave in a group, together with the queen and begin looking for a new place to settle" H. Yahya, "The miracle of the honeybee", p. 112

1 Introduction

Cooperatives worldwide have been undergoing waves of successes and failures, many cooperatives cease to exist, others restructure, many have demutualized, while organizational innovations have emerged (Chaddad and Cook 2004; Galor 2008; Fulton and Hueth 2009). The trend in the 1980s and 1990s has been for cooperatives to strive for economies of scale through mergers and acquisitions.

K. Hakelius (✉) • K. Karantininis • L. Feng
Department of Economics, Swedish University of Agricultural Sciences, PO Box 7013, 750 07 Uppsala, Sweden
e-mail: Karin.Hakelius@slu.se; Karantininis.Konstantinos@slu.se; Li.Feng@slu.se

T. Ehrmann et al. (eds.), *Network Governance*, Contributions to Management Science,
DOI 10.1007/978-3-7908-2867-2_8,

Some farmers, however, have been following the opposite path: instead these farmers de-associate themselves from large cooperatives and form smaller entities, just as bees swarm out of the old crowded beehive in search for a new one. Recently, in Sweden, a number of farmers left their cooperatives and formed smaller cooperatives. We document two cases in this chapter, one dairy and one beef cooperative. In both cases the founding members of these cooperatives were members of very large cooperatives, *Arla Foods* and *Swedish Meats*, respectively.

What is fascinating for the organizational economist is the resilience of the cooperative form of organization. Why do farmers choose the cooperative form for the new business? What drives farmers away from large cooperatives? What are the factors of success of the new "beehives"? These are the questions in focus in this chapter. The inability of large cooperatives to deal with specialized products, and the control problem, i.e. the general dissatisfaction of cooperative members with management (Cook 1995; Porter and Scully 1987) are also central here. The choice of the cooperative as the governance structure for the new entity has been discussed earlier, e.g. by Chaddad and Cook (2004) who refer to the "stickiness" of the cooperative form. They attribute this to the cooperative's advantage to deal with transaction costs and property rights issues. Conditions under which the cooperative form is more efficient for financial decisions are derived by Hendrikse and Bijman (2002). Similarly, more stylized formulations show the advantages and disadvantages of cooperatives versus IOFs (Albaek and Schultz 1998; Karantininis and Zago 2001).

Cooperative beehiving is not a new phenomenon. Hendrikse and Bijman (2002) and Bijman and Hendrikse (2003) document a number of cases in the Dutch agri-food sector where producers gradually abandoned the cooperatively owned auction and formed specialized growers' associations. The emergence of New Generation Cooperatives in the 1980s in the USA constitutes also a similar case (Fulton and Hueth 2009). Many reasons for cooperative failure and consecutive cooperative restructuring have been sited. Financial constraints are a common reason. Cooperative organizations have not been very successful in raising capital for investing in product and market innovations, especially in the highly differentiated modern agri-food system, mainly due to not properly defined property rights (Cook 1995; Hendrikse and Bijman 2002). Based on similar argumentation, Hendrikse and Veerman (2001) argue that large multi-purpose cooperatives are ill-fit to invest in specialized assets, such as brand name capital and specialized processing and distribution systems. Smaller, more specialized cooperatives increase producers' countervailing power, although often adverse selection may be a problem. The trade-off between the production economies of size of the large cooperatives, their slow response and reluctance to involve in product differentiation on one hand, and the transaction costs and diseconomies of scale of the smaller cooperatives on the other, drive the final outcome. This choice is the subject investigated in this chapter.

This study contributes to the development of the cooperative theory by identifying and defining the concept of cooperative beehiving, and outlining a theoretical framework to analyse the beehiving phenomenon. To capture more information of interest in details and depth rather than data points, we use the case study methodology to investigate this recent phenomenon in Sweden within its real-life context (e.g. following the methodology outlined by Dul and Hak 2008). We find that when the large cooperative is reluctant to pursue opportunities for product differentiation at small scale and to pay qualified farmers higher prices, some farmers will choose to exit. Those who are more involved in the administration of their old cooperative will be the leaders in the new entity since they are the carriers of the organizational know-how. Also, the availability of an outside existing facility, and their ability to outsource and achieve economies of size, are key factors of success.

First, we start with the historical background of the Swedish agricultural cooperative sector, with a focus on dairy and slaughter. Second, the theoretical framework drawing from cooperative theories, and transaction cost economics, will provide hypotheses as to the economic and other reasons for cooperative beehiving. Third, a detailed analysis of two cases—one dairy and one meat cooperative from Sweden—will be presented. Finally, we summarize and conclude and present implications.

2 Historical Background

Influenced by German and British experiences, the farmers of Sweden founded cooperative associations, starting in the end of the nineteenth and the beginning of the twentieth centuries. Typically, the Swedish cooperatives were focusing on one line of production within agriculture, for example dairy, slaughter, grain and crop[1] (Nilsson 1997; Johansson 1985, 1994; Lindahl 2004). Somewhat later, during the 1930s, forest owner cooperatives were also formed (see Johansson 1985, 1994) (Table 1).

In the middle of the twentieth century, farmer cooperatives had reached a strong position in Swedish economy and society—judging from the share of the population they represented at that time, i.e., 20 % (Svenska Institutet 22t Qd). The reason was that the farmer cooperatives had helped greatly with the domestic food supply during the two world wars—being in direct contact with farmers and covering the entire nation. Hence, the Swedish government felt that the farmers should be protected against future competition, having as the main goal to keep an agricultural production leading to self-sufficiency of food products. A system of border protection, export subsidies, and pre-set domestic prices for agricultural inputs and agricultural products, was gradually developed (Johansson 1985, 1994; Fakta om svenskt jordbruk 1996; Lindahl 2004). In order to serve as a strong counterpart to the government, in the recurrent negotiations, a federated system was developed

[1] As can be seen in Table 2, however, two cooperatives (i.e., NNP and Norrmejerier) were both active in dairy and slaughter, mainly due to that these two cooperatives were active in the North of Sweden where the farm density is low.

Table 1 Number of cooperative associations by agricultural sector in Sweden

Coop. sector	1900	1910	1920	1930	1940	1950	1960	1970	1980
Dairy	430	550	565	715	726	375	233	46	24
Slaughter	1	3	21	25	37	25	24	19	10
Forest owners				No data	30	29	23	12	9
Grain & crop	No data	850	1,353	662	795	619	347	144	86

Source: Johansson (1994, p. 75)

and in 1971 the *Federation of Swedish Farmers* (*LRF*) was formed, having both producer cooperatives, such as dairy, slaughter, and banking, as well as individual farmers as members.[2]

The farmer cooperatives reached high market shares under these circumstances: in the dairy sector, they represented 99 % of the market and about 80 % in the slaughter, grain and crop sectors, and 70 % in the poultry sector in 1997 (Bager 1997; Nilsson 1997). Typically, regional cooperatives covered a specific geographical area and no or very little domestic competition took place.

The cooperatives gradually grew larger through mergers, and the farms also became fewer and larger due to a combination of older farmers retiring and the remaining farmers focusing on developing more efficient production processes on their farms (Lantbrukskooperativ årsbok 1996, 1998). Hence, one common feature in the agricultural sector has been that many farmers have bought or leased neighboring farms in order to become more efficient and profitable. In the case of the dairy sector, the largest cooperative in the end of the 1980s was *Arla*, having 64 % of the domestic market in 1991 (Lantbrukskooperativ årsbok 1996, 1998). The dairy cooperatives were members of the national organization *Svensk Mjölk* (*Swedish Milk*). Table 2 summarizes some key figures on the dairy cooperatives in 1991, 1994, and 1997. As can be seen there, *Arla's* number of members decreased, while the turnover increased during this period. Table 2 also shows that *Arla's* size—both in terms of number of members and turnover—exceeded the other dairy cooperatives by far. *Skånemejerier*, being on second place in size only corresponded to 17 %, and 20 % of *Arla's* membership and turnover in 1997, respectively. *Arla* attempted to merge with a small dairy cooperative, which would have given *Arla* a 66 % market share. This attempt, however, was eventually stopped by the *Swedish Competition Authority* in 1992. After this, *Arla* started to look for collaboration possibilities both in Finland and Denmark, an effort that was intensified once Sweden joined the European Union in 1995. Finally, a merger with Danish *MD Foods*, forming *Arla Foods*, took place in April 2000.

In 1971, there were 18 regional slaughter cooperatives, and one mixed (dairy and slaughter), joined together at the national level into one organization called *Slakteriförbundet* (*The Slaughterhouse Association*). There were several mergers during the 1980s and 1990s, leading to five slaughter cooperatives by 1993

[2] Today, the LRF has 29 cooperative organizations and 170,000 individuals as members. In addition, the LRF has eight subsidiaries, for example working with insurance, financial consulting services, and media. Source: LRFs homepage.

Table 2 Membership and turnover of Swedish dairy cooperatives 1991, 1994, 1997

	1991		1994		1997	
Cooperative dairy	Members	Turnover (MSEK)	Members	Turnover (MSEK)	Members	Turnover (MSEK)
Arla	15,710	11,086.8	11,628	11,689.9	9,385	13,298.0
Falköpings Mejeri	427	364.4	414	296.9	358	280.7
Gefleortens Mejeriförening	377	3.8	341	265.7	243	266.9
Gäseneost	205	58.2	163	93.6	147	96.5
Milko	2,176	1,373.0	2,075	1,466.7	1,793	1,391.6
NNP	5,385	2,048.3	4,444	2,061.0	3,515	1,973.7
Norrmejerier	1,746	1,079.2	2,319	1,536.5	1,870	1,349.4
Skånemejerier	2,184	1,533.0	2,063	2,478.9	1,641	2,596.3

Note: the NNP and Norrmejerier cooperative were active in both dairy and slaughter. *Source*: Lantbrukskooperativ årsbok (Yearbook of farmer cooperatives) (1996, 1998)

Table 3 Membership and turnover of Swedish slaughter cooperatives 1993, 1995, 1997

	1993		1995		1997	
Cooperative slaughterhouse	Members	Turnover (MSEK)	Members	Turnover (MSEK)	Members	Turnover (MSEK)
Scan Farmek	34,709	6,521.4	33,651	7,913.8	31,600	8,291.6
Scan KLS	4,810	972.0	4,329	867.9	3,955	880.9
Scan Norrland	8,340	784.5	7,266	687.2	6,864	607.2
Skanek	16,394	5,167.6	12,595	3,377.3	11,154	3,157.3

Source: Lantbrukskooperativ årsbok (Yearbook of farmer cooperatives) (1996, 1998)

(Lantbrukskooperativ årsbok 1996, 1998). The situation in 1993, 1995, and 1997 is summarized in Table 3, showing the largest slaughter cooperative (*Farmek*) with more than 30,000 members, followed by *Skanek*, having less than half the size of *Farmek's* membership (Lantbrukskooperativ årsbok 1996, 1998). In 1998, *Slakteriförbundet* changed name to *Swedish Meats*, and an attempt was made to merge all existing regional associations to *Swedish Meats*, but this attempt was unsuccessful. Eventually, the Finnish *HK Ruokatalo* bought what is now known as *HK Scan AB*[3] in the end of 2006. It is important to stress, however, that both *HK Ruokatalo* and *HK Scan AB* are so-called farmer controlled businesses (FCBs), as the majority of the votes are in the hands of farmer cooperatives: the *LSO* in Finland, and *Sveriges Djurbönder* (*Swedish Animal Farmers*) in Sweden. The number of members of *Sveriges djurbönder* has decreased from c 20,600 in the beginning of 2008 to about 16,200 in the beginning of 2011, i.e., a decrease of 21 % (Sveriges Djurbönder's Annual Report 2010).

In the opening quotation, we refer to the beehive analogy and the increasing population of bees that causes bees swarm out in search of a new beehive. Increased membership in the case of cooperatives is not the actual cause of the beehiving

[3] AB = IOF, or joint-stock company/corporation.

phenomenon, but rather increased and differentiated volume and the associated diseconomies of organization.

In order to explore what factors may lead to the fairly new phenomena on the Swedish scene of farmers leaving large, established, farmer cooperatives, to form new small-scale cooperatives, and what the factors of success for the new cooperatives might be, we have studied two cases in Sweden: *Sju Gårdar* (*Seven Farms*, referring to the number of founding members—today there are six members who deliver milk) and *Upplandsbondens* (*Uppland Farmers*). *Upplandsbondens* was formed in the fall of 2006, while *Sju Gårdar* in 2008. Both cooperatives are active in the region of *Uppland* (see map in Fig. 1) and both are producing organic products: the former is a dairy cooperative, consisting of six active dairy producers; the latter is a slaughter cooperative, having 86 members. The empirical data about these cooperatives is partly from official documents and the internet,[4] and partly from two interviews with two directors of these small-scale cooperatives.

The board secretary of *Sju Gårdar*, Mrs *Elisabeth Gauffin*, is an agronomist with a specialization in animal husbandry. She runs her farm together with her husband and two children. She was sitting on the board of *Arla*, when she decided in 2008 to exit and start a small-scale cooperative together with a few farmers in the same geographical area.

In 2006, Mrs *Inger Gauffin Carlsson*, also running her farm with her husband and two children, decided to form a new small-scale cooperative that they named *Upplandsbondens*. In the beginning, the production on her farm included dairy cows, but since large investments were required and *Inger* wanted to devote a lot of time to board-work, it was decided that they were to focus on meat production.

As mentioned earlier, both beehive cooperatives are focusing on organic production and they have limited their activities to the region *Uppland*. The main reason for leaving the large cooperative was that their requirement of premium prices for their organic products was not respected. In the case of the meat producing farmers, they also objected against the extensive transportation of the animals: up to 275 km—or 3 h—to the closest slaughterhouse in the city of *Linköping*. The new beehives consist of members with similar production orientation and potential. *Elisabeth* described those leaving *Arla* as having different educational background, being risk tolerant, willing to work together with others, tired of the old system and therefore willing to try something new. Albeit it being a big step, both could always return to the old cooperative (*Elisabeth* has to pay a new member fee of SEK 18,000[5] in order to re-enter as a member of *Arla*, while *Inger* has stayed as a passive member of *HK Scan AB*).

These two cases show examples of why members leave their cooperative, but further analysis is needed to understand issues such as: under what conditions a member can afford to leave the mother cooperative and form a new one and even compete with the original cooperative; what challenges does such an exit imply?;

[4] See homepage for *Sju Gårdar* and *Upplandsbondens* in Swedish.

[5] 1 SEK is about 0.15 USD, or 0.11 EUR (April 5, 2012).

Fig. 1 Map of Sweden and Uppland

what are the factors leading to the success of the new endeavour?; and why is the new entity also a cooperative?

3 Theoretical Framework

Below, the theoretical framework, based on cooperative theory, organization and transaction cost theory, will be presented. The theoretical framework leads to the development of hypotheses.

3.1 *Theory of Cooperatives*

A producer cooperative is an enterprise collectively owned by many independent suppliers. It involves both a horizontal arrangement between members and a vertical coordination mechanism between the upstream members and the downstream processor.

What distinguishes a cooperative from an IOF (investor owned firm) with a single focus on profit maximization is members' plurality of interests (Trifon 1961). So the guiding principle regarding understanding a cooperative is that members advance the interests of their own farm portfolio through a cooperative. They place the cooperative between themselves and a market they must deal with (Fulton 1988). However, agricultural markets are showing the tendency to become more heterogeneous on the demand as well as the supply side. Consumers demand more variety and higher quality; producers respond to intensified competition from globalisation and saturated markets by developing and marketing a broader range of new products (Hendrikse 2011). The developments in agricultural markets highlight the importance of specific assets at the downstream stage of production, which puts pressure on the upstream oriented cooperatives. Wirenga et al. (1997, p. 53) state that a "drawback of co-operatives is that their locus of power (and perspective), even if they have integrated processing and distribution facilities, is close to primary production and far moved from the market. This does not make them very suitable for taking the guiding role in an agrifood value-added partnership, the very purpose of which is to derive competitive advantage from adding those values that consumers want." The implication seems to be to abandon the cooperative structure. The trend towards differentiation and innovation has resulted in changes in the governance of marketing channels, like horizontal mergers, associations falling apart, emergence of dual distribution and heterogeneous cooperatives (Fulton and Hueth 2009). What is very curious, still, is the persistence of the cooperative organizational form, especially when it comes to the governance of innovations and entrepreneurship.

Opportunities for successful innovations are by all means important to an enterprise. The free riding problem as the result of team production impairs cooperatives' incentives for differentiation and innovation. Helmberger (1966) raises the question "… how can an individual with an entrepreneurial flair be rewarded for his talents by the creation of a cooperative?" (p. 1430). A member who perceives and seizes an opportunity has to share the surplus or residual in proportion to his patronage, not according to his contribution of good judgment and business acumen. Another internal factor, according to Helmberger (1966), that confines the growth and development of cooperatives is the "single origin constraint" imposed by the special interest group that form the organization in the first place. "The cooperative … may need to pass up many good prospects that are incompatible with its life blood" (p. 1431). Furthermore, the leadership paradox of members leads to the tendency that cooperatives are often poorly managed. "To the extent that farmers participate in the leadership role, they may contribute to poor decisions and hamstring management; to the extent that they don't, ownership is separated from control" (p. 1431).

Cooperatives, especially those large complex ones, suffer also from the collective decision making process entailed by the democratic nature of the organizations. The cost of group decision making is likely to increase with the size and diversity of the cooperative. The decision making procedure in cooperatives is usually much slower than in IOFs. Because many of the decisions affect the distribution of income among the members, cooperative members are more likely than their IOF counterparts to seek involvement (e.g., via the board) in deciding a broad range of issues that are considered merely strategic in an IOF (Staatz 1987). When multiple principals engage in an entrepreneurial exercise, the challenge is how to combine the institutional frameworks of investor-driven shareholder firms and patron-driven forms of collective action (Cook and Plunkett 2006). Of interest to this study is the observation that cooperative entrepreneurs often choose to re-organize a cooperative entity, even after they break up from the original cooperative.

Despite of its resilience, the cooperative is not the dominant organization form, and we observe a lot of cooperative failures. To explain this, Cook (1995) suggests a life cycle theory, where the dominance of the cooperative form rises and declines through time. Thurow (2001) points to the economic history of the U.S. as evidence for a decline of most forms of cooperative organizations, attributing this to factors related to "social capital". Thurow describes the ability to get organized as one of the first elements of social capital. The success of this ability is to a large extent a function of the willingness to have leaders and follow them. He underlines the importance of the nature of organization, namely whether they have a cooperative or competing outlook. These observations are key to the success and failure of the cooperative and are crucial to the formation of beehive cooperatives. In the next section we elaborate more on cooperative failures.

3.2 Cooperative Failures

We have seen that there are fundamental differences between a cooperative and an IOF. It is useful to view the differences among these two business entities in terms of property rights (Hansmann 1996). "The residual claimants to the income generated by the cooperative are its users, whereas in an IOF the capital owners are the residual claimants" (Fulton 1995, p. 1146). It turns out that this fundamental difference creates several problems for the cooperative resulting from the conflict over residual claims: the horizon problem, the non-transferability problem, and the control problem.

1. *The horizon problem* is created when the claims on an asset are shorter than the life of the asset. It is argued that this is the case for producer cooperatives, where members' claims last as long as they are users which is usually shorter than the productive life of most assets. The horizon problem may be one of the main obstacles of capital acquisition by the cooperatives (Harris et al. 1996).
2. *The non-transferability* refers to the fact that members' claims on the cooperative's cash flow are contingent on patronage and are not marketable. This further creates what Jensen and Meckling (1979) call "the portfolio problem": because cooperative claims cannot be bought or sold (no such market exists), the members' ability to diversify or concentrate their asset portfolios is limited. In turn, they will pressure the cooperative management to re-arrange the cooperative's investment portfolio to fit their needs. Hence, one may expect the level and pattern of diversification of a cooperative to differ from that of an IOF. Caves and Petersen (1986) suggest that cooperatives will diversify more than IOFs. Their argument mainly derives from political theory and theory of clubs. The non-transferability problem may cause also some free rider problem since new members of a cooperative are entitled to the same level of price and residual claims as existing members (Cook 1995).
3. *The control problem* is the typical principal-agent problem arising in any firm where there exists separation of ownership and management: the agents being the cooperative managers, whereas the principals are the members. It is argued that this problem is more severe in a cooperative than in an IOF and will result in

scale inefficiency because monitoring costs increase as the number of members increases (Porter and Scully 1987).

The empirical evidence on the impact of these problems on cooperative performance is rather conflicting. Porter and Scully (1987) and Ferrier and Porter (1991) attribute allocative as well as X-inefficiency found in cooperatives in their sample, to these factors. On the other hand, Parliament and Taitt (1989) and Sexton and Iskow (1993), fail to accept most of these hypotheses.

Cooperative failures may take different forms. Fulton and Hueth (2009) document 13 cases of cooperative failures in U.S. and Canada. Broadly, they categorize them in three groups: (1) those that went into bankruptcy or converted to an IOF because of poor financial performance; (2) those that converted to an IOF because of a need to acquire additional capital or a desire to access market value; and (3) those that were in the process of forming or were re-engaging in the market (for example, after bankruptcy). In this chapter, we distinguish a fourth type, which is not actually a failure but rather a re-birth of another cooperative through the process of beehiving. In order to understand further this process it is important to delve further into the organizational aspect of this governance structure.

To understand the process of failure-rebirth through beehiving, we need one more piece in the theoretical puzzle: transaction cost economics (TCE).

3.3 Transaction Costs

The cooperative is one form of governance to deal with vertical integration, forward or lateral, between firms. Perry (1987) sites three broad categories of determinants of vertical integration: (1) technological economies; (2) transactional economies; and (3) market imperfections. While the traditional view was founded on technological economies of scale (Stigler 1951), it is broadly recognized today that this technological argument does not hold, unless we assume absence of transaction costs (Coase 1937; Williamson 1975).

Since TCE have been presented repeatedly in a multitude of publications, it is not necessary to replicate this theory here.[6] However, in a nut-shell we should point out that one of the key elements of TCE is that contracts are incomplete. This incompleteness can distort ex ante investment incentives, and can weaken the efficiency of ex post performance and the adaptation to unforeseen changes in supply and demand conditions. These problems surface themselves when the parties involved in the contractual arrangements are locked-in, especially due to specific investments (asset specificity). Other factors that contribute to the potential contractual hazards are uncertainty, information asymmetries, and bounded rationality.

[6] See for example a recent treaty on TCE by Tadelis and Williamson (2012); also Joskow (2005).

In brief, the principal hypothesis of the transaction cost theory can be summarized in three parts: (a) transactions differ in their attributes; (b) transactions are aligned with governance structures which differ in their cost competences; (c) governance structures are chosen by minimization of the sum of production plus transaction costs. Furthermore, the three pervasive attributes (dimensions) of transactions are (1) asset specificity; (2) uncertainty; and (3) frequency. Asset specificity can take several forms, such as physical, human, site, brand name capital, etc., and measures the degree to which an asset is redeployable outside the transaction. Uncertainty is important because it results into imperfect contracting and maladaptation of the transaction process. Frequency of the transaction "...bears both on the efficacy of reputation effects in the market and the incentive to incur the setup cost of specialized governance" (Williamson 2004).

The analysis following the transaction cost approach is dynamic; it covers the transaction in its entirety, and can be divided into two stages: ex ante and ex post. Ex ante costs refer to those costs of searching, haggling, writing the contract, etc. Ex post costs are associated with motivating, managing and monitoring to deal with opportunistic behaviour. While the emphasis of the property rights and agency theories is on ex post costs, transaction cost theory focuses on ex ante costs.

4 Hypotheses

While large farm cooperatives often involve themselves into product differentiation in large scale, for example, by introducing new types of processed products, yogurts, cheeses, etc., they are reluctant to introduce product differentiation that necessitates differentiation among producers. Organic production and products with denomination of origin are common examples of this phenomenon. One reason being a significant amount of remorse emerging when members receive different prices.

Let us take two dairy farms, farm Type-I and farm Type-II. We assume that Type-I farm has higher production costs and the farmer-owner is heavily involved in the cooperative board. One important reason for the higher production costs for Type-I farmers is that these farmers will have to have more employees at the farm, in order to be able to devote time to the cooperative board. The farmer-owner in the Type-II farm is more efficient in terms of production, and is not involved in the cooperative besides its business part. The underlying assumption is that a Type-I farm has a comparative advantage in "organizational technology", while farm Type-II has a comparative advantage in "production technology"—and vice versa. The farms of Type-I are relatively productively inefficient, while Type-II farms are inefficient when it comes to organization and cooperation with other farms. In fact, not both assumptions are necessary, our arguments and results would work as long as some farms are relatively better "organizers" and better "co-operators" than others.

Let there be a potential for a differentiated product, say with a local brand name which can fetch a higher price than the conventional products of the cooperative. This is potentially feasible with a sunk investment in marketing, promotion, building the brand name, etc. Let us assume that only a small group of producers would qualify for such a program de facto—for example because of geographic location, or some previous sunk investment (for example in some technology, say organic production, new variety or breed, etc.). However, the cooperative management is often reluctant to engage in such a process due to the control problem, or potential conflict between those who are eligible and those who are not, since the former group would receive a higher price. Hence we can put forward the first hypothesis:

H1. When an opportunity for product differentiation at small scale arises the large cooperative is reluctant to pursue it and pay qualified farmers higher prices.

Several of those frustrated potentially eligible producers are tempted to search for ways to reap these benefits, however there are two barriers: production economies of size, and transaction costs. The industry (say dairy, or meat processing) is characterised by very large economies of size in procurement, packaging, distribution, processing, etc. The potential differentiated market is very small relative to the scale economies in this industry and as a result the potential production cost of the differentiated product can be very high if it is produced at a small scale. When this cost is internalised by the member farms it raises their costs at such levels that the new business is uneconomical since costs exceed the expected price. How can the farmers that want to produce the differentiated product resolve this problem? The farms may outsource the processing and distribution operation, however, they have to resolve two more burdens: sunk costs and transaction costs.

4.1 Sunk Costs and Subsidies

There is no easy way to start up a new business, especially when it involves a new brand. Start-ups require investments in lawyers, brand name, market research, etc., and these costs are sunk and often not insignificant. It is a heavy burden for a small number of farmers to undertake. These farmers have then the following alternatives: one solution is to spread this over a large number of farmers; however this is by definition impossible because we assumed the differentiated product is a small scale. A second alternative is to seek capital from outside investors or to borrow. "Going public" is not an easy endeavour for a small number of farmers entering a marginal business in the food industry. This is a model more suited to young start-ups in the Silicon Valley—not for farmers with dirty boots in the farm lands, producing bulk products. Borrowing is of course always an option but, again, it puts an extra burden which will have to be paid eventually and is very risky. Finally, there is a start-up subsidy from the state or elsewhere—for example from EU funds. This is not uncommon and as we shall see in our case studies, this was successfully used by both of the newly formed cooperatives, primarily by *Upplandsbondens*. Since the start-up investment is often too large they require a large scale. Neither of the two types of farms, I or II, can afford to make the investment on their own. Hence any form of subsidy may act as a catalyst. This brings us to the second hypothesis:

H2. In the absence of other sources of funding, a start-up assistance plan may help the break-up farmers to undertake the necessary setup sunk costs.

What is required for a successful absorption of such outside funding is a serious plan, which is an outcome of a coherent and solid organization. To this we turn soon, but before this, let us look at the other piece of the puzzle: outsourcing.

4.2 Outsourcing: A Source of Transaction Costs

Given the very large economies of size, a newly formed small entity must rely on another large existing facility with excess capacity which is willing to serve the new *beehive* for a fee. This is a *sine qua non*—a necessary condition-for the success of the quest for the new beehive.

This requires a contract loaded with transaction costs and leads to the "make or buy" decision, i.e. will the new cooperative build (make) a new facility or simply outsource (buy) from an existing one? If there exists such a facility, the new cooperative will find it cheaper to outsource the processing; otherwise it will be too expensive to build its own capacity. Hence we can put forward the third hypothesis:

H3. An existing facility with excess capacity for outsource processing is necessary for the success of the new beehive.

4.3 The New Beehive Organization: Why a Cooperative?

The new beehive cooperative must deal with these transaction costs described above. The choice of governance is the key question at this stage. The farmers at the beginning attempt to achieve their goals through their original cooperative [what Hirschman (1970) calls "voice"]. When this is not achieved, a small number of them consider splitting ["exiting", using the terminology of Hirschman (1970)]. Their alternatives are to (a) join another cooperative—if it exists; (b) sell in the spot market, i.e. to an existing IOF; (c) form a new IOF; (d) form a new cooperative. We will rule out options (a) and (b), in our case: (a) because no other cooperatives exist in the market; (b) because their idea of the differentiated production is new, and no IOF will be willing to invest on this, besides if the IOF does make any long-term investment benefits will be taken by the IOF and not by farmers. Hence the farmers are limited in options (c) and (d). We will see why farmers chose option (d): to form a new cooperative.

We put forward two reasons why the choice of the cooperative form is appropriate in this case: one reason has foundations in agency theory and the second in transaction cost economics.

Agency problems: first, there exist information rents due to asymmetry of information between outside investors and farmers—especially Type-I, who are better informed than any outsider about the quality characteristics and the potential of their products. As a result, an outside IOF will incur higher costs and bear larger risks than the Type-I farmers. The second reason is similar to the franchising problem: ex post, after the launch of the differentiated product proves successful there will be a brand name capital created, having characteristics of public good for all participants, and with the potential to be appropriated by, for example the IOF. In order to safeguard the ownership of this brand name capital the cooperative is an appropriate governance structure (Hansmann 1996; Holmström 1999). We can then propose Hypothesis 4:

H4. Internalizing information costs and safeguarding the brand name of the new entity is best done with a cooperative structure.

One may argue that size is a necessary condition for building brand name capital, since it requires significant investments. However, as we have seen above (**H2**) this problem in our two cases was solved by outside funding through subsidised start-up funds.

The hypothesis put forward by transaction cost theory is that the choice of governance structure is in a discrete cost-minimizing manner. So far, we have illustrated two types of production costs: the sunk costs and the processing costs. It is our hypothesis that both of them are outsourced: sunk costs from grants (**H2**), processing from existing facilities with excess capacity (**H3**). What about the transaction costs?

First, as we have seen above, due to low frequency there exist high set up costs of specialized governance. New organizations do not come for free. There is a large amount of time and resources required in negotiations, building trust, creating a common understanding, besides the "ink" costs of forming the new entity legally. It is our hypothesis that the farmers who choose to split and form the new beehive are Type-I farmers when their comparative advantage in "organizational technology" is significant to offset their inefficiency in production. Being close to decision making and corporate information, they are the first to discover and spot the new opportunities. Also, having the experience of meetings and organizational matters they are better equipped to set up the new entity at a lower transaction cost than their Type-II colleagues.[7] Hence, just like in the bee colonies a group of bees swarms around the queen and leaves to form a new beehive:

H5. The Type-I farmers will form the new beehive as a cooperative.

[7] Hendrikse and Bijman (2002) propose slightly different results. However, according to our definition Type I farmers are farmers efficient in organization technology and not in producing higher quality product like the Hendrikse-Bijman Grower 1.

5 Empirical Findings

In this Section we illustrate the hypotheses presented above with the two case studies *Sju Gårdar* and *Upplandsbondens*. The information concerning these two cases has been collected from interviews, from official documents, various internet sources,[8] and a lecture[9] delivered by Mrs. *Elisabeth Gauffin* (Gauffin, 2011a). Two independent interviews were conducted with mrs *Elisabeth Gauffin*, chairwoman of *Sju Gårdar* (Gauffin, 2011b), and Inger *Gauffin Carlsson*, chairwoman of *Upplandsbondens* (Carlsson, 2011). The interviews took place at the interviewees own farms on April 6, and October 5, 2011 respectively. Both interviews were administered by the authors and lasted approximately 2 hours each. The interviewees are sister-in-laws.

5.1 *Production Costs*

As mentioned in Sect. 2, the key reason for leaving the large cooperative and starting a new, small one was that the established cooperative did not adhere to price premium requirements. In both cases, voice was tried first, but failed. It was not only the farmers that eventually broke loose from their large cooperatives who practiced voice, many others did too, but only some took the step to exit. Producing organic milk and meat is more costly than producing conventionally. All feed has to be organically produced and this leads to higher feed costs. In addition, the per-animal volumes produced are lower than in conventional production, adding to the unit costs. Hence, not receiving a premium price means economic difficulties on farm-level and this spurred some farmers to take the step and leave the large cooperative.

In both cases, the costs for legal assistance, development of trademarks, and performing market analyses—what we refer to as sunk costs above (Sect. 4.1)—was highly underestimated. In addition, especially in the case of *Sju Gårdar*, it has turned out that “organic” was not the most important label—“locally produced” was, however. Therefore, the necessary investments have been putting great pressure on the financial situation of both beehive cooperatives. In the case of *Upplandsbondens*, a great part of these costs have been covered by EU-support money (from the rural development program) and they are worried what will happen if that money ceases to come.

The second part of the production costs, is the processing costs. Being small-scale organizations, it is not feasible to invest in processing facilities of their own. Instead, they have to rely on outsourcing the processing of their products. In the case of *Sju Gårdar*, a medium-sized established cooperative (*Gefleortens mejeriförening*[10]) proved to have excess capacity and they managed to reach an agreement, which

[8] See homepage for *Sju Gårdar* and *Upplandsbondens* in Swedish.

[9] In the course “Cooperatives and Other Agri-Food Systems”, held at the Swedish University of Agricultural Sciences (SLU).

[10] The Dairy Association of Gävle (i.e., a city north of Stockholm, on the east coast).

seems to work smoothly. *Upplandsbondens* has a less stable situation when it comes to outsourcing their processing. They use a skilled butcher and have invested in processing equipment, but the impression we got is that this collaboration is somewhat less stable than between *Sju Gårdar* and *Gefleortens mejeriförening*. In addition, *Sju Gårdar* has so far been more successful in creating a well-known brand as well as finding channels to reach the consumer.

In conclusion, we find that Hypotheses **H1**–**H3** hold—especially when it comes to handling the processing costs. By succeeding well with this, *Sju Gårdar* has been better at bearing the sunk costs related to becoming an established market actor themselves, while the *Upplandsbondens* has had to rely on EU-subsidies in order to manage their market entry.

Previous studies have shown that cooperatives in other countries, for example The Netherlands (Hendrikse and Bijman 2002) have adjusted by involving into product differentiation and offering members price premiums, etc.

5.2 Type-I Versus Type-II Farmers

A closer look at the individuals leaving a large cooperative in order to start a beehiving cooperative reveals interesting facts. First, the sizes of the two small-scale cooperatives in our study are quite different: *Sju Gårdar* consists of seven dairy farmers (six are delivering milk today), and *Upplandsbondens* has 86 farmers as members. Naturally, this implies that the costs for collaborating are greater in the latter. Also, it was stressed by the interviewees that it was crucial for success that members took their part of the responsibility and contributed to the cooperative—not only by sending their milk and meat to it. It is clear that both *Elisabeth* and *Inger* have an enormous responsibility in their cooperatives—they do a lot of work-hours and are well-informed about operation details. In the case of *Sju Gårdar*, all members can sit on the board, which deters free-riding behaviour. In *Inger*'s cooperative, seven members sit on the board as well, but the size of the body of members leads to that they have some problems with free-riding.

Elisabeth described the members of her cooperative as being tired of the old cooperative (*Arla*) and willing and capable to start a new one. As mentioned above, many farmers used voice in the old cooperative, but only a handful took the step to exit. Traits she mentioned were "risk tolerant", and "willingness to work together—no 'lonely wolves'". She also believed it to be a strength having differences in age and educational backgrounds, and that the farmers in *Sju Gårdar* were entrepreneurs to a higher degree than the average *Arla*-member. *Inger* expressed the same farmer-traits, in combination with a feeling in the large cooperative of disappointment, due to the failed merger of *Swedish Meats* in the end of the 1990s. Farmers were dissatisfied with the general development of the slaughter sector. This supports our Hypothesis **H5**. In addition, both interviewees have extensive board-experience,[11] which further confirms the hypothesis.

[11] *Inger* even sat on the board of Arla during the time she was an active dairy farmer.

5.3 Outsourcing

Both small-scale cooperatives have outsourced the processing of their products, since they do not have the sufficient economic muscles for investing in processing plants of their own. Investing into a dairy or a slaughter house imply investments "in the 8-digit range", at least, and hence this is not possible. In addition, in both cases, there exists enough excess capacity in the region of Uppland, so finding external processing capacity was not that difficult, albeit somewhat "shaky" in the case of *Upplandsbondens*. Both interviewees said that it would not be possible to start their beehive cooperative, should the possibility to outsource been unavailable. Here, it is important to outsource, both when it comes to processing the products, but also getting a chance to take part in the distribution channels as well as—to some extent—the contacts with the retailers. Put short, outsourcing leads to fairly low transaction costs. Hence, Hypothesis **H3** holds.

5.4 Why a Cooperative?

In the case of *Sju Gårdar*, Elisabeth stressed that a critical factor in closing a deal with the cooperative dairy that processes their milk was that *Sju Gårdar* also was a cooperative: "Had we not been a cooperative, *Gefleortens mejeriförening* would not have been interested in helping us." Elisabeth also mentioned that working in a cooperative was "a tradition" among farmers, and that the traditional cooperative organizational form ensured that all members were treated equally—something important to the involved farmers.

The main reason for *Upplandsbondens* being a cooperative is that the members "feel at home" in this organizational form, a phenomenon also mentioned by *Elisabeth*. Since both interviewees have experience in sitting on cooperative boards—being Type-I farmers—and being the prime movers in founding the small-scale cooperative, the choice of organizational form was not difficult, they both say. They all agreed on that it was worthwhile to accept the time-consuming decision process of a cooperative. Our belief is that the fact that the collaboration between the two cooperatives *Seven Farms* and *Gefleortens mejeriförening* works well, compared to the collaboration between *Upplandsbondens* and its trading partners, could partly be explained by the fact that the latter is not collaborating with a cooperative. Doing business between a cooperative and an IOF often implies problems, due to the fact that the actors do not understand each other's business rationales, etc. Hence, we found support for Hypothesis **H5**. They both stressed that it was essential to safeguard that all future benefits created by the cooperative were distributed to the members and not to external investors. The main reason for this being that they want to get a payoff from exposing themselves to the risks associated with exiting the large cooperative and starting the beehive cooperative. This is an indication of Hypothesis **H4** being supported by the two cases studied here.

Our interviewees also indicated that "luck and timing" are important factors in leaving an established cooperative to create a new one, oftentimes competing with the cooperative they exited from. It is essential that the ones exiting really are willing to make sacrifices in order to safeguard success for the beehive cooperative.

6 Summary and Conclusions

In this paper we have set to analyse the phenomenon of *cooperative beehiving*, when cooperative members leave their cooperative to form a new one. We examine the reasons for leaving the mother cooperative and the choice of the new cooperative organizational form. We present two cases from Sweden, one dairy and one beef beehive cooperative.

We set out five hypotheses which we verify with the two case studies. We find that one main factor leading some farmers to leave their cooperative is the discovery of a demand for differentiated product where the large cooperative is reluctant to engage. A group of entrepreneurial, risk-taking farmers decides to exit and cease this business opportunity. We find that the leaders are usually cooperative members engaged in the leadership of the mother cooperative. This gives them a double advantage. On one hand, they have inside information of the opportunities and the weaknesses of the large cooperative; on the other hand, they can put their previous leadership experience to use in the formation of the new cooperative. The success of the new entity depends also on two key cost-related factors: (a) it is crucial that some processing capacity can be outsourced, otherwise the new coop cannot afford to process on their own, since the scale economies are very large; (b) the set-up sunk costs are also very large for the size of the new entity. Both cases found these costs too high, and both relied on some government and EU subsidy cover up the marketing and other costs related to promotion and branding.

The two cases in Sweden shed light onto the resilience of the cooperative form of organization, which resembles the beehiving process of the bee colonies. This differs somewhat from previous literature on cooperative re-structuring and cooperative failure, where dissatisfied members either dismantle the cooperative entirely, or they often choose other governance forms, such as bilateral or collective contracts (Hendrikse and Bijman 2002; Bijman and Hendrikse 2003; Fulton and Hueth 2009).

Although a complete theoretical model is not presented here, the theoretical underpinnings and the methodology are founded on transaction cost theory. The existing cooperative fails to capture the opportunity of a new differentiated product market. Those who leave choose a governance form that minimizes the total of transaction and production costs. The transaction costs of organization are minimized due to the experience and organizational know-how of the farmers leading the initiative who were previously engaged in the administration and leadership of the large cooperative. Asset specificity is dealt with in two ways. The sunk set-up promotion and branding costs are to a great extend subsidised, whereas the large investment and economies of scale in processing are avoided through outsourcing, under contract with an existing processing facility. If the organizational structure of the outsourcing entity is also a cooperative that makes the transaction smoother, as we find in one of the two cases (dairy).

Cooperative beehiving was interpreted as a natural process by those involved. There was no animosity between the mother cooperative and the new beehive cooperative. One of the interviewees stressed that it had been a conscious strategy not to talk in negative terms about the mother cooperative. She believed that this had contributed to the positive development of the new beehive.

The study of cooperative beehiving contributes to the deeper understanding of the cooperative firm. Where does a large cooperative fail and why a cooperative is chosen again by dissatisfied members are the key research questions. Further research is required. First, we need to investigate and document other cases of cooperative beehiving in other sectors in other countries, in order to put the hypotheses developed in this chapter to further scrutiny. Second, more theoretical work is needed to formulate the conceptual framework presented here. Transaction cost economics and agency theory provide a solid background against which this theory can be advanced.

References

Written Sources

Albaek S, Schultz C (1998) On the relative advantage of cooperatives. Econ Lett 59(3):397–401

Bager T (1997) Institutional and organisational change in the European food sector: a meso-level perspective. In: Nilsson J, van Dijk G (eds) Strategies and structures in the agro-food industries. Van Gorcum, Assen, pp 3–19

Bijman WJJ, Hendrikse GWJ (2003) Co-operatives in chains: institutional restructuring in the Dutch fruit and vegetables industry. J Chain Netw Sci 3(2):95–107

Caves RE, Petersen BC (1986) Cooperatives' shares in farm industries: organizational and policy factors. Agribusiness 2(1):1–19

Chaddad FR, Cook ML (2004) Understanding new cooperative models: an ownership-control rights typology. Rev Agric Econ 26(3):348–360

Coase RH (1937) The nature of the firm. Economica 4(16):386–405

Cook ML (1995) The future of U.S. agricultural cooperatives: a neo-institutional approach. Am J Agric Econ 77:1153–1159

Cook ML, Plunkett B (2006) Collective entrepreneurship: an emerging phenomenon in producer-owned organizations. J Agric Appl Econ 38(2):421–428

Dul J, Hak T (2008) Case study methodology in business research. Butterworth-Heinemann, Oxford. ISBN 978-0-7506-8196-4

Fakta om svenskt jordbruk (Facts on Swedish agriculture) (1996) Jordbruksverket, Jönköping

Ferrier GD, Porter PK (1991) The productive efficiency of US milk processing co-operatives. J Agric Econ 42(2):161–173

Fulton M (1988) Cooperative organizations in Western Canada. Occasional Paper 88.03, Center for the Study of Cooperatives, University of Saskatchewan

Fulton M (1995) The future of cooperatives in Canada: a property rights approach. Am J Agric Econ 77(5):1144–1152
Fulton M, Hueth B (2009) Cooperative conversions, failures and restructurings: case studies and lessons from U.S. and Canadian agriculture. Center for the Study of Cooperatives, University of Saskatchewan, Saskatoon
Galor Z (2008) Demutualization of cooperatives: reasons and perspectives. http://www.coopgalor.com/i_publications.html
Hansmann H (1996) Ownership of enterprise. The Belknap Press of Harvard University Press, Cambridge
Harris A, Stefanson B, Fulton M (1996) New generation cooperatives and cooperative theory. J Coop 11:15–28
Helmberger P (1966) Future roles for agricultural cooperatives. J Farm Econ 48(5):1427–1435
Hendrikse GWJ (2011) Pooling, access, and countervailing power in channel governance. Manage Sci 57(9):1692–1702
Hendrikse GWJ, Bijman WJJ (2002) On the emergence of new growers' associations: self-selection versus countervailing power. Eur Rev Agric Econ 29(2):255–269
Hendrikse GWJ, Veerman CP (2001) Marketing co-operatives: an incomplete contracting perspective. J Agric Econ 52(1):53–64
Hirschman AO (1970) Exit, voice, and loyalty. Responses to decline in firms, organizations, and states. Harvard University Press, Cambridge
Holmström BR (1999) Future of cooperatives: a corporate perspective. Finnish J Bus Econ 4:404–417
Jensen MC, Meckling WH (1979) Rights and production functions: an application to labor-managed firms and codetermination. J Bus 52(4):469–506
Johansson T (1985) Lantbrukarnas föreningsrörelse – framväxt och utveckling. (The farmer cooperative associations – emergence and development). LRF, Helsingborg
Johansson T (1994) Bondesamverkan i Sverige. (Farmer cooperation in Sweden). LRF, Stockholm: Media Nova
Joskow PL (2005) Vertical integration. In: Ménard C, Shirley MM (eds) Handbook of new institutional economics. Springer, Berlin, pp 319–348
Karantininis K, Zago A (2001) Cooperatives and membership commitment: endogenous membership in mixed duopsonies. Am J Agric Econ 83(5):1266–1272
Lantbrukskooperativ årsbok 1996. (Yearbook of farmer cooperatives). LRF, Stockholm
Lantbrukskooperativ årsbok 1998. (Yearbook of farmer cooperatives). LRF, Stockholm
Lindahl B (2004) Svensk lantbrukskooperation, jordbrukspolitik och nationalekonomi (Swedish farmer cooperatives, agricultural policy, and economics). In: Rydén R (ed) Jordbrukarnas kooperativa föreningar och intresseorganisationer i ett historiskt perspektiv (The farmers' cooperative associations and interest organizations in a historical perspective). KSLA, Stockholm, pp 255–267
Nilsson J (1997) Agricultural co-operatives in the 15 EU member states: Sweden. In: van Bekkum OF, Van Dijk G (eds) Agricultural co-operatives in the European Union, trends and issues on the eve of the 21st century. van Gorcum, Assen, pp 145–155
Parliament C, Taitt J (1989) Mergers, consolidations, acquisitions: effect on performance of agricultural cooperatives, staff paper no 89-37: 1–24
Perry C (1987) Union corporate campaigns. University of Pennsylvania, Philadelphia
Porter PK, Scully GW (1987) Economic-efficiency in cooperatives. J Law Econ 30(2):489–512
Sexton RJ, Iskow J (1993) The competitive role of cooperatives in market-oriented economies: a policy analysis. In: Csaki C, Kislev YH (eds) Agricultural co-operatives in transition. Westview Press, Boulder, pp 55–83
Staatz J (1987) Farmers' incentives to take collective action via cooperatives: a transaction cost approach. In: Royer J (ed) Cooperative theory: new approaches, agricultural cooperative service. USDA, Washington, pp 87–107, Service Report 18

Stigler G (1951) The division of labor is limited by the extent of the market. J Polit Econ 59 (3):185–193
Svenska Institutet (The Swedish Institute). Fact Sheets About Sweden, 22t Qd
Sveriges Djurbönders (Swedish Animal Farmer's) Annual Report, 2010
Tadelis S, Williamson OE (2012) Transaction cost economics. In: Gibbons, Roberts (eds) Handbook of organizational economics. Princeton University Press, pp 159–191
Thurow L (2001) Social and antisocial capital. The Luxembourg Institute for European and Institutional Studies (LIEIS). Organized at the Harvard Faculty Club
Trifon R (1961) The economics of cooperative ventures: further comments. J Farm Econ 43:215–235
Williamson OE (1975) Markets and hierarchies: analysis and antitrust implications. Free Press, New York
Williamson OE (2004) Transaction cost economics and agriculture – an excursion. In: Huylenbroeck GV, Verbeke W, Lauwers L (eds) Role of institutions in rural policies and agricultural markets. Elsevier Science, Amsterdam, pp 19–42
Wirenga B, Tilburg AV, Grunert KG (1997) Agricultural marketing and consumer behavior in a changing world. Springer, Berlin

Internet

LRFs homepage: www.lrf.se
Sju Gårdar (Sju gårdars) homepage: www.sjugardar.se
Upplandsbondens (Upplandsbondens) homepage: www.upplandsbondens.se

Personal Interviews

Carlsson I (2011) Interview, October 5
Gauffin E (2011a) Lecture, January 31
Gauffin E (2011b) Interview, April 6

Collective Entrepreneurship and Rural Development: Comparing Two Types of Producers' Organizations in the Ethiopian Honey Sector

Clarietta Chagwiza, Roldan Muradian, Ruerd Ruben, and Worku Tessema

Abstract This paper deals with the comparison of two types of honey producers' enterprises in the Masha district, western Ethiopia. Cooperatives and private limited companies (PLCs) are both collectively owned by a group of farmers, but the former do not face a legal restriction regarding the number of members, are strongly regulated by the government, and their shares are not tradable. We argue that the collective entrepreneurial capacity varies significantly among the two types of organizations. We found that members of PLCs have higher productivity and income derived from honey, are more prone to adopt new technologies, as well as receive higher dividends and price per kilo of honey. Additionally, the incentive scheme exercised by the PLCs was more market oriented. Furthermore, as compared to cooperatives, PLC members market a higher proportion of honey through their organizations. These results are relevant for the design of development interventions aiming at enhancement of market integration of small farmers in Ethiopia.

Keywords Collective action • Collective entrepreneurship • Cooperatives • Ethiopia • Honey • Private limited companies • Value chain development

C. Chagwiza (✉) • R. Muradian • R. Ruben
Centre for International Development Issues Nijmegen, Radboud University Nijmegen, Thomas van Aquinostraat 4, 6525 Nijmegen, The Netherlands
e-mail: clarrytee@yahoo.co.uk

W. Tessema
Department of Agricultural Economics, Hawassa University, P.O. Box 05, Awassa, Ethiopia

T. Ehrmann et al. (eds.), *Network Governance*, Contributions to Management Science,
DOI 10.1007/978-3-7908-2867-2_9, © Springer-Verlag Berlin Heidelberg 2013

1 Introduction

The way farmers groups function and relate with downstream value chain players (processors, traders, retailing, etc.) are important determinants of economic rents, and therefore they influence the prospects of rural economic development. In agricultural markets dominated by small-scale farmers, producers' groups might play a critical role both reducing transaction costs for the downstream players, as well as enhancing the market opportunities of growers through economies of scale (Mangus and Piters 2010). However, the fact of being "organized" is not enough. The manner in which farmers are organized is also critical. In the same market, the type/form of producer organization might have a strong influence on the economic performance of members. In this study, we seek to assess the relationship between the form of organization and the capacity to seize wealth creation opportunities by members. For doing so, we compare two types of collective enterprises; cooperatives and "private limited companies" (PLCs) involved in the production and commercialization of honey in Ethiopia. The overall objective of this study is therefore to compare the levels of collective entrepreneurship between cooperatives and PLCs, and to discuss how such variation is related to the institutional and structural differences between these producers' groups.

In the study area, both, cooperatives and PLCs are producers-owned and run enterprises, with a formal legal status. Both have a board of directors, which is elected by the farmers and both market the largest part of their honey production through a single (the same) processor. Just like cooperatives, a PLC is formed and collectively owned by a group of farmers to pursue their economic goals. The main differences between these two groups are however: (1) *the size of membership*: primary cooperatives can have several hundreds of members while the PLCs are allowed to have a maximum of 50 members; (2) *the level of external regulation*: the cooperatives are strongly regulated by the government through district cooperative offices, while PLCs hold a higher degree of freedom from the government in their operation; (3) *organizational layers*: primary cooperatives are organized into unions, while the latter level of organization does not exist in the case of PLCs; (4) *the ownership structure*: PLC members can buy shares, while shares in cooperatives are not tradable. It is important to note that a PLC is also a type of producers' group formed as an alternative to the cooperatives, and the latter have operated in the area for a much longer period of time; (5) *the functions*: the cooperatives considered for this study are multi-purpose. In addition to honey, they also market other commodities such as peas, beans and spices, while the considered PLCs specialize in the marketing of honey.

In the following section we define collective entrepreneurship and discuss its determinants. Section 3 provides a brief background of the Ethiopian honey sector. Section 4 presents the sampling techniques, type of data collected and the methods used for data analysis. Empirical results and the discussion of results are presented in Sect. 5. Finally, conclusion and suggestions for further research are elaborated in Sect. 6.

2 Collective Entrepreneurship

Collective entrepreneurship (CE) is the process through which the organizational and governance structure as well as the attitudes of members are translated into economic performance and benefits. Yan and Sorenson (2003) also defined CE as a process by which agents are able to identify and seize economic opportunities by means of collective action. CE therefore is determined by social norms, values, and networks for the production of goods or services (Connell 1999) and the ability to take collective risks (Trompenaars and Hampden-Turner 2002). Stewart (1989) suggests that collective entrepreneurship might result in an increase in the ability of each member of the group to create and reap opportunities for economic development, as compared to agents that operate by their own. CE changes market conditions by means of building and modifying the organization's resources, competences, and organizational architecture to respond to opportunities and influence market relations (Bratnicki 2005). Cook and Plunkett (2006) point out that for any form of a collective organization to achieve a higher level of performance, members' decisions about their own (in this case on-farm) activities and investments should be aligned with the cooperative. CE can also be defined as the ability to align these two levels of decision making.

CE is a property of collective enterprises, reflected in their ability to exercise efficiency and accrue rents whenever they are faced with opportunities. The performance of producers' groups depends to a large extent on their level of collective entrepreneurship. In this type of organizations, individual skills and attitudes are integrated into the group in order to achieve a common economic goal (Dana and Dana 2007). Collective entrepreneurship is a property of the group, which however is determined by individual behavior. That is, CE results from the interaction between individuals when they face a common economic dilemma (collective action situation). We argue that institutional and structural differences between groups may cause differences in the way in which producer groups react to opportunities and innovations.

A vast literature on management of natural resources has already addressed the factors that are likely to affect collective action, and this body of literature offers numerous lessons that can be applied to collective action in marketing (Markelova and Meinzen-Dick 2009). Social and economic heterogeneity, group size, and the level of autonomy in setting the rules have been highlighted as important variables determining the ability of groups to solve social dilemmas (Poteete and Ostrom 2004; Agrawal 2000). Group size has been identified as a key factor influencing the performance of groups (Olson 1965; Agrawal and Goyal 2001; Hussi et al. 1993). The effects of size on performance have been often explained from the perspective of transaction costs. Olson (1965) hypothesized that "unless the number of individuals in a group is small, rational, self-interested individuals will not act to achieve their common or group interests unless certain conditions are present". In order to solve this free-riding problem, Olson (1971) proposes to create incentives that will induce individuals to contribute to a collective good as a by-product of

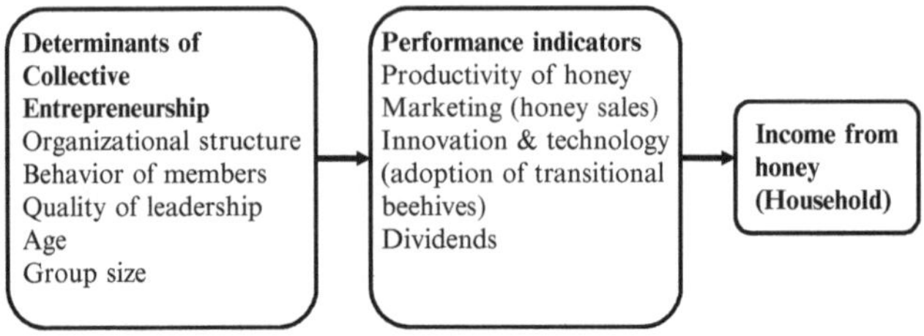

Fig. 1 Operationalization of collective entrepreneurship

their pursue of individual interest. The free-riding problem generates monitoring costs. Agrawal and Goyal (2001) and Hussi et al. (1993) argue that the cost of monitoring rise more than proportionately as group size increases. In the same vein, Bandiera et al. (2005) argue that the institutional features that make collective action successful, such as monitoring, are more easily accrued in small groups. Due to the higher monitoring costs, the possibilities of free-riding are higher in large groups. In the same line, Hardin (1982) argues that the larger the number of people who must be coordinated, the higher the costs of organizing them to an effective level. However, the relationship between group size and performance is not as straight-forward as we might expect according to the transaction costs literature. The advantages offered by economies of scale in large groups might compensate higher transaction costs.

Apart from size, other factors such as resource characteristics and diversity among members influence the transaction costs of collective action (Ostrom 1994). Furthermore, social norms and values, such as trust and loyalty, can play a role in the economic performance of collective enterprises, and mediating the relationship between size and group performance. In groups with high levels of social capital, members will forego opportunistic behavior, thereby lowering transaction costs and increasing the group and individual returns (Kirsten 2004; O'Brien et al. 2005). In addition to the factors discussed above, Aiken and Hage (1971) identified age as another important variable that can hinder innovation in a collective firm or organization. According to these authors, the older the organization, the more bureaucratic and the less receptive it is to innovation. Younger organizations are believed to be in a better position to embrace new technologies and to be more willing to innovate than older organizations. In the current study, we refer to technological innovation as improvements in the way commodities are produced or transformed (Devaux et al. 2007).

The conceptual framework is presented in Fig. 1. We consider the organizational structure (rules and regulations; degree of autonomy), group size, the behavior of members (social capital), the quality of leadership and age as the main determinants of CE. The level of CE is reflected in performance indicators, both at the individual and group levels. We considered honey productivity, honey sales, dividends provided, and the level of innovation (adoption of transitional beehives) as the most important performance indicators for comparing members of cooperatives and PLCs. These factors, we argue, have a significant influence on the income derived from honey, and therefore on the prospects of rural economic development in the study area. Based on the theoretical considerations summarized above, we adopted the following working hypotheses: (1) PLCs hold a higher level of collective entrepreneurship than cooperatives and (2) honey producers in PLCs reap higher income from honey, as compared to their peers in cooperatives.

3 The Ethiopian Honey Sector

Beekeeping is a traditional and important farming activity in Ethiopia (Agonafir 2005). Ethiopia's total honey production is approximately 39,700 tons per year (GDS 2009). The country is one of the five biggest wax exporters, with an average annual export estimated at 3,000 tons (EEPD 2006). Ethiopia is one of the leading honey producers in Africa and one of the ten largest honey-producing countries in the world. However, honey exports have started only recently, facilitated by interventions of the international cooperation. Currently, the main importers of Ethiopian honey are the USA, Japan and the EU.

Different stakeholders (the government, non-governmental organizations, etc.) have initiated development interventions in the country as a whole and the study area (Masha district) in particular in order to promote the production and export of honey. The government has given attention to the promotion of improved hives (transitional and modern), which have been provided at subsidized prices through the Ministry of Agriculture and Rural Development. This policy has triggered the participation of women in beekeeping activities since the management of these types of hives requires less physical efforts (they can be placed in the backyard instead of hanging on trees as the traditional hives use to be). The NGOs have facilitated the adoption and use of low-cost and appropriate hive technologies and have provided training to the beekeepers.

Small-scale producers are the most important honey producers in Ethiopia. The main buyers for the honey produced in Masha are private traders (local merchants), local *Tej* (Ethiopian traditional honey liquor) brewers, and the lead firm Bezamar, a honey processing, trading and exporting company. A lead firm can be described as a firm that has forward and backward commercial linkages with a number of small-medium enterprises within the value chain and holds a significant market share and power in the sector. Contracting relations between the private sector and honey producers and their organizations (e.g. farmers' groups) is considered essential to effectively align production, processing, and the specific demands and standards of the international market. Thus, in order to satisfy the market requirements on quality and volumes producers (suppliers) and buyers (processors) need to closely coordinate their activities. As a result, their degree of interdependence is increasing. The owner and manager of Bezamar is one of the key entrepreneurs who have facilitated the transformation of the honey sector in Ethiopia. The Dutch development agency SNV promoted a mutually beneficial relationship between the producer groups and the lead firm through the provision of grants for training on quality, technology transfer; and business development services.

Table 1 Average yield potential per each hive type

Hive type	Farmer's average yield (kg/hive)	Research center[a] yield (kg/hive)
Traditional	5.0–7.0	NA
Transitional (intermediate)	15.0–25.0	25
Framed (boxed)	30.0–45.0	40

Source: Global Development Solutions, LLC (2009)
[a]Research centre refers to the centre built for scientific research

Beekeeping requires techniques that can be easily managed and it does not require investment to acquire big land size, which is often a constraint for the poor rural dwellers (Debela 2010). In Ethiopia, there are three types of bee husbandry systems namely traditional, transitional and modern beekeeping. In the traditional way of beekeeping, the hives are made out of logs, bark, reeds, gourds and clay pots. The hive has to be hanged on top of a tree (in the forest). The number of traditional beehives a household might handle is very high (up to 200) but yield per hive is very low. Kerealem et al. (2009) report that about 95 % of bees are still kept in traditional hives. The term "transitional beehive" refers to a hive technology that is between the traditional and the modern one, and it is managed at the backyard. Transitional hives are made of local wood, and they have typically a higher honey yield, compared to the traditional hives. They provide also a mechanism for monitoring the maturity of honey, thus enabling harvest at optimal time. Finally, modern hives are created from rectangular and square boxes of better quality wood. These include Langsroth and Top Bar hives. The modern hives are more complex and difficult to build but they are easily transportable and generate greater quantities of better quality honey, which will command higher prices (Mehari 2007).

The three types of bee husbandry systems described above have different costs, harvesting techniques and productivity expectations (GDS 2009). By adopting the transitional and framed (modern) types of hives, alongside with proper training on management of the honeybees, producers can harvest higher yields. The average potential yield for each type hive is shown in Table 1.

4 Methodology

This section describes the study area and the sampling strategy applied in the selection of the respondents, as well as the procedure for data collection, source of data, the kind of data collected and the methods used for data analysis are also described.

4.1 Description of the Study Area

Fieldwork was carried out in the Masha district, South west part of Ethiopia. Masha is one of the 77 woredas in the Southern Nations, Nationalities and Peoples' Region of Ethiopia (SNNPRs). Different nationalities are found in this zone, including

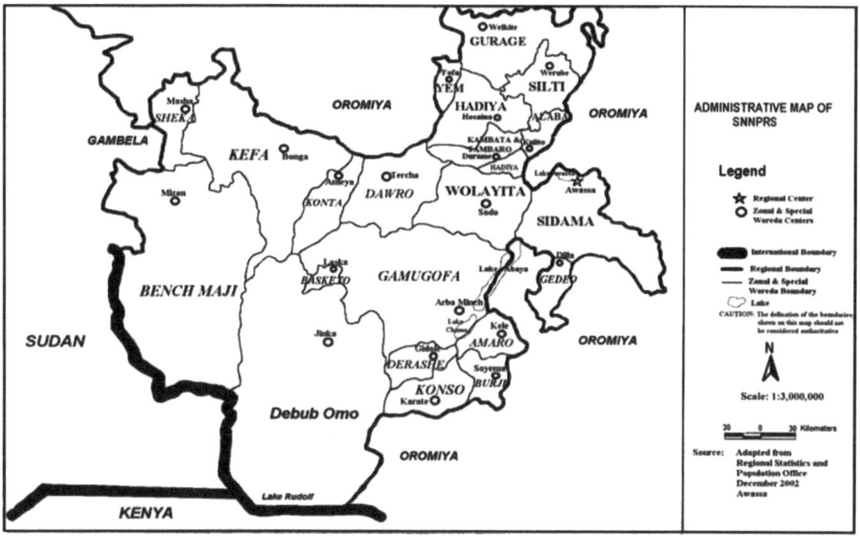

Fig. 2 Map showing the study area Masha-Sheka Zone (Abbute 2003)

Shekicho, Sheko, Megzenger, Keficho, Amhara, Oromo and Guraghe. This woreda falls under the Sheka Administrative Zone (see Fig. 2). More specifically, Masha is the administration center of Sheka Zone and is located 676 km south west from Addis Ababa, along the Addis-Jimma road. Geographically, the zone lies between 7°24′–7°52′ N latitude and 35°13′–35°35′ E longitude.

The altitudinal range of the areas in the zone is between 900 and 2,700 m above sea level, and it receives a high amount of rainfall, with an average of 2.000 mm annually. This woreda is notable for its relatively high forest cover as compared to other parts of Ethiopia. In general, the area is characterized by dense forests and woodlands containing diverse plant species that provide nectar and pollen to foraging bees. However, this important attribute is threatened by the high rate of deforestation in the area, which has aggravated in recent years due to increased conversion to monoculture plantations such as coffee and tea.

4.2 *Sampling Strategy and Data Collection*

Out of the five cooperatives (one per kebele)[1] producing honey in the district, three primary cooperatives (Genobay, Akach and Degele) were selected for the study. Accessibility was taken into consideration in selecting these primary cooperatives. Sixty producers (20 from each cooperative) were selected randomly from a list

[1] A *kebele* is defined in Ethiopia as the smallest administrative unit, below the municipality-district level.

Table 2 Number of members in the producers' groups that participated in the study

Organization	Total no. of members	No. of participants/group
Gada PLC	14	12
Chiefdale PLC	17	15
Shatto PLC	19	16
Akach primary cooperative	445	19
Degele primary cooperative	270	19
Genobay primary cooperative	451	20

obtained from the cooperative marketing office. Fifty-eight producers participated in the survey and the remaining two could not participate due to various reasons. All the members from the three operational PLCs in the district (Chiefdale, Gada and Shatto) were included in the survey, since the smaller membership size of the PLCs allows interviewing all the members. In total, 43 PLC members were interviewed. The total membership (group size) and number of members of the collective enterprises that participated in the survey are shown in Table 2.

Data was collected from March to April 2011, through a household survey applied to 101 households, as indicated above. A semi-structured questionnaire was designed to generate information in the following fields: duration of membership, main source of income, dividend, annual income from honey, number and type of beehive owned, constraints of beekeeping, perceptions of the producers on the transitional beehives, amount of honey harvested, price received per kilogram of honey, year of adoption of transitional beehives and beekeeping experience. Face-to-face interviews were also conducted with the management teams of the producers' organizations (cooperatives and PLCs), private traders, and the processor (Bezamar). More relevant secondary information and data were gathered from various institutions, including the zonal and woreda sector offices. Documents from NGOs (SNV and NTFP) active in the study area were also consulted and reviewed.

Data was analyzed using the Statistical Package for Social Scientists (SPSS) 19.0 for Windows. T-test and χ^2 were employed to test the significance of differences between groups for continuous and discrete variables, respectively. We ran a regression on income from honey sales, in order to assess the explanatory power of different independent variables.

5 Empirical Results

The presentation of results on the comparison of the two producers' groups (cooperatives and PLCs) is guided by the hypotheses proposed in the previous section.

Table 3 Distribution of respondents according to age and education

	PLCs (43)		Coops (58)	
Variable	Count	%	Count	%
Gender				
Female	9	20.9	2	3.4
Male	34	79.1	56	96.6
Total	43	100	58	100
Age groups				
24 & below	1	2.3	1	1.7
25–34	9	20.9	7	12.1
35–44	14	32.6	21	36.2
45–54	14	32.6	13	22.4
55–64	5	11.6	15	25.9
65 & above	0	0	1	1.7
Total	43	100	58	100
Education level				
Never been to school	10	23.3	12	20.7
Primary school	8	18.6	15	25.9
Secondary school	19	44.2	24	41.4
High school	6	14.0	7	12.1
Total	43	100	58	100

Sources of income	Count		%	Count		%
First choice	Honey	25	56.8	Kocho	22	62.9
Second choice	Kocho	13	46.4	Honey	19	61.3
Third choice	Livestock	18	51.5	Livestock	17	48.6

5.1 Socio-Economic Characteristics of Respondents

The frequencies of several socio-economic indicators are summarized in Table 3. Most of the households interviewed were male-headed. Across the two types of producers' group (coop and PLCs), most beekeepers (83 %) were 34 years or older. 78.3 % of respondents went to school, and the majority has completed the primary education. About 23 % of respondents indicated that they have never been to school. We did not find significant differences in education levels between members of cooperatives and PLCs.

About 72 % of all the respondents report beekeeping experience of 20 years or more. The minimum and maximum numbers of years of experience indicated by respondents were 2 and 48, respectively. Forty-nine percent of the respondents have been members of their organizations for 8 years or more. The longest time of membership by cooperatives members was 19 years. PLC members ranked honey as their main source of income, followed by kocho (banana-like tree whose stems are edible) and lastly livestock. For cooperative members, their main source of income was kocho, followed by honey and then livestock.

Table 4 Summary of independent-samples t-test results

	Cooperatives (58)		PLCs (43)		(Differences)	
Variables	Mean	SD	Mean	SD	T-stat	*P*-value
Socio-economic variable						
Beekeeping experience (years)	26.29	11.06	26.23	10.15	−0.028	0.978
Duration of membership (years)	13.03	4.35	6.12	1.12	−11.610	0.000**
Production and income variable						
Number of transitional beehives in 2007 (unit)	0.31	1.08	2.09	3.06	3.654	0.001*
Number of transitional beehives in 2010 (unit)	1.48	1.64	4.56	3.71	5.080	0.000**
Production transitional 2007 (kg)	3.28	13.81	21.26	35.97	3.112	0.003*
Production transitional 2010 (kg)	12.47	21.05	53.51	56.38	4.545	0.000**
Production change transitional (kg)	9.10	17.50	32.26	62.01	2.37	0.022*
Productivity of honey (kg/hive)	2.47	1.83	3.48	2.14	2.560	0.012*
Income honey sales 2008 (Birr)	1,075.84	863.63	2,542.59	2,014.24	4.480	0.000**
Income honey sales 2010 (Birr)	1,615.50	1,229.49	4,060.21	2,740.78	5.456	0.000**
Dividend paid (second payment) (Birr)	14.93	46.18	276.95	308.72	5.520	0.000**

*significant at 5 %; **significant at 1 %

5.2 *Performance at Household and Group Level*

5.2.1 Level of Adoption of Transitional Beehives

In the study area, the use of transitional beehives was low across the sample. However, we found that PLC members owned a significant higher number of transitional beehives in 2007 and 2010, as compared to members of cooperatives (see Table 4). The beekeepers were asked whether they were willing to give up traditional beehives and focus only on transitional beehives. Eighty-nine percent of all the respondents answered “Yes”. The results reveal that there is no significant difference between the two groups with regards to their willingness to give traditional hives up for transitional hives (see Table 5).

5.2.2 Dividend for Members and Incentives to the Management Team

Our results show that PLC members receive a significant higher amount of dividends (second payment), compared to members of coops (see Table 5). In addition, we found a statistically significant difference between the two groups in relation to the proportion of farmers that received dividends (see Table 6). Producers highlighted a number of reasons why they did not receive dividend from their organizations. The information was generated only from the members who indicated that they did not receive dividends (coop = 40, PLC = 21).

Table 5 Summary of Pearson χ^2 test results

Organization variables	Coop (58) Yes	PLC (43) Yes	χ^2 value	Asymp. Sig (2-sided)
Dividend	18	**22**	4.183	0.041*
Marketing channels				
(1) Own organization	43	**42**	10.26	0.001*
(2) Private traders	**37**	8	20.41	0.000**
(3) Neighboring coops	0	2	2.75	0.097
Willingness to give up traditional beehives	52	38	0.042	0.838
Advance payment from organization	0	**5**	7.095	0.008*
Receive equipment	10	**16**	5.150	0.023*

The underlined and bold figures are just there to show the group (PLC or coop) where there are many respondents who answered YES to the questions we asked during the interviews
*significant at 5 %; **significant at 1 %

Table 6 Marketing constraints as reported by the beekeepers

	Challenge	Number of respondents (coop = 58)	%
1	Low price of honey	15	26
2	Lack of access to credit	9	16
3	Lack of support from the union	8	14
4	Private trader cheats on price and weight	6	10
5	Lack of capital for organization to buy all our honey	5	9
6	Transport problem	5	9
7	Fewer buyers	3	5
8	Unable to get timely information	2	3
9	The organization does not buy honey on time	2	3
10	The coop cheat when weighing honey	2	3
		Number of respondents (PLC = 43)	%
1	Low price of honey	16	37
2	Lack of access to credit	7	16
3	Few buyers	6	14
4	Lack of capital for organization to buy all our honey	4	9
5	Unable to get timely information	3	7
6	Private trader cheats on price and weight	2	5
7	The organization does not buy honey on time	2	5
8	Transport problem	1	2

The most important reason given by members of both types of groups was the need to re-invest for the expansion of the business.

By incentives we refer to the payment or compensation given to the board members of an organization for their managerial tasks. The board members of the PLCs are entitled to 10 % of the net profits made by the organization. However, at the moment fieldwork was conducted they have agreed not to claim it, but to invest it back into the organization. During the interviews, unlike in PLCs, all cooperative board members revealed that they are not entitled to receive incentives; rather they receive perdiems if they attend meetings or workshops. They mentioned as their main motivation to be a board member their willingness to contribute to societal goals and the common good.

5.2.3 Honey Productivity

Overall honey productivity was calculated as follows:

$$\text{Honey productivity} = \frac{\text{Quantity of honey (kg)}}{\text{Number of beehives (traditional + transitional)}}$$

The mean annual honey yield from transitional beehives (for the whole sample) in 2010 was 10.7 kg/hive/annum. The yield was far below the expected yield from transitional beehives: 15.0–25.0 kg/hive/annum (GDS 2009). For the traditional beehives, the mean annual yield was 2.1 kg/hive/annum, which is also below the expected yield of 5.0–7.0 kg/hive/annum (GDS 2009). Cooperative members reported significant lower quantities of honey produced from transitional beehives in 2007 and 2010, as compared to PLC members.

Our findings suggest that members of PLCs have higher honey productivity, as compared to members of cooperatives (see Table 5). In addition, we used the quantile method for classifying the households according to their productivity level. The data was arranged in descending order. The 101 households were grouped into 3 classes; high, medium and low, according to their productivity performance. The results from the comparison show that all the three PLCs (Gada, Chiefdale and Shatto) have higher percentages of their members in the higher ranks (33, 27 and 50, respectively) as compared to the three cooperatives (Akach, Degele and Genoby), with 21, 10 and 20, respectively.

5.2.4 Honey Prices

Figure 3 presents how prices paid by cooperatives, PLCs and private traders have evolved across time. Producers delivering their honey to PLCs are consistently receiving better prices than those delivering to cooperatives and private traders. More interestingly, the figure shows that, except for 2007, the private traders are offering higher prices than cooperatives but less than PLCs.

In addition, an independent-sample t-test was used to compare means of the income obtained from honey for the years 2008 and 2010. Table 4 clearly shows that PLC members obtained significantly higher incomes from honey sales as compared to members of cooperatives.

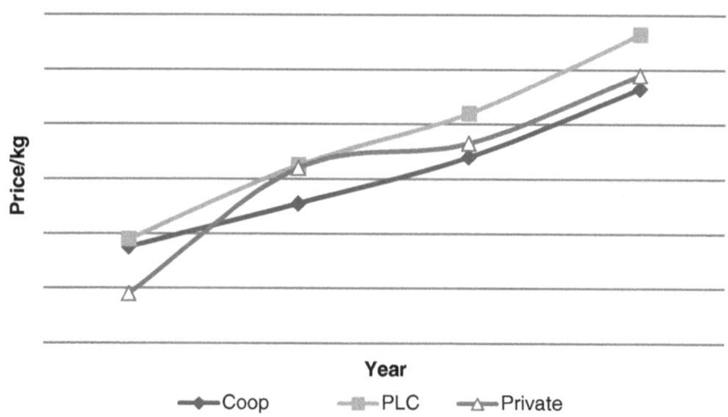

Fig. 3 Price variation among buyers across years

5.2.5 Perceptions About the Effect of Improved Hives and Training on Income

Respondents were asked to respond to the statement "I am very satisfied with my income over the past three years as a result of the training on beekeeping". Answers were ranked on a 5 point-Likert scale from strongly disagree to strongly agree. The results indicate that significant differences ($p < 0.05$) were observed between the responses from members of PLCs and cooperatives. Forty-nine percent of the PLC members tend to strongly agree with the statement as compared to only 24 % of cooperative members ($p = 0.01$).

The respondents were also asked to give their responses on the statement "As a result of the training and access to modern technologies on beekeeping, I would confidently say that my household income has..." For answers we used a 3 point-Likert scale with the following options; increased, slightly increased and remained the same. Across the two groups, about 65% of the whole sample indicated that their income has increased. However, significant differences ($p < 0.05$) were again observed between the means of PLC' and cooperative' producers.

5.3 *Marketing*

Three marketing channels were identified among respondents: own organization (cooperative or PLC), private traders and neighboring cooperatives. Choice and utilization of marketing channel varies significantly ($p < 0.05$) across producer organization (coop or PLC). Sixty-four percent of cooperative members are marketing a proportion of their honey through private traders, whereas only 19 % of PLCs sell part of their production through this channel (see Table 6). Bezamar (honey processing and exporting company) is the main buyer of honey from both the cooperatives and PLCs. Producer groups buy honey from their suppliers (members), bulk it and sell to the processor. The honey sales reported by both PLCs and cooperatives from 2007 to 2010 are shown in Figs. 4 and 5, respectively.

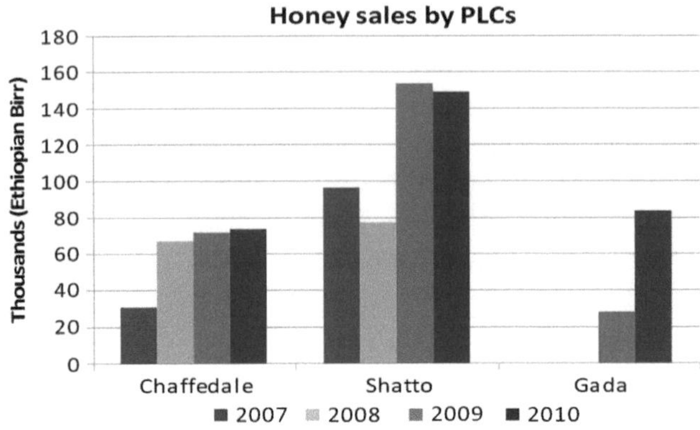

Fig. 4 Honey sales by PLCs

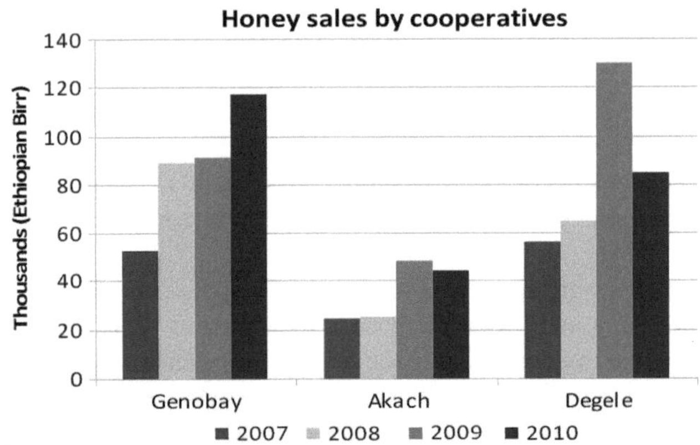

Fig. 5 Honey sales by cooperatives

5.3.1 Marketing Constraints

Despite all the benefits that honey can bring to the beekeepers in the area, the producers are confronted with a number of challenges and constraints that can potentially hamper the future of honey production and the economic contribution it brings to their livelihoods.

We present separately the constraints identified by cooperatives and PLC producers (see Table 6). Beekeepers of the two types of groups ranked low price, and lack of access to credit as the most important constraints. As the third most important constraint, cooperative members indicated that they lack support from the union, while the members of PLC stated that they would like to have more buyers.

Table 7 Variables and their expected signs

Dependent variable	Explanatory variables	Expected sign	Description of relationship
Income from honey	Education level	(+)	Knowledge and skills in production and marketing should influence positively income from honey
	Age	(+)	More experience in production and marketing is expected to translate into higher income from honey
	Gender (0 = female, 1 = male)	(+)	Males are expected to have better access to market information
	Duration of membership	(+)	Producers with more years of membership are expected to be more knowledgeable and experienced about the market, and hence able to reap higher income from honey
	Total number of transitional beehives	(+)	More transitional beehives translate into higher productivity, thereby increasing income from honey
	Productivity	(+)	Higher productivity results in an increase in income from honey
	Organization code (0 = PLC, 1 = coop)	(−)	PLC members are expected to have higher income from honey sales

5.4 Regression Results

In order to identify the determinants of income from honey sales, we ran a regression analysis (OLS) taking some household and organization characteristics as independent variables, according to the following model:

$$Y = a + b_1X_1 + b_2X_2 + b_3X_3 + b_4X_4 + e \quad (1)$$

where Y is the dependent variable and X_1, X_2, X_3.... are the explanatory variables. We present the dependent variables, explanatory variables, expected signs and the description of the expected relationships (see Table 7).

For this analysis, we used the income from honey sales for 2010 as the dependent variable. Our regression results show that 30 % of the variation in income from honey is explained by the considered explanatory variables (see Table 8). We checked for collinearity using the variance inflator factor (VIF) and the Durbin–Watson test.

Significant variables include organization code and productivity. Organization code (0 = PLC, 1 = coop) is negatively related to the income from honey sales. Productivity of honey has a positive relationship with the income derived from honey sales.

Table 8 Factors influencing income from honey sales (regression analysis)

Model	B	Std. error	Beta	t	Sig.
(Constant)	2,783.373	983.715		2.829	0.006**
Education level	−50.021	232.288	−0.021	−0.215	0.830
Total number of transitional beehives	37.984	26.826	0.145	1.416	0.160
Duration of membership	69.921	66.011	0.143	1.059	0.292
Gender of the respondent	145.067	676.945	0.019	0.214	0.831
Age	−3.507	225.978	−0.002	−0.016	0.988
Organization code	−2,194.744	671.394	−0.465	−3.269	0.002**
Productivity	107.271	37.015	0.262	2.898	0.005**

Dependent variable: income from honey sales (2010)
Durbin–Watson = 1.914
**significant at 5 %

5.5 *Discussion of Results*

In this section we will address two main overarching questions derived from the results presented above:

1. Why do PLCs hold in a higher level of collective entrepreneurship?
2. Why do cooperatives remain operational in the area?

We evaluated CE in two types of farmers' organizations by means of using a number of parameters characterizing the performance of collective enterprises, including, but not limited to, the rate of adoption of new technologies (process upgrading), member loyalty (in supplying output to his/her own organization), dividends and incentive schemes for board members. Based on these indicators, our main overall finding is that PLCs hold a higher level of collective entrepreneurship as compared to cooperatives. The difference in the level of collective entrepreneurship is expected to be influenced by a number of factors. As mentioned earlier on (see Sect. 2), group size is one variable that has been mentioned in the literature as an important determinant of group performance when they deal with collective action situations. Smaller groups tend to realize lower transaction costs as compared to larger groups. Institutional economists have argued long ago that transaction costs are a key element explaining the performance and survival of collective endeavors (North 1990). In addition, social capital features related to the performance of groups such as trust, commitment, participation and loyalty might be affected by the group size. For instance, Nilsson et al. (2009) report an inverse relationship between group size and membership satisfaction and trust in leadership among traditional cooperatives in Sweden. Furthermore, Jones (2004) argues that individuals in small groups can expect personal action to prove "significant" (for example, to affect the probability that others will contribute). In addition, previous studies (see Sykuta and Cook 2001; Chaddad and Cook 2004) have shown that a major problem associated with collective action in cooperatives is that members' property and decision rights tend to be vaguely defined in this type of enterprises. This causes social tensions amongst members. This is more likely in larger groups, since they are more likely to experience higher levels of agency costs, free riding or apathy in terms of monitoring (Chambers 2007). Vague property rights are more pronounced in larger groups because of higher degree of heterogeneity of economic interests among members. In addition, as membership becomes more heterogeneous the degree of vaguely defined property rights increases (Chaddad and Cook 2004). In our case, PLCs, being smaller and more homogenous, have clearer property and decision rights (it is less costly to take collective decisions). In addition, property rights are flexible (shares can be bought).

However, contrary to the previous argumentation, Agrawal (2000) found a positive relationship between group size and success in raising resources needed to hire a guard for protecting forest resources (a typical collective action situation). In a similar vein, Agrawal and Goyal (2001) found medium-sized groups to be more effective than smaller and larger groups, in the management of common-pool resources. Nonetheless, Barham and Chitemi (2009) found no evidence of any relationship between the group size and performance of the group in Tanzania. The downside of small groups is that they often lack economies of scale, a particular advantage in marketing and for achieving efficiency (Markelova et al. 2009). As the studies mentioned above show, the relationship between group size and collective action is not always straightforward. Ostrom (1997) argues that the impact of group size on collective action is usually mediated by a variety of other variables. We deal with some of them below.

There are several institutional factors that influence the performance of groups when dealing with collective action problems. North (1990) defined institutions as humanly devised constraints, formal (rules, laws, constitutions) and informal (norms of behavior, conventions and self-imposed codes of conduct) that structure human interactions, and their enforcement characteristics. An important set of institutional factors have to do with the type of rules that, according to Coleman (2009), are used to effectively manage long-term collective endeavors. Moreover, the rules that govern an organization influence the extent to which collective entrepreneurship is realized. Absence of rules or poor monitoring of rules is consistently associated with poor performance of common-pool resources (Ostrom and Nagendra 2006). For instance, Coleman (2009) highlights that issuing harvesting rights to local users provides incentives to invest in the common forest and results in good forest condition. Well structured incentive schemes and the right to buy shares within the PLCs are some of the rules that can facilitate collective entrepreneurship. Unlike PLCs, all cooperative board members do not receive monetary incentives for the work they do and during the interviews they justified their unpaid managerial tasks as a contribution to the common good and a social duty. Furthermore, the possibility to buy shares gives PLC members an incentive to be loyal to their organization, to increase honey sales and consequently to reap higher economic benefits derived from collective action. In this line, Chambers (2007) states that loyalty may be a form of selective social incentive or social coercion that maintains a collective business.

Livelihood strategies might be another important factor influencing the level of collective entrepreneurship. Cramb (2000) suggests that farmers in the same environment may have different objectives and livelihood strategies, and so respond differently to a given technology or innovation. The fact that PLC members consider honey to be their highest source of income (whereas kocho is the main source of income for cooperative members) might explain the differences in the rate of adoption and use of transitional beehives. PLC members have clearer incentives for their adoption. Another factor that might contribute to explain the differences between PLCs and cooperatives is the level of external influence on the organizations. There is a much higher level of external influence (government control) on cooperatives, as compared to PLCs. During our interviews, some PLC members pointed this as one of the advantages of their organizations. In this regard, Coleman (2009) argues that externally imposed rules and monitoring institutions have often failed in inducing effective management of common pool resources.

Nevertheless, the differences we have found in the level of collective entrepreneurship might also be due to selection bias. It could be the case that more productive and entrepreneurial producers tend to prefer join PLCs. However, our data does not allow us to assess the importance of selection bias in explaining the results. In any case, it is likely that a better performance attracts more entrepreneurial farmers, inducing virtuous cycles among PLCs.

If PLCs hold a higher degree of collective entrepreneurship and seem to be dominating in the supply of honey, one may wonder then why the cooperatives still remain operational in the area. There are several reasons that can explain this phenomenon. Firstly, one possible explanation has to do with the multipurpose nature of cooperatives. Besides marketing of honey, the cooperatives also engage in trading of other agricultural commodities like peas, beans and spices. During our interviews, the cooperative members highlighted this multipurpose nature as an important feature of cooperatives. The farmers are able to market their various commodities at one place thereby reducing transaction costs that arise from searching for buyers and transportation. The lower level of efficiency in the marketing of honey might be offset by the other services offered by the cooperative.

Secondly, there might be an information gap/inadequate information amongst the cooperatives members about the performance of other collective enterprises that they could join. For example, cooperative members might lack information about how PLCs are functioning and performing. The third and final reason has to do with the high start-up costs. The initial investments required to set up a collective enterprise in the study area are high. PLCs were heavily subsidized (financially, as well as through capacity building and other services) during their establishment by the international cooperation. It is likely that without this financial support it is extremely difficult to set up new collective enterprises.

Our findings shed light on the importance of organizational features, such as group size, rules and incentives, for the successful implementation of rural development interventions. Some policy recommendations can be derived from our study. For instance, the division of large cooperatives into smaller subgroups might facilitate collective entrepreneurship. Additionally, very likely cooperatives will reap efficiency gains if they change their incentive mechanisms. For example, they could adopt a system where board members benefit from the organization's profits, as a way of motivating them to invest in the common good. They could also allow members to trade shares. However, we should be careful about generalizations. The conclusion that PLCs are more entrepreneurial than cooperatives should not be generalized across all sectors and regions. The relationship between organization type, size and performance among collective enterprises run by farmers is a subject that requires much further research in Ethiopia before we can arrive to robust policy recommendations.

6 Conclusions

Producers groups in the Ethiopian honey sector have the potential to promote exports of honey from the area (capitalizing on supplying organic and forest honey) and improve the livelihoods of the rural households through increased income from honey sales. This has attracted the attention of the government, non-governmental organizations and private players in the sector. However, the way farmers are organized (organizational type) influences the performance of these collective enterprises. We addressed empirically these issues by comparing the levels of collective entrepreneurship between cooperatives and PLCs, and discussed how such variation is related to the institutional and structural differences between the two types of groups. The analysis shows that PLCs demonstrated a higher level of collective entrepreneurship as a result of some qualities (specialization, member loyalty, incentives, social capital with buyer; and group size) that they possess over their counterparts, cooperatives.

References

Abbute WS (2003) Resettlement as a response to food insecurity: the case of southern nations, nationalities, and peoples' region (SNNPR). United Nations. http://www.ocha-eth.org/Archive/DownloadableReports/resettlementSNNPR0603.pdf. Accessed 27 October 2011

Agonafir J (2005) Strategic intervention plan on honey & Beeswax value chains. SNV support to business organizations and their access to markets (BOAM). Strategic Intervention Plan, SNV/Ethiopia. Available online http://www.beesfordevelopment.org/uploads/Ethiopia_vc-honey-beeswax2005.pdf. Accessed 25 October 2011. Addis Ababa

Agrawal A (2000) Small is beautiful, but is larger better? Forest-management institutions in the Kumaon Himalaya, India. In: Gibson CC, McKean M, Ostrom E (eds) People and forests: communities, institutions, and governance. MIT Press, Cambridge, pp 57–85

Agrawal A, Goyal S (2001) Group size and collective action: third-party monitoring in common-pool resources. Comp Political Stud 34(1):63–93

Aiken M, Hage J (1971) The organic organization and innovation. Sociology 5(1):63–82

Bandiera O, Barankay I, Rasul I (2005) Cooperation in collective action. Econ Transition 13 (3):473–498

Barham J, Chitemi C (2009) Collective action initiatives to improve marketing performance: lessons from farmer groups in Tanzania. Food Policy 34:53–59

Bratnicki M (2005) Organizational entrepreneurship: theoretical background, some empirical test and directions for the future research. Hum Factors Ergon Manuf 15(1):15–33

Chaddad FR, Cook ML (2004) Understanding new cooperative models: an ownership-control rights typology. Rev Agric Econ 26(3):348–360

Chambers M (2007) Informing the theory of collective entrepreneurship: investment choice. Published Doctorate Dissertation. University of Missouri-Columbia. https://mospace.umsystem.edu/xmlui/bitstream/handle/10355/4881/research.pdf

Coleman EA (2009) Institutional factors affecting biophysical outcomes in forest management. J Policy Anal Manage 28(1):122–146

Connell DJ (1999) Collective entrepreneurship: in search of meaning. http://www.djconnell.ca/papers.html. Accessed 27 October 2011

Cook M, Plunkett B (2006) Collective entrepreneurship: an emerging phenomenon in producer-owned organizations. J Agric Appl Econ 38(2):421–428

Cramb RA (2000) Processes influencing the successful adoption of new technologies by smallholders. In: Stür WW, Horne PM, Hacker JB, Kerridge PC (eds) Working with farmers: the key to the adoption of forage technologies. ACIAR Proceedings No. 95: 11–22. Australian Centre for International Agricultural Research, Canberra

Dana LP, Dana TE (2007) Collective entrepreneurship in a Mennonite community in Paraguay. Latin Am Bus Rev 8(4):82–96

Debela S (2010) SNV support to Capacity Development for the Ethiopian honey Value Chain. Case Report IOB-SNV Capacity Development Evaluation

Devaux A, Velasco C, López G, Bernet T, Ordinola M (2007) Collective action for innovation and small farmer market access: the Papa Andina experience. Collective Action and Property Rights (CAPRi) Working Paper 68

Ethiopian Export Promotion Department (EEPD) (2006) Exports of honey and beeswax. Draft Report. EEPD, Ministry of Trade and Industry, Addis Ababa

Global Development Solutions (GDS) (2009) Integrated value chain analyses for honey and beeswax production in Ethiopia and prospects for exports. The Netherlands Development Organization (SNV)

Hardin R (1982) Collective action. John Hopkins University Press, Baltimore

Hussi P, Murphy J, Lindberg O, Breneeman L (1993) The development of cooperatives and other rural organizations. Technical Paper No 199. The World Bank, Washington

Jones P (2004) 'All for One and One for All': transactions cost and collective action. Political Stud 52:450–468

Kerealem E, Tilahun G, Preston TR (2009) Constraints and prospects for apiculture research and development in Amhara region. Livestock Res Rural Dev 21 (Article #172). Retrieved December 23, 2012 from http://www.lrrd.org/lrrd21/10/ejig21172.htm

Kirsten J (2004) Expanding the frontiers of agricultural economics to meet the future challenges of agricultural development. Paper presented for African Association of Agricultural Economists, proceedings of the inaugural symposium, Nairobi

Mangus E, Piters BS (2010) Dealing with small scale producers: linking buyers and producers. KIT (Royal Tropical Institute) Publishers, Amsterdam

Markelova H, Meinzen-Dick R (2009) Collective action for smallholder market access: the case for collective action in smallholder marketing. Policy Brief 6

Markelova H, Meinzen-Dick R, Hellin J, Dohrn S (2009) Collective action for smallholder market access. Food Policy 34:1–7

Mehari G (2007) Impact of beekeeping on household income and food security: the case of Atsbi Wemberta and Kilte Awlailo Woredas of Eastern Tigray, Ethiopia. M.Sc. Thesis, Mekelle University

Nilsson J, Kihlén A, Norell L (2009) Are traditional cooperatives an endangered species? About shrinking satisfaction, involvement and trust. Int Food Agribus Manage Rev 12(4):103–123

North D (1990) Institutions, institutional change and economic performance. Cambridge University Press, Cambridge

O'Brien DJ, Phillips JL, Patsiorkovsky VV (2005) Linking indigenous bonding and bridging social capital. Reg Stud 39(8):1041–1051
Olson M (1965) The logic of collective action: public goods and the theory of groups. Harvard University Press, Cambridge
Olson M (1971) The logic of collective action: public goods and the theory of groups. Harvard University Press, Cambridge
Ostrom E (1994) Constituting social capital and collective action. J Theor Politics 6(4):527–562
Ostrom E (1997) Self-governance of common-pool resources (w97-2, workshop in political theory and policy analysis). Indiana University, Bloomington
Ostrom E, Nagendra H (2006) Insights on linking forests, trees, and people from the air, on the ground, and in the laboratory. Proc Natl Acad Sci USA 103(51):19224–19231
Poteete AR, Ostrom E (2004) Heterogeneity, group size and collective action: the role of institutions in forest management. Dev Change 35:435
Stewart A (1989) Team entrepreneurship. Sage, Newbury Park
Sykuta ME, Cook ML (2001) Cooperative and membership commitment: a new institutional economics approach to contracts and cooperatives. Am J Agric Econ 83:1273–1279
Trompenaars F, Hampden-Turner C (2002) 21 Leaders for the 21st century: how innovation leaders manage in the digital age. McGraw-Hill, New York. September 30, 2001.
Yan J, Sorenson RL (2003) Collective entrepreneurship in family firms: the influence of leaders' attitudes and behaviors. N Engl J Entrepren 6(2):37–51

CEO Incentive Provision in Cooperatives: The Impact of Membership Size and Heterogeneity

Li Feng and George Hendrikse

Abstract A multi-task principal-agent model is formulated to capture the effect of membership size and heterogeneity on the incentive provision of the CEO in a cooperative. An increase in membership size as well as an increase in member heterogeneity decreases the optimal incentive intensity of the CEO.

Keywords Cooperatives • Governance • Performance measurement

1 Introduction

A cooperative is an enterprise collectively owned by many independent suppliers. It involves both a horizontal arrangement among many independent farmers and a vertical coordination mechanism between the upstream members and the downstream processor. Members of a cooperative have two roles. On the one hand, a member is a patron, implying a transaction relationship with the enterprise by providing inputs. On the other hand, a member is an owner. Members collectively possess the residual rights over the cooperative and take decisions regarding it. Many cooperative researchers considers the cooperative as an inherently less efficient governance form when compared to IOF (investor owned firm), mainly due to a number of property rights constraints (Cook 1995; Fulton 1995).

L. Feng (✉)
Department of Economics, Swedish University of Agricultural Sciences, PO Box 7013, 750 07 Uppsala, Sweden
e-mail: li.feng@slu.se

G. Hendrikse
Rotterdam School of Management, Erasmus University, PO Box 1738, Office T08-56, 3000, Rotterdam, The Netherlands
e-mail: ghendrikse@rsm.nl

T. Ehrmann et al. (eds.), *Network Governance*, Contributions to Management Science, DOI 10.1007/978-3-7908-2867-2_10,

We compare these two governance structures from the perspective of managerial incentive provision. The members-CEO relationship in cooperatives is similar to the investors-CEO relationship in IOFs to the extent that the members exercise their decision rights mainly by critically following the policies of the management, rather than by giving it directions (Trifon 1961).

Despite of the similarities, the issue of incentive provision in cooperatives is more complex than a standard principal-agent relationship (Feng and Hendrikse 2012). First, a managerial incentive contract is based on a performance measurement system, creating incentives that align the goal of the agent with that of the organization. However, there are no simple indicators of cooperative managerial performance or automatic incentive systems (such as a stock price) to close the gap in interests. Giving a CEO equity, a common way to tie the CEO's wealth to firm performance and thus to alleviate the interests conflict in IOFs, is not feasible in cooperatives. The reason is that a cooperative CEO is not eligible to hold equity in the business and receives only limited benefits from such ownership given the fact that most cooperative stocks do not appreciate in value (Trechter et al. 1997). Trechter et al. (1997) document various CEO compensation schemes in cooperatives. Some use pre-set performance-based bonuses, some allow for bonuses paid on past performance, and others do not use bonuses.

Second, there is a group of principals whose interests differ. What distinguishes a cooperative from an IOF with a single locus of profit maximization is members' plurality of interests (Trifon 1961). The guiding principle regarding understanding a cooperative is that members advance the interests of their own farm portfolios through a cooperative (Fulton 1988). Members differ from each other in terms of size, location, risk aversion, attitude towards innovation, growth potential, member involvement, and financial contribution to the cooperative. When colliding interests exist among principals, the agent's tasks involve devising workable compromises and acting as a neutral guardian of everybody's priorities (Trifon 1961).

Trechter et al. (1997) conduct a series of studies on the executive compensation practices in US cooperatives and identify some alternative sources of information revealing the performance of cooperatives. Some use patronage refunds per member as a factor of the financial performance measure. Others tie the CEO bonuses to some accounting measures (such as accounts receivable). Hueth and Marcoul (2008) investigate how the unique features of cooperatives influence the managerial incentives and information asymmetry between the CEO and the owners. They suggest subjective performance assessment as another source of information, based on the stable long-run relationships between owners and the management and the fact that the patrons are in a privileged position to observe and monitor managerial operations. We model the cooperative CEO's incentive provision based on these observations and focus on a special feature of this governance structure, i.e. the multiplicity of owners, and its impact on the incentive intensity.

This article is organized as follows. The next section formulates a multi-task principal-agent model and tailors it to the differences between cooperatives and IOFs. The impact of the membership size and heterogeneity on the cooperative CEO's incentives are analysed in Sect. 3. Finally, Sect. 4 concludes.

2 Model

A multi-task principal-agent model (Gibbons 1998) consisting of a two-stage non-cooperative game is presented in this section. In the first stage, the principal chooses the strength of incentives while the agent's optimal choice of activities is determined in the second stage. Assume that the CEO in governance structure i (C for cooperative and F for IOF) can take two actions: a_{1i} denoting the action to advance the downstream value, and a_{2i} denoting the action adding value to the upstream producers. The CEO's total contribution to firm value is denoted by y_i. The marginal product of action a_{ji} is f_{ji}. The production function is $y_i = f_{1i}a_{1i} + f_{2i}a_{2i} + \varepsilon$, where ε is a stochastic variable with expected value of 0, representing the noise in the production process that is beyond the agent's control.[1]

Since it is difficult to measure the overall effect of the CEO's actions on firm value, no compensation contract based on y_i can be enforced in court. An alternative performance measure p_i becomes therefore necessary. Suppose the technology of performance measurement is $p_i = g_{1i}a_{1i} + g_{2i}a_{2i} + \phi$, where g_{ji} denotes the performance measurement parameter, i.e. the weight attached to a_{ji}, and ϕ denotes the noise in performance measurement with expected value of 0.

The compensation contract in governance structure i specifies the wage w_i paid to the CEO as a linear function of p_i, i.e. $w_i = s_i + b_i p_i$, where s_i stands for the salary and b_i for the bonus rate. The CEO's payoff is the difference between the wage and the cost of actions: $U_i = w_i - c_i(a_{1i}, a_{2i})$. We assume that the cost function takes the form $c_i(a_{1i}, a_{2i}) = \frac{a_{1i}^2}{2} + \frac{a_{2i}^2}{2}$. The principal receives the difference between the CEO's total contribution to firm value and the CEO's wage: $\pi_i = y_i - w_i$. Notice that with this specification, the CEO's incentives are to produce a high value of p_i, not of y_i, whereas the principal does not directly benefit from increased realizations of measured performance p_i, rather, he/she benefits from increased realizations of the CEO's total contribution y_i. As a result, the incentives may be distorted. To minimize the distortion the principal wants to minimize the divergence between p_i and y_i.

The game is solved by backward induction. The CEO's optimal action in the second stage is determined by maximizing his/her expected utility, i.e. $\max_{a_{1i}, a_{2i}} E(U_i)$, where $E(U_i) = E[w_i - c_i(a_{1i}, a_{2i})] = s_i + b_i(g_{1i}a_{1i} + g_{2i}a_{2i}) - c_i(a_{1i}, a_{2i})$. The first

[1] We assume the actions taken by the CEO only have consequences for the principal, which excludes the possibility for tunneling and the CEO directly benefiting from acting against the interests of the principal.

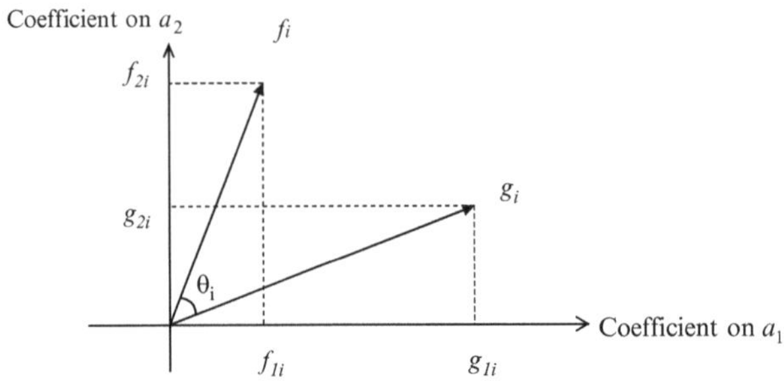

Fig. 1 The scale and alignment effect of the performance measure

order condition $b_i g_{ji} = \frac{\partial c_i}{\partial a_{ji}}$, $j = 1{,}2$, characterizes the CEO's equilibrium actions $a^*_{ji}(b_i) = b_i g_{ji}$. The payoff-maximizing reply in the second stage is anticipated in the first stage when the principal determines the efficient intensity of incentives. Maximizing the expected total surplus $\max_{b_i} E(\pi_i + U_i)$, where $E(\pi_i + U_i) = E[y_i - c_i(a_{1i}, a_{2i})] = f_{1i}a^*_{1i} + f_{2i}a^*_{2i} - c_i(a^*_{1i}, a^*_{2i})$ results in the efficient bonus rate $b^*_i = \frac{f_{1i}g_{1i}+f_{2i}g_{2i}}{g^2_{1i}+g^2_{2i}} = \frac{\sqrt{f^2_{1i}+f^2_{2i}}}{\sqrt{g^2_{1i}+g^2_{2i}}} \cos(\theta_i)$, where θ_i is the angle between the vectors $f_i \equiv (f_{1i}, f_{2i})$ and $g_i \equiv (g_{1i}, g_{2i})$ as depicted in Fig. 1.

There are two important features in the expression of efficient bonus rate, scale and alignment. More specifically, $\frac{\sqrt{f^2_{1i}+f^2_{2i}}}{\sqrt{g^2_{1i}+g^2_{2i}}}$ reflects the relative scale of f_i and g_i. A high $\frac{\sqrt{f^2_{1i}+f^2_{2i}}}{\sqrt{g^2_{1i}+g^2_{2i}}}$ indicates that the weights of actions are higher in the production function compared to those in the performance measure. As a result, the firm will optimally increase the incentive intensity based on such a performance measure. Cos (θ_i) captures the alignment effect. To the extent that the performance measure induces CEO's actions less aligned with firm value, θ will increase, and the performance measure will distort incentives more (Baker 2000). As a result, the firm will optimally reduce the incentive intensity.

Next we identify the differences between a cooperative and an IOF in terms of the parameters in the production function and performance measure. First, the CEO's contribution to firm value depends on organizational form. In cooperatives, it is equivalent to the change in total member value. Members want to bring both upstream farms and the downstream cooperative to value, i.e. $f_{1C}>0, f_{2C}>0$. Investors of an IOF processor care only about value added to the downstream stage, i.e. $f_{1F}>0, f_{2F} = 0$. Second, the performance measures of IOFs and cooperatives differ. It is common in IOFs that the CEO's bonus is paid in the form of firm shares, i.e. $g_{1F}>0$, $g_{2F} = 0$. This instrument is lacking in cooperatives and we

Table 1 Marginal product and performance measure parameters

	F	C
f_{1i}	>0	>0
f_{2i}	0	>0
g_{1i}	>0	0
g_{2i}	0	>0

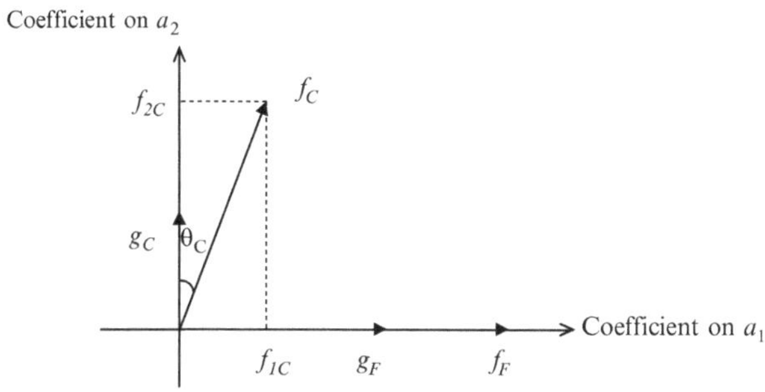

Fig. 2 Scale and alignment differences between a cooperative and an IOF

capture this by $g_{1C} = 0$.[2] However, member interests are usually present in the incentive scheme for a cooperative CEO, e.g. by benchmarking the transfer price or production volume. This results in $g_{2C}>0$. To wrap up, members' plurality of interests is represented by $f_{2C}>0$, while the absence of patron-members, and therefore serving their interests, in an IOF is represented by $g_{2F} = 0$. The absence of public listing of a cooperative is embodied by $g_{1C} = 0$, while the use of the stock price in an IOF's performance measure is captured by $g_{1F}>0$. The distinct features of two governance structures are displayed in Table 1 and Fig. 2.

Plugging these parameter values in the expressions of efficient bonus rates results in $b_F^* = f_{1F}/g_{1F}, b_C^* = f_{2C}/g_{2C}$. Subsequently, the CEO's equilibrium actions are determined $a_{1F}^* = f_{1F}, a_{2F}^* = 0$ and $a_{1C}^* = 0, a_{2C}^* = f_{2C}$. As shown in Fig. 2, the production function and performance measure are perfectly aligned in an IOF, while they are not in a cooperative. In equilibrium, an IOF CEO has incentives to undertake only a_{1F} because the investors care only about a_{1F} and make the CEO's pay dependent only on a_{1F}. Members of cooperatives, however, appreciate the CEO's actions on both dimensions but are able to compensate only for a_{2C}. Thus, only an incentive to increase a_{2C} is created and no incentive for a_{1C} exists even though it contributes to the firm value. In other words, when an action increases the member value without simultaneously increasing the performance measure, the CEO has no incentives to do it. When the available performance measures are incomplete, the incentive contract will lead to distortion, or 'the folly of rewarding A while hoping for B' (Kerr 1975). With the complex and sometimes ambiguous

[2] We are not stating that a cooperative has no information at all about the downstream activities, but our model will focus on the impact of lacking certain information.

goals of cooperatives, the incentive contact may provide only a partial representation of its objectives. The misalignment between the performance measure and the production function persuades the CEO to pay unbalanced attention to actions that positively affect their scores on the performance measures, neglecting areas for which performance is not assessed.

3 Society of Members: Size and Heterogeneity

The above model refers to members in general. The results can be best understood as the extent to which the CEO's interest accords with the average member' interest. Now we turn to explore the impact of membership size and member heterogeneity on the incentives provision. In the standard principal-agent model, the agent is usually assumed to be risk averse whereas the principal is assumed to be risk neutral. The assumption that the principal is risk neutral will now be relaxed. Members' risk attitude are different from that of the investors of an IOF because the latter could diversify their portfolio to spread risks. Due to the immobility of cooperative capital, members usually exhibit financial commitment to a particular line of business, having all their eggs in one basket (Staatz 1987). But we maintain that the agent is more risk averse than the principal.

Suppose there are n identical members in the cooperative. The CEO's contribution to member q and to the society of members are $y_{C(q)} = \frac{1}{n} f_{1C} a_{1C} + \frac{1}{n} f_{2C} a_{2C} + \varepsilon$ and $\sum_n y_{C(q)} = f_{1C} a_{1C} + f_{2C} a_{2C} + n\varepsilon$, respectively. Assuming that errors are independent and all members will agree on a single way of evaluating the CEO, the performance measure remains $p_C = g_{1C} a_{1C} + g_{2C} a_{2C} + \phi$. Let r denote the CEO's risk aversion, R the risk aversion of each member, R_0 the risk aversion of the group of members, v' the variance of ε, and v the variance of ϕ. When members act collusively and pool risks, $\frac{1}{R_0} = \sum_n \frac{1}{R} = \frac{n}{R}$, i.e. the existence of multiple members decreases members' risk aversion. The efficient bonus rate is now $b_C{}^* = \frac{f_{2C} g_{2C}}{g_{2C}^2 + v(r - R/n)}$, which is smaller than f_{2C}/g_{2C} given that r > R. This is in line with the results in the standard principal-agent problem regarding risk-aversion. If the agent is more risk-averse, the equilibrium compensation scheme specifies a lower incentive intensity and higher base wage. The expression also indicates that the efficient bonus rate decreases with n. That is, a larger society of members decreases the efficient bonus rate. The increasing ability of a larger membership to bear risks further widens the gap between the risk aversion of the CEO and that of the members.

Proposition 1. The managerial incentive intensity decreases with the number of members.

Next we relax the assumption of member homogeneity and keep the size of the membership fixed. Hansmann (1996) stresses the importance of a homogeneous membership for the efficiency of decision-making. However, cooperative members often differ in various dimensions, like age, location, size, investment portfolio, amount of capital investment, attitude towards risk, and being an active or retired member. The result is that members will have different preferences regarding the decisions made by the cooperative. For example, good performance for the inactive or over-invested member is associated with the amount of returned equity, but good performance for the under-invested or new member means the competitiveness of current prices or services (Cook 1994).

The investor and owner roles of members entail that they share the same goal of bringing the downstream stage of production to value in order to receive dividends. That is, they value the CEO's action a_{1C} in the same way. As independent farmers, each of them derives individual benefits from the cooperative and its services and therefore may value the CEO's action a_{2C} in different ways. Suppose n cooperative members differ regarding risk aversion and valuation of a_{2C}. The individual benefit of member q is $y_{C(q)} = \frac{1}{n} f_{1C} a_{1C} + \frac{1}{n} f_{2C(q)} a_{2C} + \varepsilon_{(q)}$, where $f_{2C(q)}$ denotes the value member q assigns to a_{2C}. Consequently the total benefits of the society of members is $\sum_{n} y_{C(q)} = f_{1C} a_{1C} + f_{2C} a_{2C} + \sum_{n} \varepsilon_{(q)}$, where $f_{2C} = \sum_{q} f_{2C(q)}$. Now the joint risk aversion of the members R_0 becomes $\frac{1}{R_0} = \sum_{n} \frac{1}{R_{(q)}}$, where R_q denotes the risk aversion of member q. The efficient bonus rate becomes $b_C{}^* = \frac{f_{2C} g_{2C}}{g_{2C}^2 + v(r - R_0)}$. It can be shown that if the sum of all members' risk aversions is fixed, R_0, and subsequently the efficient bonus rate, takes highest possible value when members have identical risk aversions. In other words, the heterogeneity of members' risk aversions leads to lower joint risk aversion and consequently a lower efficient bonus rate.

Proposition 2. The heterogeneity in the members' risk aversions decreases the incentive intensity of a cooperative CEO.

This proposition provides an explanation for the phenomenon that, compared with investors of an IOF, members of a cooperative usually are more homogeneous with regard to their social backgrounds, investment portfolios, attitudes towards risk, and so on. This finding suggests that the negative relationship between member heterogeneity and the strength of CEO incentives might be one of the considerations regarding the evolution of membership heterogeneity in the course of time. The membership may be quite heterogeneous at the founding stage of a cooperative, but the development of cooperatives are geared towards attracting more homogeneous members and encouraging heterogeneous members to leave in subsequent stages. This reduces the impact of member heterogeneity on the managerial incentive intensity.

4 Conclusion

Cooperatives have been, and are, formed by many small producers to build countervailing power (Galbraith 1952) in order to mitigate the adverse effects of a few powerful buyers as well as to exploit power on their own. The growth of membership, however, brings also challenges to the development of cooperatives. How to reduce the heterogeneity of cooperative membership has drawn considerable attention from cooperative researchers. This can be justified for a number of reasons, for example, to decreases the costs of voice and collective decision making (Hansmann 1996), and to facilitate the pooling arrangement in cooperatives. By presenting a model that highlights the principal-agent tension between members and the cooperative CEO, we formulate results regarding the sensitivity of the optimal incentive intensity to the membership size and composition. We have shown that the managerial incentive in a cooperative decreases as the cooperative membership grows larger and more diverse.

References

Baker G (2000) The use of performance measures in incentive contracting. Am Econ Rev 90 (2):415–420

Cook ML (1994) The role of management behavior in agricultural cooperatives. J Agric Cooperatives 9:42–58

Cook ML (1995) The future of U.S. agricultural cooperatives: a neo-institutional approach. Am J Agric Econ 77:1153–1159

Feng L, Hendrikse GWJ (2012) Chain interdependencies, measurement problems, and efficient governance structure: cooperatives versus publicly listed firms. Eur Rev Agric Econ 39 (2):241–255

Fulton M (1988) Cooperative Organizations in Western Canada. Occasional Paper 88.03. Center for the Study of Cooperatives, University of Saskatchewan, Saskatchewan

Fulton M (1995) The future of cooperatives in Canada: a property rights approach. Am J Agric Econ 77(5):1144–1152

Galbraith JK (1952) American capitalism: the concept of countervailing power. Houghton-Mifflin, Boston

Gibbons R (1998) Incentives in organizations. J Econ Perspect 12:115–132

Hansmann H (1996) The ownership of enterprise. The Belknap Press of Harvard University Press, Cambridge

Hueth B, Marcoul P (2008) Incentive pay for CEOs in cooperative firms. Am J Agric Econ 91 (5):1218–1223

Kerr S (1975) On the folly of rewarding for A while hoping for B. Acad Manage J, 18(4):769–83

Staatz JM (1987) The structural characteristics of farmer cooperatives and their behavioral consequences. In: Royer J (ed) Cooperative theory: new approach. ACS Service Report no. 18. USDA, Washington, pp 33–60

Trechter DD, King RP, Cobia DW, Hartell JG (1997) Case studies of executive compensation in agricultural cooperatives. Rev Agric Econ 19(2):492–503

Trifon R (1961) The economics of cooperative ventures: further comments. J Farm Econ 43:215–235

Uncertainties and Governance Structure in Incentives Provision for Product Quality

Wendong Deng and George Hendrikse

Abstract This paper compares the product quality provision of cooperatives and investor owned firms (IOFs) by highlighting the impacts of uncertainties in agricultural production and marketing, and farmers' risk aversion. In a principal-agent model, we show that the linear contract can shift the risk of market uncertainty from farmers to processors, and pooling can share the risk of production uncertainty among cooperative members. Complete pooling places the cooperative at a disadvantage relative to the IOF in a quality-differentiated market due to the loss of free-riding dominating the gain of risk-sharing. Product quality of cooperatives decreases when the membership size increases. Cooperatives can overcome this disadvantage by partial pooling. Product quality of cooperatives will be equivalent to that of IOFs when an optimal income rights structure with partial pooling is adopted.

Keywords Cooperative • Investor owned firm • Pooling • Quality

1 Introduction

In the organizational economics literature, cooperatives are commonly considered as less efficient in terms of delivering high-quality products to the market. Saitone and Sexton (2009, p. 1224) list a number of disadvantages of cooperatives in the provision of product quality, including: "(i) revenue pooling, which in quality-differentiated markets is generally regarded as disadvantageous due to the potential for adverse selection; (ii) patronage-based financing, which leads to the horizon problem and underinvestment in long-term strategies that can enhance objective or perceived product quality; (iii) providing a 'home' for member production, which is

W. Deng (✉) • G. Hendrikse
Rotterdam School of Management, Office T8-56, Erasmus University, Burgermeester Oudlaan 50, 3062PA Rotterdam, The Netherlands
e-mail: wdeng@rsm.nl; ghendrikse@rsm.nl

T. Ehrmann et al. (eds.), *Network Governance*, Contributions to Management Science,
DOI 10.1007/978-3-7908-2867-2_11,

problematic both with respect to product quality and the potential to glut niche markets; (iv) difficulties in terminating 'marginal' members; and (v) limitations on procuring product from nonmember sources." These considerations have led to the pessimistic prospect of cooperatives' future regarding their ability to compete and survive in the modern agricultural markets (Fulton 1995; Cook 1995).

However, nowadays there are many large cooperatives active in the market, competing with IOFs for market share by delivering products with superior quality. For example, in the Brazilian broiler industry, Cechin et al. (2012) find that suppliers delivering to a cooperative are performing better in terms of quality than suppliers delivering to an IOF. Another empirical observation raising doubts about the prospects for cooperatives is that in many sectors they coexist with IOFs. Mérel et al. (2009) posit that several particular characteristics of cooperatives, such as the preference of consumers for cooperative products, better communication, insurance function of pooling, and etc., have the potential to counterbalance the disadvantages of cooperatives in the provision of product quality. Other cooperative researchers consider the components of social capital in cooperatives, such as identification and trust, as cooperatives' comparative advantage in the competition with IOFs (Uzea and Fulton 2009; Feng et al. 2011).

This paper formulates a principal-agent model regarding the provision of product quality by different governance structures, including self-processing, cooperative and IOF. The impacts on product quality of three aspects are highlighted: farmers' risk-aversion, uncertainties, and (partial) pooling. First, although farmers are usually regarded as risk-averse (Staatz 1987), in most of the conceptual models analysing the decision-making of farmers, their characteristic of risk aversion is not explicitly captured. Second, agricultural production and marketing are subject to different types of risks, including biological risk, price risk and institutional risk (Bogetoft and Olesen 2004). We highlight two types of risk in agribusiness: the risk of market uncertainty and production uncertainty (Knoeber and Thurman 1995). Third, a pooling policy is often adopted by cooperatives (Hendrikse 2011). We show that a cooperative with a complete pooling policy will have lower product quality than an IOF. The growth of the cooperative membership will hamper the cooperative's provision of product quality. Cooperative researchers have pointed out that large cooperatives have to adopt the strong incentive structure by paying a "quality-specific price" to the members with high product quality (Hendrikse 2011). We investigate how a large cooperative can maintain an optimal product quality level by designing an optimal income rights structure with partial pooling. Partial pooling is effective because it provides on the one hand insurance to risk-averse farmers and on the other hand incentive for quality.

The remainder of this paper is organized as follows. In Sect. 2, we formulate the model. Section 3 compares the quality provision of different governance structures. In Sect. 4, we investigate the design of a cooperative's income rights structure, and determine the optimal quality provision for large cooperatives. Section 5 discusses the findings and the last section concludes.

2 Model

This section develops a non-cooperative game with the upstream farmers and a downstream processor. Assume that there are N identical farmers in a region producing a certain raw commodity that needs to be processed before reaching the final market. The farmers each produce only one unit of the raw produce and individually make the decisions regarding the quality of their produce. The cost related to the product quality provision of farmer i, where $i = 1, 2, \ldots, N$, is:

$$C(q_i) = \frac{1}{2} c q_i^2.$$

The quality provision cost coefficient c is identical for all farmers and is treated as a constant. Without loss of generality, the production costs of the raw produce, and the processing costs and valued-added of the final product are sunk and will not enter into the analysis. We also assume that one unit of the raw produce will be processed into one unit of the final product. We refer to the difference in the quality as in the realm of vertical product differentiation (Mérel et al. 2009). The quality of the raw produce determines the quality of the final product, and the processing itself cannot change the product quality.

The farmers are risk-averse, their von Neumann-Morgenstern utility function of an uncertain economic payoff π_i $(i = 1, 2, \ldots, N)$ is:

$$U_i = -\exp(-r\pi_i).$$

Parameter r, which is assumed to be identical for all farmers, is the farmers' coefficient of absolute risk aversion, i.e., the higher r is, the more risk averse the farmers are.

We highlight two types of risk in our model. The first type of risk is from the market uncertainty. Because we want to investigate the provision of product quality, in current model we only try to capture the uncertainty in the market's preference of product quality. Assuming that the market is competitive and quality-differentiated, the final product's unit price in the market is linearly increasing with the product quality, nevertheless, with a certain level of uncertainty:

$$P_m = (b + \varepsilon_1)q.$$

q (>0) denotes the quality of the final product sold, which is determined by the quality of the raw produce, and the coefficient b (>0) denotes the marginal market price with respect to the product quality. b can also be understood as the market's aggregate "taste parameter" in the model of Mussa and Rosen (1978). The utility that the market derives from consuming one unit of the product with the quality of q is bq, and it pays the equivalent price bq to the product seller. The market prefers higher ranked quality by paying a higher price as $b>0$. However, the preference or taste of the market is uncertain. Therefore, there is a normally distributed random noise term ε_1 in the marginal market price, with mean zero and variance σ_1^2.

The second type of risk is from the uncertainty in agricultural production per se. Assuming that farmer i's planned quality for his production is q_i. However, the farmer's realized product quality after harvest is $q_i + \varepsilon_{2i}$, where ε_{2i} is a normally distributed random noise term, with mean zero and variance σ_{2i}^2, representing the uncertainty in his production. We assume that production uncertainty is common for all farmers, i.e., $\varepsilon_{2i} = \varepsilon_2$, $\sigma_{2i}^2 = \sigma_2^2$. The variances σ_1^2 and σ_2^2 represent the objective risk of the market and production, respectively.

Three governance structures regarding the processing of a farmer's produce will be considered: self-processing, cooperative and IOF. In the following, we will analyse how the risk of market and production uncertainty affects the farmers' utility when they trade with different processors, and the consequence on the provision of product quality.

2.1 Self-processing

Consider the situation where a farmer processes the raw produce into the final product by himself, and then sells the final product in the market directly. In self-processing, a farmer, also as the processor, sells the product in the market and will receive the price P_m from the market according to his product quality q_i. There is indeed no contract between the farmer and processor. Farmer i's economic payoff is:

$$\pi_i = (b + \varepsilon_1)(q_i + \varepsilon_2) - \frac{1}{2} c q_i^2.$$

The farmer's utility function is:

$$U_i = -\exp\left\{-r\left[(b + \varepsilon_1)(q_i + \varepsilon_2) - \frac{1}{2} c q_i^2\right]\right\}.$$

The farmer's certainty equivalent payoff is:

$$CE_i = bq_i - \frac{1}{2} c q_i^2 - \frac{1}{2} k_1 q_i^2 - \frac{1}{2} k_2 b^2 + CE(\varepsilon_1 \varepsilon_2),$$

where $k_1 = r\sigma_1^2$, $k_2 = r\sigma_2^2$ denote the farmer's subjective risk toward the market and production uncertainty, respectively. Each term of subjective risk is the corresponding objective risk scaled by the farmer's degree of aversion (see Bolton and Dewatripont 2005, Chap. 4). The term $\frac{1}{2} k_1 q_i^2$ and $\frac{1}{2} k_2 b^2$ are risk premiums, which are the disutility of risk. $CE(\varepsilon_1\varepsilon_2)$ is the risk premium of the joint contribution of the market and production uncertainty. It is negative and decreases when the market and production uncertainty increases.

The farmers can also sell the raw produce to an enterprise processor. The enterprise processor has one of the two governance structures: an open-membership cooperative or an IOF. We model the transactions between the enterprise processor and the farmers in a principal-agent framework (Holmström 1979). The processor acts as a principal, and the farmers are agents who are rewarded by the outcome of their efforts invested in the product quality. The efforts per se are not observable, but the quality q of the delivered raw produce from the farmers to the processor is contractible. The processor offers the farmers a linear contract stating the payment formula as:

$$P = \alpha + \beta q.$$

P is the unit price of the raw produce that the processor will pay for. α (≥ 0) is the base (guarantee) price and β (≥ 0) is the incentive regarding the product quality or the quality premium. An important function of the linear contract between the principal and agent is to balance the costs of risk bearing against the incentive gains (Holmström 1979). This form of contract is commonly used in agribusiness, whether the processor is an IOF or a cooperative (Gow et al. 2000; Cechin et al. 2012).

2.2 Cooperative

Confronted with the market and production uncertainty, the individual farmers who used to process individually and trade directly in the market may have the incentive to form a marketing cooperative with an open-membership policy. The members of the cooperative jointly own the processor, but the farmers remain independent regarding their quality decisions. We assume that the cooperative adopts the traditional principle of complete pooling policy. This assumption will be relaxed later. The marketing contract between the cooperative and the members has the payment formula as follows:

$$P_c = \alpha_c + \beta_c Q_c.$$

$$Q_c = \frac{1}{n} \sum_{i}^{n} (q_i + \varepsilon_{2i}).$$

The price consists of a fixed base price α_c and a quality-incentive β_c. In complete pooling, the cooperative enacts a single pool for all products with various qualities, and the members share equitably on a per-unit basis in the revenue stream that has been created (Saitone and Sexton 2009). This equality principle distributes the net revenue to members based on the delivered volume, regardless the quality of the product. Therefore, in the payment formula of the marketing contract, the quality incentive is related to the pooled or aggregate quality Q_c instead of the individual product quality q_i, since the cooperative will also receive revenues from the market based on Q_c. A member i's economic payoff is:

$$\pi_i = \alpha_c + \beta_c Q_c - \frac{1}{2} c q_i^2.$$

and the member's utility function is:

$$U_i = -\exp\{-r[\alpha_c + \frac{\beta_c}{n} \sum_i^n (q_i + \varepsilon_{2i}) - \frac{1}{2} c q_i^2]\}.$$

Different from the utility function of the self-processing farmers who trade directly in the market, the subjective risk toward market uncertainty k_1 is not in the member's utility function. This is because the members are not selling their products to the final market directly. Instead, they sell to the market via the cooperative they own and they are insulated from the market uncertainty ε_1 and risk σ_1^2. Their projected income is then decided by the payment formula P_c. A member's certainty equivalent payoff is:

$$CE_i = \alpha_c + \frac{\beta_c(q_i + Q_{-i})}{n} - \frac{1}{2} c q_i^2 - \frac{1}{2} k_2 \frac{\beta_c^2}{n}.$$

Notice that Q_{-i} is the sum of the quality decisions of the other members besides member i and $\sum_i^n \varepsilon_{2i}$ has a normal distribution with variance $n\sigma_2^2$. We can see that complete pooling reduces the member's risk premium term $\frac{1}{2} k_2 \frac{\beta_c^2}{n}$ related to the production uncertainty by a factor of $\frac{1}{n}$. It captures the risk sharing function of pooling.

2.3 IOF

An IOF procures the raw produce of the farmers and sells the processed products in the same final market. The contract the IOF offers to the farmers is:

$$P_f = \alpha_f + \beta_f q_i.$$

Similarly, the price consists of a fixed base price α_f and an individualized quality-incentive β_f, and with this contract the farmers are not faced with the

Table 1 Decisions of the farmers and the processor

	Self-processing	IOF	Cooperative
Processor	–	α_f, β_f	α_c, β_c
Farmer i ($i = 1, 2, \ldots, N$)	q_i	q_i	q_i

market uncertainty ε_1 and risk σ_1^2 directly. Differently, the quality incentive is now based on the individual instead of pooled product quality. The economic payoff of a farmer i, who trades with the IOF, is:

$$\pi_i = \alpha_f + \beta_f q_i - \frac{1}{2} c q_i^2.$$

His utility function is:

$$U_i = -\exp\left\{-r\left[\alpha_f + \beta_f(q_i + \varepsilon_{2i}) - \frac{1}{2} c q_i^2\right]\right\}.$$

and his certainty equivalent payoff is:

$$CE_i = \alpha_f + \beta_f q_i - \frac{1}{2} c q_i^2 - \frac{1}{2} k_2 \beta_f^2.$$

Different from the certainty payoff of the cooperative members, in farmer i's certainty equivalent payoff, both the quality incentive and risk premium of the production uncertainty are individualized.

Table 1 lists the players' decisions in three different governance structures.

Assume that the farmers' coefficient of absolute risk aversion, quality provision cost coefficient, production uncertainty, and the market's preference and uncertainty are common knowledge. The product quality is perfectly measurable. The timing of the two-stage game is as follows: (1) the principal (processor) chooses the α and β of the payment formula; (2) the agents (farmers) choose the product quality to maximize their certainty equivalent payoff. This game will be solved by backward deduction.

3 Equilibrium

In this section, we derive the equilibrium product quality in different governance structures and compare the farmers' certainty equivalent payoff.

3.1 Self-processing

The self-processing farmer's decision regarding product quality is obtained by the FOC of his certainty equivalent payoff:

$$\frac{\partial CE_i}{\partial q_i} = b - k_1 q_i - c q_i = 0.$$

$$q_i^* = \frac{b}{c + k_1}.$$

The result entails that if the farmer trades directly in the market, the quality of the product is determined by his subjective risk toward the market uncertainty k_1. The product quality will be reduced if the farmer's subjective risk toward the market uncertainty is high. The production uncertainty doesn't play a role in the quality decision because it is intrinsic and the farmer cannot change the disutility from the production uncertainty by choosing his product quality. However, it determines whether the farmer will participate in the market. The farmer's certainty equivalent payoff is:

$$CE_i^* = \frac{b^2}{2}\left(\frac{1}{c + k_1} - k_2\right) + CE(\varepsilon_1 \varepsilon_2).$$

Assuming that the farmer's reservation certainty payoff is zero, when their subjective risk toward the market uncertainty k_1 and product measurement uncertainty k_2 is so large that $CE_i^* < 0$, the farmer is not willing to participate in the market.

3.2 Cooperative

The cooperative members make their decisions individually. Member i's decision on his product quality is obtained by the FOC of his certainty equivalent payoff:

$$\frac{\partial CE_i}{\partial q_i} = \frac{\beta_c}{n} - c q_i = 0.$$

$$q_i^* = \frac{\beta_c}{nc}.$$

While the subjective risk toward market uncertainty k_1 doesn't play a role in members' decisions now, the cooperative's membership size n and quality incentive β_c jointly determine the member's decision regarding product quality. As the cooperative becomes large in terms of n, while the arrangement of complete pooling attenuates the production risk of individual members by risk sharing, it also causes an offsetting impact—the members will free ride on other members' efforts in product quality improvement. The members have little incentive to supply high-quality product, since each member's share is relatively insensitive to his effort level in a large organization. When any individual effort will not be directly rewarded, it gives rise to free riding. Because all members are identical, the cooperative's aggregate product quality is:

$$Q_c^* = \frac{\beta_c}{nc} + \frac{1}{n}\sum_i^n \varepsilon_{2i}.$$

For the members, the risk of market uncertainty is now placed on the cooperative processor. In fact, this risk cannot be completely shifted from the members to the processor because the members are also the decision makers and residual claimants of the processor. They are actually the same people. Therefore, farmer cooperatives are usually regarded as risk averse in decision making (Staatz 1987; Vitaliano 1983). However, the risk-sharing is still possible between the members and processor because the equity in the cooperative can be used as a buffer to absorb temporary fluctuations in profits (Bogetoft and Olesen 2004). In current model, we assume that the cooperative can execute this buffering function and treat the cooperative processor as risk-neutral regarding the decision of the payment formula. Assuming that the processing costs and valued-added of the cooperative processor is sunk, the processor retains no earnings and its objective is to maximize the joint certainty equivalent payoff of the processor and members, which is:

$$\pi_c = E\left[nbQ_c^* - \frac{n}{2}cq_i^{*2} - \frac{n}{2}k_2\frac{\beta_c^2}{n}\right] = n\left(\frac{b\beta_c}{cn} - \frac{1}{2}\frac{\beta_c^2}{cn^2} - \frac{1}{2}k_2\frac{\beta_c^2}{n}\right).$$

Following the FOC regarding β_c $(0 \le \beta_c \le b)$:

$$\frac{\partial \pi_c}{\partial \beta_c} = \frac{b}{cn} - \frac{\beta_c}{cn^2} - \frac{k_2\beta_c}{n} = 0.$$

$$\beta_c^* = \frac{b}{\frac{1}{n} + ck_2}.$$

It entails that the cooperative's quality incentive payment β_c^* should increase with membership size but decrease with the members' subjective risk toward production uncertainty. The cooperative's expected aggregate product quality is:

$$Q_c = E\left[\frac{\beta_c^*}{nc} + \frac{1}{n}\sum_i^n \varepsilon_{2i}\right] = \frac{b}{c(k_2nc + 1)}.$$

The member's certainty equivalent payoff is:

$$CE_i^* = \frac{b^2}{2c(k_2nc+1)}.$$

The result shows that, in comparison with self-processing, the members of the cooperative with a complete pooling policy can always obtain a positive certainty equity payoff because the cooperative processor bears the risk of market uncertainty for its members. However, the cooperative's aggregate product quality and members' certainty equivalent payoff decreases when its membership size increases, due to the increasing free-riding problem.

3.3 IOF

The farmer trading with the IOF makes the decision of the product quality based on the FOC of his certainty equivalent payoff:

$$\frac{\partial CE_i}{\partial q_i} = \beta_f - cq_i = 0.$$

$$q_i^* = \frac{\beta_f}{c}.$$

Owned by investors who can hold diversified portfolios, the IOF is modeled as risk-neutral. Assuming that the processing costs and valued-added of the IOF processor is sunk, it will maximize its total economic payoff subject to the farmers' participation constraint. The farmers' reservation certainty equivalent payoff R is assumed to be equal to the certainty equivalent payoff of the cooperative members:

$$R = \frac{b^2}{2c(k_2nc+1)}.$$

The participation constraint of the farmers to deliver his raw produce to the IOF is:

$$CE_i^* = \alpha_f + \beta_f q_i^* - \frac{1}{2}cq_i^{*2} - \frac{1}{2}k_2\beta_i^2 \geq R.$$

The IOF will simply pay the lowest possible fixed payment so that the farmers are just willing to deliver:

$$\alpha_f^* = R - \frac{\beta_f^2}{2c} + \frac{1}{2}k_2\beta_f^2.$$

The total expected payoff of the IOF is:

$$\pi_f = E\left[nbq_i^* - n\left(\alpha_f^* + \beta_f q_i\right)\right] = n\left[\frac{b\beta_f}{c} - \frac{1}{2}k_2\beta_f^2 - \frac{\beta_f^2}{2c} - R\right].$$

The IOF maximizes its payoff by choosing β_f $(0 \leq \beta_f \leq b)$:

$$\frac{\partial \pi_f}{\partial \beta_f} = n\left[\frac{b}{c} - k_2\beta_f - \frac{\beta_f}{c}\right] = 0.$$

$$\beta_f^* = \frac{b}{1 + ck_2}.$$

Given the contract offered by the IOF, the farmer's decision on the product quality can be obtained. As all farmers are identical, and assuming that there are m farmers supplying the IOF, the expected aggregate product quality of the IOF is:

$$Q_f = E\left[\frac{\beta_f^*}{c} + \frac{1}{m}\sum_i^m \varepsilon_{2i}\right] = \frac{b}{c(1 + ck_2)}.$$

From the equation above we see that the farmers' subjective risk toward production uncertainty k_2 determines the IOF's product quality. The farmers' certainty equivalent payoff is equal to his reservation certainty payoff and the IOF keeps the remaining part of the certainty payoff for each unit of the product. The IOF exists because by offering the contracts to the non-member farmers, it also insures them from the market uncertainty and elicits supply.

3.4 Comparison

Table 2 presents the comparison of the product quality, farmers' certainty equivalent payoff and the processor's payoff per unit of product in different governance structures. When the farmers process individually and trade in the market directly, the product quality is merely decided by their subjective risk toward the market uncertainty. However, when the farmers' subjective risks toward the market and production uncertainty are so large to produce a negative certainty equivalent payoff, they will not participate in the market. By contrast, when the farmers trade with a (enterprise) processor, the risk of market uncertainty is shifted from the farmers to the processor through the contract. This result is supported by empirical findings (Knoeber and Thurman 1995), and it may justify the trend that, fewer and fewer products are traded on open markets and production contracts are

Table 2 Product quality and certainty equivalent payoff in the three governance structures

	Self-processing	IOF	Cooperative
Product quality	$\frac{b}{c+k_1}$	$\frac{b}{c(1+ck_2)}$	$\frac{b}{c(1+nk_2c)}$
Farmers' CE payoff	$\frac{b^2}{2}\left(\frac{1}{c+k_1}-k_2\right)+CE(\varepsilon_1\varepsilon_2)$	$\frac{b^2}{2c(1+nk_2c)}$	$\frac{b^2}{2c(1+nk_2c)}$
Processor's payoff	n.a.	$\frac{b^2}{2c(1+ck_2)}-\frac{b^2}{2c(1+nk_2c)}$	0

CE certainty equivalent

more and more common (Bogetoft and Olesen 2004). However, the farmers still face the risk of production uncertainty. The product quality will thus be determined by the contract offered by the processor, which balances the production risk bearing and incentive provision.

The cooperative processor has the advantage over self-processing, since the risk-averse farmers can always earn a positive certainty equivalent payoff as members of the cooperative in an uncertain market. This provides a justification for the formation of agricultural cooperatives. However, the complete pooling policy is problematic. Although the complete pooling policy can bring the benefits of risk sharing that supports the quality provision of the cooperative, it goes at the detriment of its members' incentive in quality improvement. With the complete pooling policy, when a new member joins the cooperative, the loss from the free-riding dominates the benefit of risk sharing. In addition, the cooperative is not able to provide sufficient incentives for the provision of product quality. As a consequence, the cooperative's product quality will continuously decrease as its membership size increases.

Instead of using the quality incentive based on pooled quality, the IOF processor offers the farmers quality incentive based on individual product quality. Without pooling, the individualized incentive will expose the farmers more to the production uncertainty. However, the IOF processor can design an optimal contract which reaches a trade-off between providing incentives and minimizing the cost of risk. Therefore, the IOF processor is able elicit farmers to deliver products with higher quality. We can formulate the first proposition as follows:

Proposition 1. *The product quality of the cooperative with a complete pooling policy is always lower than that of the IOF.*

Figure 1 provides a graphical illustration that compares the product quality of the IOF and the cooperative. How will the farmers choose the processor? As discussed previously, if self-processing and trading directly in the market brings negative or no certainty equivalent payoff to the farmers due to the large uncertainties, i.e.

$$\frac{b^2}{2}\left(\frac{1}{c+k_1}-k_2\right)+CE(\varepsilon_1\varepsilon_2)\leq 0,$$

No farmers are not willing to participate in the market alone. They either form a cooperative or trade with the IOF. Assuming that both a cooperative and an IOF exist, they are attracting the supplies from the farmers in the same region. Figure 2

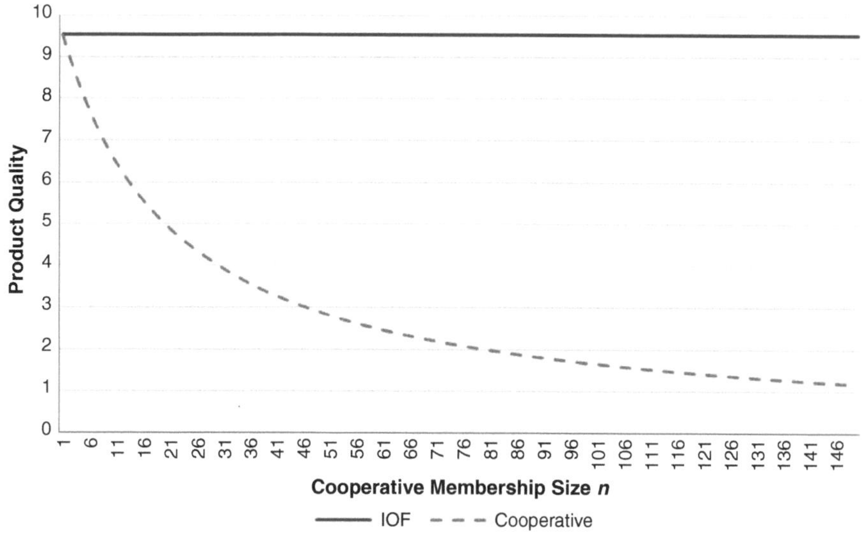

Fig. 1 Product quality

illustrates the competition between the cooperative and the IOF over raw produce supply. When the cooperative's membership size is M, each member's certainty equivalent payoff is equal to 1. The IOF processor designs the contract subject to the farmers' participation constraint, which will be equal to 1. The cooperative's membership size will no longer grow because each member's certainty equivalent payoff will decrease to below 1 if more farmers join in the cooperative. As a consequence, some members will leave and turn to the IOF. However, if the IOF processor wants to attract more suppliers, it can simply modify the contract offered to the suppliers by increasing the base payment α_f. As such, the certainty equivalent payoff of the farmers who supply the IOF will be higher than the cooperative members' certainty equivalent payoff. For example, if the certainty equivalent payoff of the farmers who supply the IOF is increased from 1 to 1.5, some cooperative members will then leave the cooperative and trade with the IOF. The cooperative membership size will decrease. With fewer members, the cooperative's product quality and members' certainty equivalent payoff will increase because the free-riding problem is relatively eased. When the membership size decreases to M', cooperative members' certainty equivalent payoff is again equal to the certainty equivalent payoff received by the farmers trading with the IOF, the members will stay in the cooperative. The membership size of the cooperative is determined by the certainty equivalent payoff that the IOF offers to its suppliers. Generally, the cooperative with a complete pooling policy is in a disadvantageous position in the competition with the IOF. The IOF can not only elicit supply with better quality, but also attract the supplies from the cooperative's members by increasing payment. The total certainty equivalent payoff (sum of farmer and processor) of each unit of

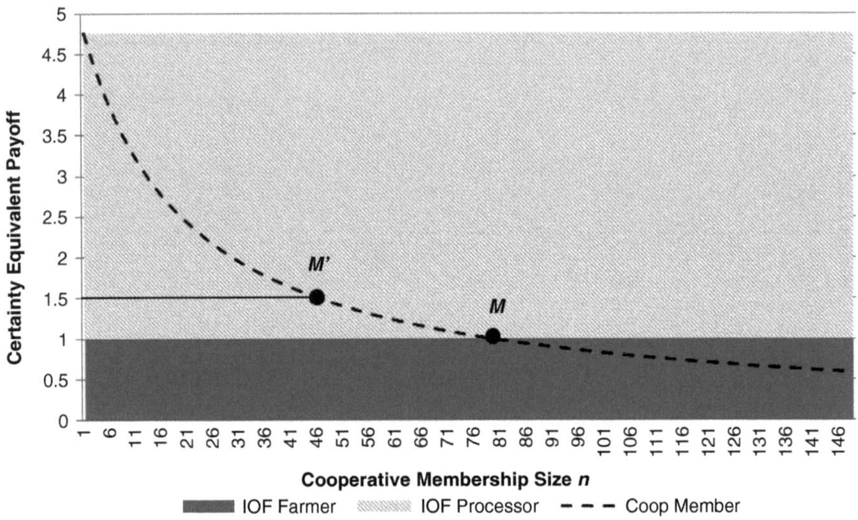

Fig. 2 Certainty equivalent payoffs

product generated by the IOF is larger than that by the cooperative, i.e. $\frac{b^2}{2c(1+ck_2)} > \frac{b^2}{2c(1+nck_2)}$. The IOF is thus more efficient than the cooperative.

Another situation we have to consider is when self-processing and trading directly in the market also can bring a positive certainty equivalent payoff to the farmers, i.e. $\frac{b^2}{2}\left(\frac{1}{c+k_1} - k_2\right) + CE(\varepsilon_1\varepsilon_2) > 0$. The curve in Fig. 3 approximates the situations where the certainty equivalent payoff of self-processing is equal to 0. The shaded area below the curve thus represents the range of k_1 and k_2, where the self-processing farmers can obtain a certain level of positive certainty equivalent payoff. This positive certainty equivalent payoff also serves as the reservation payoff of all farmers. According to Fig. 2, the certainty equivalent payoff of the cooperative members will continuously decrease when the membership size increases. When the certainty equivalent payoff of the cooperative members is equivalent to the certainty equivalent payoff of the self-processing farmers, the farmers are indifferent between self-processing and becoming members of the cooperative. The cooperative's membership size will thus no longer grow. The membership size of the cooperative is determined by the certainty equivalent payoff of the self-processing. When k_1 and k_2 increase, the certainty equivalent payoff of the self-processing will decrease, so will the reservation payoff of all farmers. The membership size of the cooperative will increase. If there exists also an IOF in the region, the IOF will design the contract subject to the reservation payoff as well and takes it as the farmers' participation constraint. As such, the farmers will be indifferent in self-processing or supplying to the cooperative or the IOF. Given that $k_1 > 0$, $k_2 > 0$ and $CE(\varepsilon_1\varepsilon_2) < 0$, through simple derivation we can obtain the result that the total

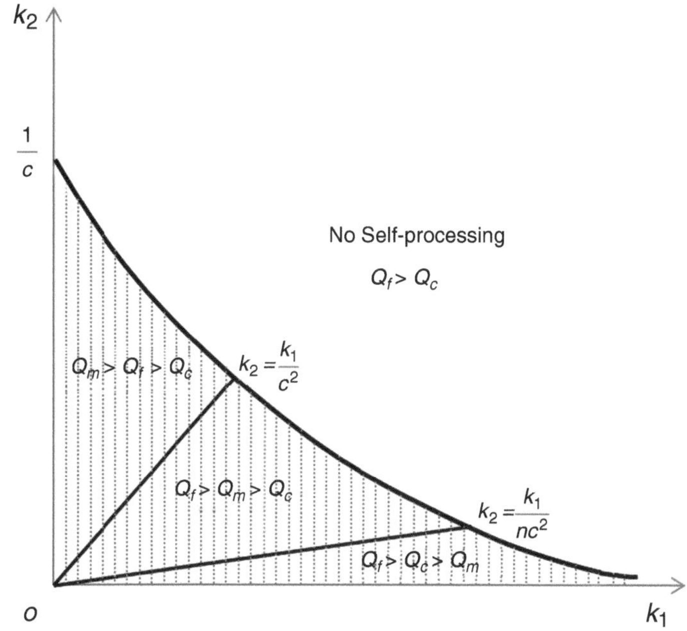

Fig. 3 Uncertainties, governance structures and product quality

certainty equivalent payoff of each unit of product generated by the IOF is larger than that by self-processing, i.e. $\frac{b^2}{2c(1+ck_2)} > \frac{b^2}{2}\left(\frac{1}{c+k_1} - k_2\right) + CE(\varepsilon_1\varepsilon_2)$. The IOF is also more efficient than self-processing.

We can compare the product quality of different processors according to the value of k_1 and k_2. First, the IOF's product quality is always higher than that of the cooperative (Proposition 1). Second, when the certainty equivalent payoff of self-processing is positive and $k_2 < \frac{k_1}{c^2}$, the IOF's product quality is higher than that of the self-processing products, i.e. $\frac{b}{c(1+ck_2)} > \frac{b}{c+k_1}$. Third, when the certainty equivalent payoff of self-processing is positive and $k_2 < \frac{k_1}{nc^2}$, the cooperative's product quality is higher than that of the self-processing product, i.e. $\frac{b}{c(1+nck_2)} > \frac{b}{c+k_1}$. Finally, when self-processing and trading directly in the market brings no positive certainty equivalent payoff to the farmers (k_1 and k_2 are in the area above the curve), there will be no self-processing. Denote the product quality of self-processing, cooperative and IOF as Q_m, Q_c and Q_f, respectively. According to the values of k_1 and k_2, the rank of product quality of the three different governance structures is illustrated Fig. 3.

4 Partial Pooling

In this section, we investigate how a cooperative can design an optimal income rights structure by adjusting its pooling policy, in order to achieve a high product quality when the membership size is large. We extend the model of the cooperative processor of Sect. 2 by the decision regarding a pooling ratio Δ $(0 \leq \Delta \leq 1)$, in addition to the decisions of the fixed payment $\alpha_c (\geq 0)$ and quality incentive β_c $(0 \leq \beta_c \leq b)$. The payment that a member will receive is modified to:

$$P_i = \alpha_c + \beta_c \Delta Q_c + \beta_c (1 - \Delta) q_i.$$

The pooling ratio Δ measures to what extent the quality incentive will be paid according to the pooled quality Q_c, whereas $1 - \Delta$ denotes the portion of a member's production that receives a quality-specific price (Saitone and Sexton 2009). Member i's economic payoff is:

$$\pi_i = \alpha_c + \beta_c \Delta Q_c + \beta_c (1 - \Delta) q_i - \frac{1}{2} c q_i^2.$$

The member' utility function is:

$$U_i = -\exp\left\{ -r \left[\alpha_c + \frac{\beta_c \Delta}{n} \sum_i^n (q_i + \varepsilon_{2i}) + \beta_c (1 - \Delta)(q_i + \varepsilon_{2i}) - \frac{1}{2} c q_i^2 \right] \right\}.$$

The member' certainty equivalent payoff is:

$$CE_i = \alpha_c + \frac{\beta_c \Delta}{n} \sum_i^n q_i + \beta_c (1 - \Delta) q_i - \frac{1}{2} c q_i^2 - \frac{1}{2} k_2 \beta_c^2 \left(\frac{\Delta^2}{n} + (1 - \Delta)^2 \right).$$

The member's decision on quality is obtained by:

$$\frac{\partial CE_i}{\partial q_i} = \frac{\beta_c \Delta}{n} + \beta_c (1 - \Delta) - c q_i = 0.$$

$$q_i^* = \frac{\beta_c \Delta}{nc} + \frac{\beta_c (1 - \Delta)}{c} = \frac{\beta_c}{c} \left(\frac{\Delta}{n} + 1 - \Delta \right).$$

The cooperative's aggregate quality is then:

$$Q_c^* = \frac{\beta_c}{c} \left(\frac{\Delta}{n} + 1 - \Delta \right) + \frac{1}{n} \sum_i^n \varepsilon_{2i}.$$

Similarly, the cooperative processor retains no earnings and maximizes the joint certainty equivalent payoff of the processor and members, which is:

$$\pi_c = E\left[nbQ_c^* - \frac{n}{2}cq_i^{*2} - \frac{n}{2}k_2\beta_c^2\left(\frac{\Delta^2}{n} + (1-\Delta)^2\right)\right]$$
$$= n\left(\frac{b\beta_c}{c}\left(\frac{\Delta}{n} + 1 - \Delta\right) - \frac{\beta_c^2}{2c}\left(\frac{\Delta}{n} + 1 - \Delta\right)^2 - \frac{1}{2}k_2\beta_c^2\left(\frac{\Delta^2}{n} + (1-\Delta)^2\right)\right).$$

The cooperative maximizes the π_c by choosing β_c $(0 \le \beta_c \le b)$ and Δ $(0 \le \Delta \le 1)$:

$$\frac{\partial \pi_c}{\partial \Delta} = n\left(\frac{b\beta_c}{c}(\frac{1}{n} - 1)\right) - \frac{n\beta_c^2}{c}\left(\frac{\Delta}{n} + 1 - \Delta\right)\left(\frac{1}{n} - 1\right)$$
$$- nk_2\beta_c^2\left(\frac{\Delta}{n} - (1-\Delta)\right)$$
$$= 0.$$

$$\Delta^* = \frac{\left(\frac{b}{c} - \frac{\beta_c}{c}\right)\left(\frac{1}{n} - 1\right) + k_2\beta_c}{\frac{\beta_c}{c}\left(\frac{1}{n} - 1\right)^2 + k_2\beta_c\left(\frac{1}{n} + 1\right)}.$$

When n is large, $\frac{1}{n} \approx 0$:

$$\Delta^* \approx \frac{ck_2\beta_c + \beta_c - b}{ck_2\beta_c + \beta_c} = 1 - \frac{1}{(1 + ck_2)\frac{\beta_c}{b}}.$$

Because $0 \le \beta_c \le b$, the pooling ratio the cooperative can choose is:

$$0 \le \Delta^* \le \frac{ck_2}{1 + ck_2}$$

And:

$$\frac{\partial \pi_c}{\partial \beta_c} = \frac{nb}{c}\left(\frac{\Delta}{n} + 1 - \Delta\right) - \frac{n\beta_c}{c}\left(\frac{\Delta}{n} + 1 - \Delta\right)^2 - nk_2\beta_c\left(\frac{\Delta^2}{n} + (1-\Delta)^2\right) = 0.$$

$$\beta_c^* = \frac{\frac{b}{c}\left(\frac{\Delta}{n} + 1 - \Delta\right)}{\frac{1}{c}\left(\frac{\Delta}{n} + 1 - \Delta\right)^2 + k_2\left(\frac{\Delta^2}{n} + (1-\Delta)^2\right)} \approx \frac{b}{(1 + ck_2)(1 - \Delta)}.$$

$$\frac{\beta_c^*}{b} = \frac{1}{(1 + ck_2)(1 - \Delta)}.$$

Because $0 \le \Delta^* \le \frac{ck_2}{1+ck_2}$:

$$\frac{1}{1+ck_2} \le \frac{\beta_c^*}{b} \le 1$$

In sum, we obtain the optimal policy of the cooperative:

$$\frac{\beta_c^*}{b}(1-\Delta^*) = \frac{1}{(1+ck_2)}.$$

Denote $\gamma^* = \frac{\beta_c^*}{b}$, $(0 \le \gamma \le 1)$, as the ratio between the quality incentive of the cooperative and the marginal market price with respect to the product quality in the market. It measures the relative strength of the cooperative's quality incentive. In sum, the optimal income rights structure S^* of the cooperative can be written as:

$$S^* = \gamma^*(1-\Delta^*) = \frac{1}{1+ck_2}, \quad \frac{1}{1+ck_2} \le \gamma^* \le 1 \; and \; 0 \le \Delta^* \le \frac{ck_2}{1+ck_2}.$$

With the optimal income rights structure, the expected aggregate quality of the cooperative is:

$$Q_c = E\left[\frac{\beta_c^*\Delta^*}{nc} + \frac{\beta_c^*(1-\Delta^*)}{c} + \frac{1}{n}\sum_i^n \varepsilon_{2i}\right] \approx \frac{b}{c(1+ck_2)}.$$

Because the cooperative operates with a zero-profit constraint, the base price can be obtained by:

$$bQ_c^* - \left[\alpha + \beta_c^*\Delta^* Q_c^* + \beta_c^*(1-\Delta^*)q_i^*\right] = 0.$$

$$\alpha_c^* = q_i^*\left(b - \beta_c^*\right).$$

Figure 4 illustrates the optimal income rights structure the cooperative can choose. Given a certain level of members' subjective risk toward the production uncertainty k_2 and quality provision cost coefficient c, the solid part of the curve represents the efficient frontier of the optimal income rights structure.

Several important implications regarding the optimal income rights structure can be drawn. First, a high pooling ratio is associated with a high relative quality incentive strength γ. While the high pooling ratio reduces the disutility of the risk premium term $\frac{1}{2}k_2\beta_c^2\frac{\Delta^2}{n}$ in the members' certainty equivalent payoff, it also reduces the members' incentive to improve product quality and boosts free-riding. Hence, a high quality incentive is needed to maintain the product quality provision from the members when the pooling ratio is high. On the other hand, when the pooling ratio is low, the relative quality incentive strength γ must decrease. When the pooling ratio is low, its risk-sharing function will decrease whereas the quality incentive will become effective due to less free-riding. The low pooling ratio individualizes

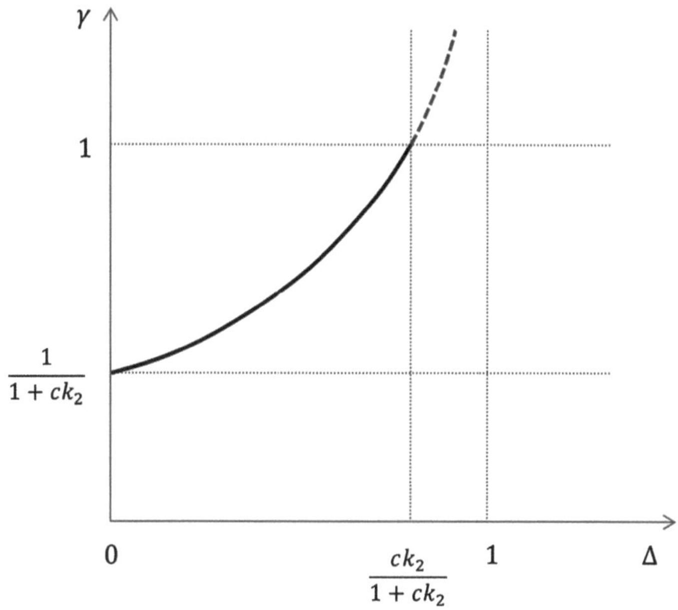

Fig. 4 The optimal income rights structure of cooperatives

not only the risk of production uncertainty but also the rewards of product quality. Therefore, with a low pooling ratio, the cooperative only needs a relative low incentive strength to support the product quality but a high base price to decrease the members' disutility from the risk of production uncertainty.

Second, the complete pooling policy, i.e. $\Delta = 1$, should by no means be adopted by the cooperative. Instead, the highest pooling ratio that the cooperative can enact is $\Delta_{max} = \frac{ck_2}{1+ck_2}$. When the cooperative chooses Δ_{max}, the base price α_c will be zero and the quality incentive β_c will be b in the corresponding optimal payment formula. It entails that the members will received no base price and the cooperative's quality incentive will be equal to the quality marginal price of the market. The relative quality incentive strength γ is then equal to 1. In other words, with the pooling ratio Δ_{max}, the cooperative does not need to pay a base payment to members to bear their risk of production uncertainty. The pooling arrangement itself has already minimized the cost of risk by risk-sharing. If the cooperative's pooling ratio is set to be higher than Δ_{max}, the cooperative has to use a relative quality incentive strength $\gamma > 1$, i.e. $\beta_c > b$, to maintain the product quality level. The reason is that a pooling ratio higher than Δ_{max} further reduces the members' incentive in quality improvement. To sustain the product quality, a more powerful incentive must be provided. However, as the cooperative operates on a zero-profit condition, choosing $\beta_c > b$ entails that $\alpha_c < 0$, i.e. the cooperative charges the members a base fee for each unit of produce they deliver. This is impractical and it also proves that the traditional principle of complete pooling, which we have modeled in Sect. 2, is not efficient. Therefore, there is an upper bound on the pooling ratio that the

Table 3 Product quality and certainty equivalent payoff in the three governance structures

	Self-processing	IOF	Cooperative
Product quality	$\frac{b}{c+k_1}$	$\frac{b}{c(ck_2+1)}$	$\frac{b}{c(ck_2+1)}$
Farmers' CE payoff	$\frac{b^2}{2}\left(\frac{1}{c+k_1}-k_2\right)+CE(\varepsilon_1\varepsilon_2)$	$\frac{b^2}{2c(ck_2+1)}$	$\frac{b^2}{2c(ck_2+1)}$
Processor's payoff	n.a.	0	0

CE certainty equivalent

cooperative can adopt. Beyond that, the cooperative will be not able to maintain its product quality with the increase of membership size. The dashed part of the curve in Fig. 3 represents the inapplicable income rights structure. Another implication of the upper bound on the pooling ratio is that Δ_{max} will decrease when the farmers' subjective risk toward production uncertainty k_2 and quality provision cost coefficient c decrease. It entails that, while agricultural modernization nowadays attenuates the production uncertainty and quality provision cost, the upper bound on the pooling ratio is continuously lowered.

Third, the cooperative can adopt the no-pooling policy i.e., $\Delta = 0$, which means that the quality incentive will be fully individualized. This is the same arrangement as in the contract of the IOF. Under this circumstance, a lowest quality incentive $\beta_c = \frac{b}{1+ck_2}$ must be chosen, otherwise the members will encounter a large disutility brought by the risk premium $\frac{1}{2}k_2\beta_c^2$, which is also fully individualized under the no-pooling policy. As the low quality incentive β_c is accompanied with a high base payment α_c, when there is no pooling to share the members' risk of production uncertainty, the highest base payment is provided to the members for bearing the risk.

In sum, the optimal income rights structure S^*, which consists of the decisions of the pooling ratio and relative quality incentive strength, provides the cooperative with optimal alignments between pooling, incentive and risk bearing, thereby supporting the quality provision of the cooperative. The cooperative's decision regarding the pooling ratio is flexible as it can choose from a range of pooling ratios. However, the cooperative may prefer a specific pooling ratio, which can bring the organization some additional benefits. We leave this topic for further research.

The second proposition is formulated as follows:

Proposition 2. *The range of the efficient equilibrium pooling ratio of the cooperative is* $\left[0, \frac{ck_2}{1+ck_2}\right]$.

Table 3 compares the product quality, farmers' certainty equivalent payoff and the processor's payoff per unit of product in different governance structures when the cooperative adopts the optimal income rights structure with partial pooling. With the optimal income rights structure, the cooperative's product quality can reach the same level as that of the IOF. Importantly, the certainty equivalent payoff the members receive increases, so does the farmers' reservation payoff. As the IOF processor is competing with the cooperative for the supplies of raw produce from the farmers in the same region, it has to increase the certainty payoff of its suppliers to the level as high as the farmers' reservation payoff. Therefore, the competition pushes the IOF processor's profit to zero and makes the farmers indifferent between supplying the cooperative and the IOF. Because $\frac{b^2}{2c(1+ck_2)} > \frac{b^2}{2}\left(\frac{1}{c+k_1} - k_2\right) + CE \times (\varepsilon_1\varepsilon_2)$, i.e. the certainty equivalent payoff of the farmers trading with an enterprise processor is strictly larger than that of self-processing farmers, no farmer will choose self-processing. The governance structure of the cooperative and the IOF are both efficient, while self-processing is inefficient. Our third proposition can be formulated as follows:

Proposition 3. *The product quality of the cooperative with an optimal income rights structure will be equivalent to that of the IOF.*

5 Discussion

We have highlighted two different types of risk in our model and investigated their impacts on the quality decisions of risk-averse farmers. Specifically, our results imply that the market uncertainty and production uncertainty both will deter the provision of product quality. If the payoff regarding the product quality is uncertain, the risk-averse farmers will be reluctant to invest efforts in quality improvement. We show that an important attribute of the production or marketing contract is to shift the risk of market uncertainty from the farmers to the processor. Via the contract, the farmers' participation can be secured. Since the farmers still face the risk of production uncertainty, another function of the contract is to balance the production risk bearing and incentive, in order to elicit the optimal product quality from the farmers. The processor thus optimizes the contract according to its objective by choosing the payment formula. It is proved that the cooperative processor with a complete pooling policy is disadvantageous in the competition with the IOF processor. The latter can elicit the optimal quality provision from the farmers by offering an efficient contract with individualized quality incentives. By contrast, the product quality of the cooperative with the complete pooling policy will be lower than that of the IOF, and will decrease when the cooperative becomes large in terms of membership size. Therefore, the cooperative must change its income rights structure and adopt a partial pooling policy.

When investigating the optimal income rights structure of the cooperative, we relax its traditional principle of complete pooling policy and the cooperative can choose a pooling ratio. This adds an additional dimension in the cooperative's approach of aligning risk-sharing and incentive of the members. One important benefit of the pooling is to share the risk of production uncertainty among members. When the risk is shared by more members, the disutility of risk is smaller for each member and the members are more willing to invest efforts in quality improvement. However, the pooling also has a negative impact on the quality provision because it will reduce the member's incentive and cause free-riding when the members make quality decisions. Under the circumstance of pooling, the more members the cooperative has, the weaker is the incentive. Therefore, the cooperative must find the applicable pooling ratios, with which the pro and con of the pooling policy can

Table 4 Effects on quality provision in the three governance structures

	Self-processing	IOF	Cooperative-complete Pooling	Cooperative-optimal structure
Shifting market risk	No	Yes	Yes	Yes
Pooling of production risk	No	No	Yes	Yes
Free-riding	No	No	Yes	Yes
Bearing production risk by α	No	Yes	Yes	Yes
Providing sufficient quality incentive by β	No	Yes	No	Yes

be balanced by the linear contract $P = \alpha + \beta q$. Based on this rationale, we derive the optimal income rights structure for the cooperative, under which the cooperative can maintain a high product quality even when the membership size is growing and large. The configuration of the optimal income rights structure is flexible. In order to reach the optimal product quality, the cooperative needs not necessarily imitate the IOF by abandoning pooling and adopting a fully individualized quality incentive. Instead, the cooperative can choose from a range of pooling ratios. When a high pooling ratio is chosen, the risk of production uncertainty is well-shared, the cooperative can choose a contract with a stronger quality incentive. When the a low pooling ratio is chosen, the incentive as well as the risk is more individualized, the cooperative thus must choose a contract with a large base payment and a weak quality incentive, which better bears the members' production risk. As such, the cooperative has more flexibility in its payment arrangements. Importantly, we emphasize that when the members have subjective risk toward production uncertainty, the pooling ratio must be lower than an upper bound. With a pooling ratio higher than this upper bound, the pro and con of pooling can no longer be balanced by a contract, and the cooperative's product quality will therefore decrease as the cooperative grows. Table 4 summarizes the effects of the institutional arrangements of different governance structures on the provision of product quality.

Our model also provides an explanation for the coexistence of IOFs and cooperatives in agricultural markets. We argue that by abandoning the complete pooling policy and adopting an optimal income rights structure, cooperatives can overcome their disadvantageous position in the competition with IOFs. Theoretically, the product quality of the IOF and the cooperative with an optimal income rights structure can both reach the same optimal level. However, in reality, the competition between the IOF and cooperative is much more dynamic. First of all, they may have different and non-precise judgments on the farmers' absolute risk aversion, quality provision cost coefficient, and the level of production uncertainty, which can lead to their different decisions regarding the payment formula. Second, the quality incentive of the cooperative is normally projected by the members as a certain promise, because the members own and control the processor, they can decide and enforce the incentive collectively. By contrast, when trading with the IOF, the farmers may have additional subjective risk toward the IOF's quality measurement and payment (Balbach 1998; Gow et al. 2000). This may distort the farmers' decisions in product quality. Third, the cooperative processor may be, to some extent, risk averse instead of risk neutral. They may thus adopt a more conservative policy regarding the quality provision. All these factors may play a role in the quality competition between cooperatives and IOFs. In addition, different processors are also competing in the quality dimension by other means. For example, they may provide the farmers with farming supplies and technical supports in order to decrease the farmers' subjective risk toward production uncertainty k_2, or help the farmers to decrease the quality provision cost coefficient c. With such measures, they are able to further increase the product quality.

6 Conclusion and Further Research

Normally an uncertain payoff is considered less valuable than the certain payoff with the same expected value. Confronted with the risk of market and product uncertainty, the risk-averse farmers' efforts of product quality provision will be deterred, especially, when they trade directly in the market individually. By forming a cooperative, the risk-averse farmers can obtain benefits given that the cooperative processor insures them from the risk of market uncertainty. Pooling also reduces the risk of product uncertainty. However, with a complete pooling policy, the members are rewarded for their product quality according to the pooled quality of the cooperative. The farmers can also trade with the IOF, which rewards the farmers' product quality on an individualized base. In a principal-agent framework with the processor as risk-neutral principal and farmers as risk-averse agents, we compare the quality provision of a cooperative and an IOF. It is shown that the traditional principle of complete pooling policy places the cooperative at a disadvantage relative to the IOF in a quality-differentiated market. The reason is that, with the complete pooling policy, when a new member joins the cooperative, the loss from the free-riding dominates the benefit of risk sharing. As a consequence, the cooperative's product quality will continuously decrease as its membership size increases. By contrast, the IOF processor can design an optimal contract that reaches a trade-off between providing incentives and minimizing the cost of risk. Therefore, the IOF processor elicits higher quality from farmers.

However, cooperatives can overcome this disadvantage by relaxing the traditional principle of complete pooling to partial pooling. We find that given the members' subjective risk toward product uncertainty k_2, the complete pooling policy should be by no means adopted by cooperatives. Instead, there is an upper bound on the pooling ratio that the cooperative can adopt. We prove that by designing an optimal income rights structure for the organization, cooperatives can maintain an optimal product quality level, which is equivalent to the product quality level of the IOF. The configuration of the optimal income rights structure is flexible. Cooperatives can choose from a range of pooling ratios, from no pooling to the upper bound pooling ratio. When a high pooling ratio is chosen, the risk of production is well-shared, the cooperative can choose a contract with a stronger quality incentive. When a low pooling ratio is chosen, the incentive as well as the risk is more individualized, the cooperative thus must choose a contract with a large base payment, which bears the members' production risk, and a weak quality incentive. As such, the cooperative also has more feasibility in payment arrangements. However, why cooperatives may choose a specific pooling ratio within the range needs to be further investigated.

In general, our model contributes to comparing the product quality provision of cooperatives and IOFs by capturing the uncertainties in agribusiness and the farmers' characteristic of risk aversion. We argue that cooperatives are able to compete with IOFs in a quality-differentiated market if an optimal income rights structure is adopted. This may provide an explanation for the coexistence of cooperatives and IOFs in many agricultural sectors.

There are various possibilities for further research by relaxing some assumptions of our model. One assumption is that the farmers are identical, with respect to both the absolute risk aversion level and quality provision efficiency. Hence, the adverse selection effect of heterogeneous farmers is not addressed in our model. Second, we don't distinguish the common and idiosyncratic production uncertainty, and just model the contract based on the absolute quality evaluation. However, the contract rewarding farmers based on the relative product quality is also commonly used in agricultural production, which shifts the common part of the production uncertainty to the processor. Third, as mentioned in Sect. 3, the cooperative processor may be risk-averse as well. However, the level of absolute risk aversion of the cooperative as a whole may be less than that of each individual member. In addition, the IOF may behave opportunistically ex-post regarding quality measurement and payment. This entails an additional risk for the farmers who trade with the IOF. Lastly, the final product market is assumed to be perfectly competitive in our model. However, in many agricultural sectors, the markets are oligopolistic. Different market settings may change the behaviors of the processors regarding contract optimization. In sum, we argue that there are several additional factors which may potentially influence the quality provision of cooperatives and the competition with IOFs.

References

Balbach JK (1998) The effect of ownership on contract structure, costs, and quality: the case of the U.S. beet sugar industry. In: Royer JS, Rogers RT (eds) The industrialization of agriculture: vertical coordination in the U.S. food system. Ashgate Publishing Ltd., Aldershot (United Kingdom), pp 155–184

Bogetoft P, Olesen HB (2004) Design of production contract: lessons from theory and agriculture. Copenhagen Business School Press, Copenhagen, Denmark

Bolton P, Dewatripont M (2005) Contract theory. MIT Press, Cambridge

Cechin A, Bijman J, Pascucci S, Zylbersztajn D, Omta O (2012) Quality in cooperatives versus investor owned firms: evidence from broiler production in Paraná, Brazil. Manage Decis Econ. Published online in Wiley Online Library (wileyonlinelibrary.com). doi:10.1002/mde.2586

Cook ML (1995) The future of U.S. agricultural cooperatives: a neo-institutional approach. Am J Agric Econ 77:1153–1159

Feng L, Nilsson J, Ollila P, Karantininis K (2011) The human values behind farmers' loyalty to their cooperatives. Paper presented at the 5th international conference on economics and management of networks, Limassol, 1–3 December 2011

Fulton M (1995) The future of canadian agricultural cooperatives: a property rights approach. Am J Agric Econ 77(5):1144–1152
Gow HR, Streeter DH, Swinnen JFM (2000) How private contract enforcement mechanisms can succeed where public institutions fail: the case of Juhocukor a.s. Agric Econ 23:253–265
Hendrikse GWJ (2011) Pooling, access, and countervailing power in channel governance. Manage Sci 57(9):1692–1702
Holmström B (1979) Moral hazard and observability. Bell J Econ 10(1):74–91
Knoeber CR, Thurman WN (1995) "Don't Count Your Chickens...": risk and risk shifting in the broiler industry. Am J Agric Econ 77(3):486–496
Mérel PR, Saitone TL, Sexton RJ (2009) Cooperatives and quality-differentiated markets: strengths, weaknesses and modelling approaches. J Rural Cooperation 37(2):201–224
Mussa M, Rosen S (1978) Monopoly and product quality. J Econ Theory 18:301–317
Saitone TL, Sexton RJ (2009) Optimal cooperative pooling in a quality-differentiated market. Am J Agric Econ 91(5):1224–1232
Staatz JM (1987) The structural characteristics of farmer cooperatives and their behavioral consequences. ACS Service Report No. 18. U.S. Department of Agriculture, Washington
Uzea FU, Fulton, M (2009) The use of identity in strategic network. In: Proceedings of the international association for research in economic psychology and society for the advancement of behavioural economics joint conference, Halifax, 7–11 July 2009
Vitaliano P (1983) Cooperative enterprise: an alternative conceptual basis for analyzing a complex institution. Am J Agric Econ 65:1078–1083

Part C
Franchising Chains

What Makes Franchisees Trust Their Franchisors?

Evelien P.M. Croonen and Maryse J. Brand

Abstract A lack of trust of franchisees in their franchisor will negatively affect franchise system performance. However, very little is known about how franchisors can create and maintain franchisee trust. This paper presents a theoretical framework of antecedents of franchisees' trust in their franchisors and franchise systems. To develop our framework we combined franchising literature with literature on trust in other organizational contexts. We argue that a franchisee's general propensity to trust together with its perception of the trustworthiness of the franchisor and franchise system determine this franchisee's level of organizational trust. We distinguish three franchise system components that each entail a set of determinants used by franchisees to evaluate the trustworthiness of the franchisor and the franchise system, the system's strategic positioning in the market, its operational management and the franchisee management.

Keywords Franchising • Relational view • System performance • Trust

1 Introduction

In business format franchising, a parent organization (the franchisor) replicates a business format—entailing a positioning in the market and internal procedures—by allowing independent small business owners (franchisees) to use this format in return for fees (Davies et al. 2011). The franchisees form part of a franchise system with units that all operate under the franchisor's business format. In many regions, business format franchising has become an increasingly important instrument for entrepreneurial wealth creation, accounting for a large share of business in a wide

E.P.M. Croonen (✉) • M.J. Brand
Faculty of Economics and Business, University of Groningen, Nettelbosje 2, 9747 AE Groningen, The Netherlands
e-mail: e.p.m.croonen@rug.nl

T. Ehrmann et al. (eds.), *Network Governance*, Contributions to Management Science,
DOI 10.1007/978-3-7908-2867-2_12, © Springer-Verlag Berlin Heidelberg 2013

range of industries, such as retailing, fast food, hospitality, construction, accounting and funeral services (Szulanski and Jensen 2008). The United States, for example, have over 1,500 franchise systems, representing more than 760,000 franchisees and almost 18 million employees (Dant 2008). In Europe the situation is similar; in the Netherlands franchising has a market share of 80% in food retailing and 71% in non-food retailing (Van Essen and Pleijster 2009).

The franchise relationship between a franchisor and a franchisee is characterized by mutual interdependence; the franchisor relies upon its franchisee to perform at expected levels and within specified guidelines while the franchisee depends upon its franchisor for support. Such mutually interdependent relationships, of which franchise relationships are a specific subset, rely on mutual trust to be successful (e.g. Morgan and Hunt 1994; Zaheer et al. 1998). Although these types of relationships are generally governed by formal contracts, a complete coverage of all possible contingencies is never guaranteed. Therefore, these relationships are governed by both formal mechanisms, such as contracts, and informal instruments, such as trust (cf. Bradach and Eccles 1989; Cochet and Garg 2008; Davies et al. 2011).

Moreover, franchise relationships are characterized by asymmetrical control; by virtue of the franchise contract the franchisor has generally more power than its individual franchisees, which renders the latter vulnerable to possible opportunistic behaviors of the franchisor (cf. Croonen 2010; Davies et al. 2011; Storholm and Scheuing 1994). Furthermore, a lack of trust of the franchisees in their franchisor and franchise system may lead to various kinds of undesired franchisee behavior, such as diminished efforts to comply with the franchise regulations or franchisees leaving the franchise system (Davies et al. 2011). These franchisee behaviors could lead to all kinds of problems, such as diminished system sales, problematic franchisee recruitment, and ultimately a decrease in and/or stagnation of the franchise system's profitability. In sum, franchisors have a large economic interest in creating and maintaining franchisee trust, since franchisees form an important ingredient in their franchise systems' success (cf. Michael and Combs 2008).

However, despite the importance of the topic, the research attention paid to franchisee trust so far has been relatively insignificant while as yet very little is known about how franchisors can create and maintain the trust of their franchisees. Even though the scientific literature has largely focused on antecedents and the consequences of trust in different organizational contexts, such as employer-employee relationships (e.g. Gillespie and Dietz 2009; Searle et al. 2011) and various types of inter-organizational relationships, for example marketing/distribution channels or buyer–supplier relationships (e.g. Anderson and Narus 1990; Gullett et al. 2009; Morgan and Hunt 1994; Lusch et al. 2003), still very little is known about antecedents of trust in franchise relationships as a specific form of inter-organizational relationship.

Since franchise relationships have specific characteristics, insights from other studies on trust cannot be directly transferred to franchisees. First, as opposed to actors in other inter-organizational relationships, franchisees generally have less freedom; franchisees operate according to a "full" business format, which includes a specific strategic positioning in the market and various internal operational procedures that franchisees are obliged to follow (Kaufmann and Eroglu 1998). With respect to evaluating (the trustworthiness of) their franchisor and its franchise system, franchisees will therefore take all aspects of the business format into account. Second, in contrast with other organizational actors (i.e. employees), franchisees are legally independent business owners who pay fees/royalties for the use of their franchisor's business format, who take the risk of investing capital in their units' assets and who are the residual claimants of these units (Ketchen et al. 2011; Sorenson and Sørensen 2001; Yin and Zajac 2004). This context may make franchisees more critical in evaluating their franchisor and franchise system as opposed to other types of actors, such as employees or professionals engaged in looser types of inter-organizational relationships. In sum, because of these particular characteristics, it is plausible that franchisees use other and stricter criteria to assess their partner organization's trustworthiness than other actors do. Existing studies, however, do not provide sufficient insights into the antecedents of franchisees' assessment of the trustworthiness of their franchisors and the franchise systems.

Most of the franchising literature has examined franchising from the franchisor's perspective (cf. Combs et al. 2011; Michael and Combs 2008), for example by explaining why franchisors use franchised units as opposed to company-owned units and how this decision affects franchise system performance (e.g. Combs et al. 2009). Much less research has taken the franchisee's perspective (exceptions are Davies et al. 2011; Kidwell et al. 2007; Michael and Combs 2008). As a result, we have a relatively limited understanding of the perceptions, motivations and behaviors of franchisees.

This lack of research on the franchisees' perspective has also resulted in a very limited understanding of antecedents and/or consequences of franchisee trust. There are a handful of studies available, most of which have focused on the consequences of franchisee trust regarding, for example, the level of franchisee compliance (cf. Davies et al. 2011), long-term orientation and satisfaction (cf. Bordonaba-Juste and Polo-Redondo 2004; Chiou et al. 2004) and performance (cf. Bordonaba-Juste and Polo-Redondo 2004; Dahlstrom and Nygaard 1995). Some authors have in fact dealt with antecedents of franchisee trust, but either in an explorative way (e.g. Croonen 2010) or by presenting only a very limited number of examples (e.g. Chiou et al. 2004; Dahlstrom and Nygaard 1995). Additionally, franchising researchers seem to consider franchisee trust mostly as a uni-dimensional construct (see Croonen 2010; Davies et al. 2011 for exceptions), whereas the literature on trust in general distinguishes different dimensions and levels of trust, each with specific antecedents and/or consequences (e.g. Searle et al. 2011; Zaheer et al. 1998).

Considering the above, our paper contributes to the literature in multiple ways. First, we have added to the franchising literature by taking the franchisee's perspective as opposed to that of the franchisor, which has until now attracted most research attention (cf. Dant 2008; Davies et al. 2011; Michael and Combs 2008). Second, given the importance of franchisee trust for franchise system performance, the lack of research on antecedents of franchisee trust represents an important knowledge gap in the franchising literature. This paper has aimed to fill this gap by developing a theoretical framework on antecedents of franchisee trust based on a multidimensional approach to defining franchisee trust. To this end, we have combined franchising literature with literature on trust in other organizational contexts.

The structure of this paper is as follows. First, we define trust in general and distinguish several forms of trust and levels of analysis. We then deal with trust in a franchise context and discuss different dimensions and levels of franchisee trust. Second, we create a theoretical framework on antecedents of franchisee trust by building on different literature streams. We conclude the paper by presenting our framework and addressing the implications for future research and practice.

2 Trust: Definitions, Dimensions and Levels of Analysis

2.1 Trust: Definitions, Dimensions and Levels of Analysis

Mayer et al. define trust as "the willingness of a party to be vulnerable to the actions of another party based on the expectation that the other will perform a particular action important to the trustor, irrespective of the ability to monitor or control that other party" (Mayer et al. 1995, p. 712). This definition has been adopted by a number of other authors who all identify two critical components of trust, viz. (1) a trustor's positive expectations regarding a trustee's intentions or behavior, and (2) a willingness to be vulnerable, thus accepting a level of risk in the relationship (cf. Rousseau et al. 1998; Six and Sorge 2008; Zaheer et al. 1998). Following this body of literature, we distinguish between a trustor (i.e. the party that has a certain degree of trust) and a trustee (i.e. the party that is trusted).

Researchers generally differentiate between two dimensions of trust (cf. Davies et al. 2011; Nooteboom 1999). The first dimension is trust in the other party's competencies to perform a certain action that is important to the trustor. This type of trust is termed competence trust (cf. Nooteboom 1999). The second dimension concerns a party's trust in the other partner's intentions or integrity, which refers to a party's expectation that the other party will demonstrate appropriate behavior. This type of trust is referred to as intentional trust (cf. Nooteboom 1999). These two trust dimensions are very similar to the three trustworthiness dimensions of Mayer et al. (1995): ability, integrity and benevolence. Ability is related to competence trust while integrity and benevolence are associated with intentional rust. Mayer et al. argue that their three trustworthiness dimensions help in understanding why party A perceives party B as trustworthy, and why party A ultimately trusts party B (see Sect. 3 for a more elaborate description of the link between trust and trustworthiness and a discussion of Mayer et al.'s three dimensions). Mayer et al.'s three-dimensional framework is a helpful instrument in understanding the different trust dimensions because it incorporates both cognition-based and affect-based trust

(Dirks and Ferrin 2002; Dunn et al. 2012). Cognition-based trust primarily concerns beliefs about the trustee's ability and integrity and affect-based trust refers to beliefs about the trustee's benevolence (Dunn et al. 2012).

In addition to distinguishing multiple *dimensions* of trust and trustworthiness, the current literature explicitly differentiates between (inter)personal and (inter) organizational *levels* of trust (e.g. Bachmann 2001; Currall and Inkpen 2002; Nooteboom 1999; Searle et al. 2011; Zaheer et al. 1998). As argued earlier, trust is based on positive expectations as regards a trustee's behavior. The trustee can be an individual or an organization. Trust in an individual is based on direct interactions with this actor, while trust in an organization is based on a trustor's image of this organization as a result of past decisions and actions. Nooteboom (1999, p. 28) argues that "organizational trust is a constellation of behavioral trust (i.e. personal trust, the authors), with organizational structure and culture acting as institutions that limit and guide behavior of staff." In other words, in organizations there are certain explicit or implicit norms for how things are generally done. In real life however, individuals within organizations may deviate from these norms because of conflicting interests. Some scholars have indeed empirically proven the relevance of this difference between personal and organizational trust. For example, in their study of 107 buyer–supplier relationships, Zaheer et al. (1998) confirmed that interpersonal and inter-organizational trust are distinct but related constructs and that they have different effects on organizational outcomes. Therefore, Zaheer et al. (1998) argue that research on trust between organizations should clearly specify the level of the trustee in order to avoid the risk of "cross-level fallacy" (i.e. attributing individual motivations and behaviors to organizations).

Zaheer et al. (1998) convincingly claim that it is not correct to say that organizations trust each other; individuals within organizations may collectively share a trust orientation toward another organization, but this is quite different from saying that an organization has trust. In other words, a trustee can be an individual or an organization, but a trustor can only be an individual. We use the term personal trust to refer to an individual's trust in another individual. In an organizational context, these individuals can belong to an organization or have an individual relationship with it. Personal trust can include, for example, the trust of one person in a colleague of the same organization (co-workers or managers, e.g. Ferrin and Dirks 2003; Gould-Williams 2003), the trust of one individual "boundary spanner" in its counterpart in a partner organization (e.g. Zaheer et al. 1998), or the trust of an individual customer in a specific representative of a supplying organization (e.g. Dahlstrom and Nygaard 1995). In contrast, organizational trust refers to an individual's trust in an organization, for example an employee's trust in the organization that he/she works for (e.g. Gillespie and Dietz 2009; Hodson 2004), an individual boundary spanner's trust in a partner organization (e.g. Zaheer et al. 1998) or an individual customer's trust in its supplying organization (e.g. Saparito et al. 2004).

In the present section, we have argued that trust is a multi-dimensional and multi-level construct. Although researchers have increasingly recognized this perspective in the literature on trust in other organizational contexts (cf. Zaheer et al. 1998), the franchising literature generally does not take this view point into account. In the next section we will therefore further discuss the different dimensions and levels of franchisee trust.

2.2 *Franchisee Trust: Definitions, Dimensions and Levels of Analysis*

Translating Mayer et al.'s widely-used trust definition to a franchising context leads to the following description of franchisee trust: "the willingness of a franchisee to be vulnerable to the actions of its franchisor based on the expectation that the franchisor will perform particular activities important to the franchisee, irrespective of the franchisee's ability to monitor or control the franchisor" (cf. Mayer et al. 1995). This definition is very similar to that of franchisee trust by Davies et al. (2011), which also emphasizes franchisees' positive expectations regarding their franchisors' behaviors and the willingness to accept vulnerability to the franchisor's actions.

The previous section has pointed at the importance of categorizing different dimensions of trust (i.e. competence and intentional trust); however, this distinction has until now not been a common one in the franchising literature. Only Davies et al. (2011) use a multi-dimensional approach to explain franchisee compliance by distinguishing between a franchisee's trust in the franchisor's competences and its integrity. Davies et al. found for example that franchisee compliance to franchisor requests is more strongly influenced by integrity trust (cf. intentional trust) than by competence trust, which empirically confirms the multi-dimensional nature of trust in a franchising context.

Neither is the multi-level approach to trust, which distinguishes between personal and organizational trust, common in the extant franchising literature. Dahlstrom and Nygaard (1995) however, do focus on antecedents and consequences of interpersonal trust in franchised channels (i.e. trust of unit owners/managers in their organizations' sales managers), but they do not explicitly distinguish this type from organizational trust. Another study by Croonen (2010) found a clear difference between a franchisee's personal trust (trust in particular representatives of the franchisor's organization) and franchise system trust (a form of organizational trust referring to the franchisee's trust in the fair and effective functioning of the franchisor's organization and its franchise system). Table 1 presents these two levels of franchisee trust. Croonen (2010) analyzed four case studies and concluded that personal trust is generally considered important by franchisees. However, if they feel too much dependent on their franchisor, personal trust is no longer sufficient. Franchisees that felt dependent on their franchisor or felt that their dependence would increase in the near future became more focused on the concept of trust in the franchisor's organization and its franchise system. Franchisees attach less

Table 1 Distinguishing between a franchisee's personal and organizational trust

		The trustee (i.e. the party that is trusted)	
		An individual representative of the franchisor's organization	The franchisor's organization and its franchise system
The trustor (i.e. the party that has a certain degree of trust)	The franchisee as an individual	*Personal trust*	*Organizational trust* (*the focus of the remainder of this paper*)

importance to trust in particular franchisor representatives, for example because these representatives could always leave the organization, which indeed frequently occurred in Croonen's case studies.

We conclude that franchisee trust is indeed a multi-dimensional and multi-level concept, which means that its different dimensions and levels may have different antecedents and consequences. The objective of this paper is to provide researchers and practitioners with new insights into how to create and/or maintain franchisee trust. Therefore, our focus is particularly on the antecedents of trust rather than on the consequences. We also chose to concentrate on the franchisee's organizational trust as opposed to the franchisee's personal trust (i.e. the right side in Table 1). We did so for several reasons. First, as we pointed out, in franchise relationships franchisees that feel dependent—which is specifically the case in centralized franchise systems (cf. Windsperger 2004)—deem trust in the franchisor and its franchise system more important than personal trust in particular franchisor representatives (see Croonen 2010). Second, organizational trust is more complex and thus more difficult to manage than personal trust. So, franchising researchers and practitioners would benefit the most from a better understanding of antecedents of franchisee trust in the franchisor and its franchise system. Probably as a result of the complexity of organizational trust there is currently very little coherent theory available on the antecedents of organizational trust and trustworthiness (cf. Gillespie and Dietz 2009 for a discussion of this issue and one of the few exceptions), while for franchise systems as a specific organizational context there is no extant theory on antecedents of trust and trustworthiness at all. The following section is a first step in building a theoretical framework.

3 Toward a Theoretical Framework

As pointed out, franchisee trust forms a relatively under-researched area in the franchising literature and its antecedents are even less well-understood than its consequences. The few studies conducted in this field have pointed to several (potentially) relevant antecedents, such as the franchisor's communication and its service assistance (Chiou et al. 2004), the organization's level of centralization and

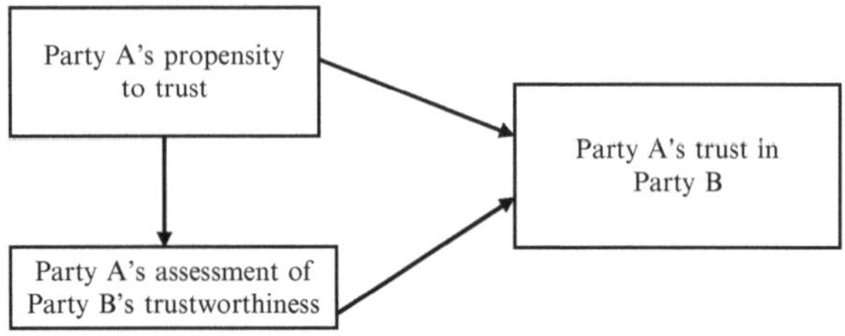

Fig. 1 General antecedents of trust

formalization (Dahlstrom and Nygaard 1995), fee structures and the establishment of a Franchise Advisory Council (FAC) (Croonen 2010). However, these sources have serious limitations since they do not distinguish between the different levels of trust and neither do they focus on personal trust (cf. Chiou et al. 2004; Dahlstrom and Nygaard 1995). Furthermore, they only explore some potentially relevant antecedents of franchise system trust (cf. Croonen 2010). The theoretical framework that we will develop in this section aims to provide a more systematic insight into antecedents of franchisees' trust in their franchisors and franchise systems. In doing so, we will build on the general literature on organizational trust and integrate this domain into the franchising literature.

It is often argued that a party A's trust in party B is the result of (1) the characteristics of party A, or more specifically party A's propensity to trust, and (2) party A's assessment of the trustworthiness of party B (e.g. Colquitt et al. 2007; Mayer et al. 1995; Mayer and Davis 1999; Schoorman et al. 2007). Figure 1 reflects this idea, which is applicable to various empirical contexts and to both personal and organizational trust. In this paper we particularly concentrate on organizational trust with franchising as the empirical context. This point of departure leads to a first proposition:

Proposition 1 A franchisee's organizational trust in its franchisor and the franchise system is a function of this franchisee's propensity to trust organizations and this party's assessment of the trustworthiness of its franchisor and the franchise system.

In the following sections we will further translate the elements presented in this figure to a franchising context and develop a number of related propositions.

3.1 A Franchisee's Propensity to Trust and Franchisee Experience Levels

Propensity to trust ("dispositional" or "generalized" trust) is an individual trait reflecting the general expectancies about the trustworthiness of others (Colquitt et al. 2007; Mayer et al. 1995; Rotter 1971); it reflects a general willingness to trust others. Typical items to measure an individual's propensity to trust are: "In dealing with strangers one is better off to be cautious until they have provided evidence that they are trustworthy", or "I believe that most people are basically well-intentioned" (cf. Mayer et al. 1995; Searle et al. 2011).

Researchers commonly assume that a trustor's trust propensity is positively related to this actor's trust in a trustee, and that this association especially applies to the early stages of a relationship, in which the trustor does not yet have much information available to assess the trustworthiness of the other party (e.g. Bigley and Pearce 1998; Colquitt et al. 2007). However, there is a discussion on whether trust propensity continues to impact trust once trustworthiness has been determined (see Colquitt et al. 2007 for a meta-analysis). Several authors have argued that trust propensity creates a "filter" which makes the trustor continue to assess the trustee's trustworthiness, even in the presence of trustworthiness information. The meta-analysis of Colquitt et al. (2007) confirms this argument, which explains the arrow between party A's propensity to trust and its assessment of party B's trustworthiness in Fig. 1.

To our best knowledge, an individual's propensity to trust is typically measured as his/her general propensity to trust other individuals (see the items above for examples). However, we propose that here a multi-level approach applies as well, and that a distinction should be made between an individual's propensity to trust other individuals and his/her propensity to trust organizations. Even though we have never seen this distinction in the literature before, it may be relevant because organizations are complex social systems in which the intentions of the individual employees may be "overruled" by the organizational formal or informal power structures. We argue that, as a result of this organizational complexity, people's propensity levels as regards trusting individuals or organizations are not necessarily identical. This means that in both our propositions and our theoretical framework we deliberately use the term "propensity" to trust organizations.

In sum, we propose that a franchisee's propensity to trust organizations may have both a direct and an indirect effect on this actor's trust in its franchisor and the franchise system, whereby the strength of the effects is dependent on the franchisee's experiences as a franchisee in this particular franchise system (cf. Bigley and Pearce 1998; Blut et al. 2011; Bordonaba-Juste and Polo-Redondo 2008; Bradach and Eccles 1989; Colquitt et al. 2007). Franchisees that have only operated in a franchise system for a short time period ("novice franchisees") have to rely more on their general propensity to trust organizations than those that have functioned within a franchise system for a longer time period ("experienced franchisees") and that have been able to assess their franchisor's and its franchise system's trustworthiness on the basis of prior exchanges and experiences in different phases of the franchise relationship.[1]

[1] This idea is similar to the view of Lewicki and Bunker (1996) that in the early stages of a relationship partners build on so-called calculus-based trust, whereas in the later stages knowledge-based trust and identification-based trust become more important.

These findings have formed the basis for the following propositions:

Proposition 2a A franchisee's propensity to trust organizations has both a positive direct and a positive indirect effect on this actor's trust in its franchisor and the franchise system, since this propensity serves as a "filter" used by this franchisee to assess the trustworthiness of its franchisor and the franchise system.

Proposition 2b The direct effect of the propensity to trust organizations is stronger for novice franchisees than for experienced franchisees.

Proposition 2c The direct effect of a franchisee's assessment of the trustworthiness of its franchisor and the franchise system on the franchisee's trust is stronger for experienced franchisees than for novice franchisees.

3.2 A Franchisee's Assessment of the Trustworthiness of Its Franchisor and the Franchise System

3.2.1 Introduction

A well-known framework for assessing an individual's or an organization's trustworthiness is formed by the three dimensions of Mayer et al. (1995): ability, integrity and benevolence. The first dimension, ability, refers to the trustor's perception of the trustee's set of skills, competencies and characteristics that are necessary to exert influence within a specific domain. It is important to recognize that the trustor's perceptions of the trustee's abilities may differ among domains (e.g. the production of a certain good or the communication with customers). The second trustworthiness dimension, integrity, concerns the trustor's perception as regards the acceptability of the set of principles used by the trustee in its business conduct. Finally, benevolence relates to the trustor's perception of the trustee's intentions with respect to the trustor's interests aside from making profit.

Although these three dimensions have already provided some preliminary insights into how organizations (i.e. franchisors) can create and/or maintain an image of trustworthiness, they are still broad constructs. The exact criteria of trustors to evaluate a trustee's trustworthiness depend on the specific organizational context. As we pointed out earlier, hardly any coherent theories or models to understand the antecedents of an organization's (un)trustworthiness have as yet been presented. The few studies available merely discuss a range of determinants of individuals' assessments of an organization's trustworthiness, such as quality assurance, interactional courtesy (cf. Caldwell and Clapham 2003; Ingenhoff and Sommer 2010), the use of certain employment practices (cf. Hodson 2004; Searle et al. 2011), and strategy, structures, policies and processes (Gillespie and Dietz 2009).

All these literature sources are focused on trustworthiness in rather general organizational contexts, such as employees' assessments of their employers' trustworthiness (cf. Hodson 2004; Searle et al. 2011). Other authors have surveyed students about what factors they considered to be important in a highly trustworthy organization (cf. Caldwell and Clapham 2003), or interviewed people about a company of their choice (cf. Ingenhoff and Sommer 2010). However, these extant models of organizational trustworthiness cannot be directly transferred to a franchising context because franchise systems form a highly specific organizational context (as pointed out in Sect. 1).

To summarize, in order to understand how franchisees assess the trustworthiness of their franchisors and the franchise systems, a new model needs to be developed. To this end, we will build on the work of Gillespie and Dietz (2009) which uses a system approach by dividing a franchise system into different "components". Each component comprises a group of determinants used by franchisees to assess the trustworthiness of their franchisor and the franchise system. Additionally, we will integrate determinants of organizational trustworthiness from other organizational contexts into this new model.

3.2.2 Introduction to the Main Components of the Trustworthiness of a Franchisor and Its Franchise System

We distinguish three main components each comprising a group of determinants used by franchisees to assess the trustworthiness of their franchisor and its franchise system: (1) strategic positioning, (2) operational management and (3) franchisee management. Since franchisees pay for the right to use the franchisor's business format, we argue that the business format forms an important component of how franchisees assess the trustworthiness of their franchisor and its franchise system. This business format entails both strategic and operational elements (cf. Croonen 2006; Kaufmann and Eroglu 1998), which we will both take into account in our model of franchise system trustworthiness.

First, the business format reflects a certain identity in a certain market by which the franchisor aims to target its "unique competitive niche" (cf. Kaufmann and Eroglu 1998, p. 71). We refer to this identity as the franchise system's strategic positioning in the market (Croonen 2006).

Second, the business format contains a wide range of operational policies and procedures that form the foundation of its effective and efficient functioning at both the individual store level and the system level ("format facilitators" in terms of Kaufmann and Eroglu 1998). These operational policies contain for example the specification of equipment, detailed operating instructions for each unit, royalty payment procedures or financial reporting requirements. Although these business format elements are not directly visible to customers, they are critical because they comprise the managerial and operational infrastructure of the entire franchise system and its units (cf. Kaufmann and Eroglu 1998). Therefore, we argue that franchisees will take these operational issues into account when assessing their franchise system's trustworthiness (cf. Croonen 2006). We refer to these issues as the franchise system's operational management, which is the second main component in our model of franchisor and franchise system trustworthiness.

The third and final main component refers to the way in which the franchisor manages its franchisees. In the model of Gillespie and Dietz (2009) this approach is reflected in the component of "structures, policies and processes". In the general organizational trustworthiness literature (cf. Hodson 2004; Searle et al. 2011), it is argued that organizations have to respect their employees' rights and interests, for example through the provision of stable and secure employment as well as adequate pay and benefits. This literature particularly focuses on how organizations can use employment or human resource management (HRM) practices to maintain or increase their trustworthiness. Although these ideas cannot be directly applied to franchise relationships since these involve independent business owners, franchisors nevertheless need to demonstrate respect for their franchisees' rights and interests (cf. Morrison 1997; Storholm and Scheuing 1994). This is why we include franchise relationship management as the third main component of franchisor and franchise system trustworthiness.

These three main components relate to Mayer et al.'s widely-used trustworthiness dimensions of ability, integrity and benevolence (Mayer et al. 1995) in a number of ways. First of all, the franchisor's ability to manage the franchise system effectively is reflected in all our three system components. The franchisor needs to be able to organize the franchise system in such a way that it can attain an organizational advantage (cf. Hodson 2004). This can be done by measures such as defining a viable strategic positioning in the market, setting up a good operational structure and selecting high-quality franchisees and keeping them satisfied. Second, a franchisor's integrity and benevolence in managing the franchise system is clearly reflected by the franchisee management component. Via their franchisee management practices, franchisors can demonstrate that they operate based on principles acceptable to the franchisees and that apart from serving their own interests, they also have those of their franchisees at heart (Searle et al. 2011).

We will now discuss the main components of our propositions as well as the groups of determinants used by franchisees to assess their franchisors' and the franchise systems' trustworthiness.

3.2.3 Component 1: The Franchise System's Strategic Positioning in the Market

The franchisor's business format ideally includes a unique strategic positioning that serves a need in a viable customer segment (cf. Kaufmann and Eroglu 1998). According to Kaufmann and Eroglu, the business format contains several elements that help in communicating the unique features of the business to the customers. The business format is the franchisor's responsibility. We thus argue that franchisees will assess the trustworthiness of the franchise system by evaluating the franchisor's ability to develop a business format with unique features and the way in which these features are communicated to the customers (cf. Kaufmann and Eroglu 1998).

We distinguish the following determinants used by franchisees in assessing their franchise system's strategic positioning in the market (cf. Croonen 2006; Kaufmann and Eroglu 1998; Sullivan and Adcock 2002):

- The "product/service deliverables" (cf. Kaufmann and Eroglu 1998). Franchisees will assess whether their franchisor has been able to compose an assortment of goods and services with unique features at a certain price level and whether the competitive niche defined is viable.
- Promotion. This determinant refers to the franchisees' assessment whether the franchisor has been able to develop promotion policies that properly communicate the unique features of the business format to the customers, attract the customers' attention and help in strengthening the format's brand name. This can for example be done through the franchise system's website, television promotion campaigns, and/or by sending out promotion materials to customers.
- Unit appearance. This determinant includes the franchisees' assessment whether the franchisor has been able to develop a unit design that clearly communicates the features of the business format, for example in terms of color schemes or materials used.

The three determinants are all related to the franchise system's overall strategic positioning in the market, which the franchisees' should assess positively. However, a major strategic and managerial issue in the franchising context is the trade-off between standardization and adaptation (cf. Bradach 1997; Kaufmann and Eroglu 1998). A high level of standardization leads to image consistency and cost minimization while it facilitates system adaptation. However, sometimes the business format may need to be adapted to the franchisee's desires and to local circumstances, especially in mature industries and in the case of experienced franchisees. We therefore argue that another criterion for franchisees in assessing their franchise system's strategic positioning in the market concerns the degree to which they are allowed to adapt elements such as the "product/service deliverables", the promotion activities and the unit appearance to their own local circumstances if necessary.

The above considerations have led to the following proposition:

Proposition 3 The franchisee's assessment of the franchise system's strategic positioning in the market positively influences this party's assessment of the trustworthiness of its franchisor and the franchise system.

3.2.4 Component 2: The Franchise System's Operational Management

In addition to a certain positioning in the market, the franchisor's business format also includes procedures and policies to support the format's effective and efficient functioning at the individual store level and at the system level (cf. Kaufmann and Eroglu 1998). These are generally referred to as the franchisor services, including for example, central purchasing, training, site selection, quality programs, sales forecasts, and ICT support (cf. Croonen 2006; Gillis and Combs 2009; Morrison 1996; Roh and Yoon 2009; Windsperger 2004). We propose that franchisees will assess the franchise system's trustworthiness by evaluating the franchisor's ability in providing a range of operational support services. On the basis of the above considerations from the franchising theory and some additions from the general organizational trustworthiness literature, we distinguish between the following determinants of trustworthiness associated with the franchisor's operational management:

- Purchasing (cf. Kaufmann and Eroglu 1998; Roh and Yoon 2009; Windsperger 2004). The franchise agreement generally contains purchasing requirements from designated suppliers, for both the goods provided to the customers and business assets such as store furniture, cars or ICT systems. Franchisees will assess their franchise systems' trustworthiness based on the conditions under which these goods and/or assets are supplied, such as the quantities of goods/assets that have to be ordered and the price levels of these items.
- Logistics (cf. Croonen 2006). This determinant refers to the actual delivery of goods and assets by the suppliers designated, involving issues such as timely delivery, flexibility in the delivery and the care with which the goods/assets are being delivered.
- ICT systems (cf. Croonen 2006; Kaufmann and Eroglu 1998; Roh and Yoon 2009; Windsperger 2004). Franchise contracts often stipulate the franchisee's obligatory use of certain ICT systems, such as accounting systems, benchmarking systems or payment systems. Franchisees will assess the quality of these systems and evaluate the level of operational support provided in case of problems with these systems.
- Site selection and sales forecasts (cf. Roh and Yoon 2009). This determinant concerns the franchisees' evaluation of the franchisor's competencies in the selection of viable sites and the composition of realistic sales forecasts.
- Quality assurance (cf. Caldwell and Clapham 2003; Ingenhoff and Sommer 2010). This element involves the franchisee's assessment of the extent to which the franchisor understands quality standards, for example in methods of operation, and adheres to these criteria on a continuous basis.
- Legal compliance (cf. Caldwell and Clapham 2003; Ingenhoff and Sommer 2010). This item entails the franchisee's assessment of the degree to which the franchisor understands and follows the laws applicable to the specific customer market.
- Overall support (cf. Morrison 1996; Roh and Yoon 2009). This element refers to the perceived support and/or assistance received by the franchisees when they specifically ask for it. Examples are support in several functional areas, such as marketing, finance, production, or human resource issues.

The above considerations have led to the following proposition:

Proposition 4 The franchisee's assessment of the franchise system's operational management positively influences this party's evaluation of its franchisor's trustworthiness and that of the franchise system.

3.2.5 Component 3: Franchisee Management

As pointed out earlier, the literature on employees' assessment of their employer's trustworthiness has included HRM practices as an important antecedent (e.g.; Gould-Williams 2003; Hodson 2004; Searle et al. 2011; Whitener 2001). HRM includes a set of practices directed at attracting, developing, and maintaining (or disposing of) a firm's human resources (cf. Lado and Wilson 1994). The main conceptual link between HRM and organizational trustworthiness is that formal HRM policies and the way in which they are enacted within an organization influence the employees' assessment of their organization's trustworthiness. HRM policies reflect the ability, integrity and benevolence of an organization.

The same line of reasoning can be applied to franchise relationship management: the way in which franchisors attract, develop and maintain their franchisees gives the franchisees an indication of the franchise system's trustworthiness. As Castrogiovanni and Kidwell (2010) point out, franchisees are the key human resources of franchisors, and an HRM perspective—of course translated to a franchising context—can therefore provide a valuable contribution to understanding franchise relationship management. However, such an approach has as yet only rarely been used; the article of Castrogiovanni and Kidwell is the only exception that we know of. Their conceptual article discusses the differences between franchisees and company managers in terms of three HRM practices: "recruitment and selection", "training and development", and "rewards". However, there have so far been no studies using an HRM perspective to investigate how franchise relationships are managed and how this impacts franchisee trust. In this paper we will take a conceptual step in that direction.

In recent years the HRM research has focused on so-called "high performance", "high involvement" or "high commitment" HR practices (cf. Evans and Davis 2005; Huselid 1995; Gould-Williams 2003; Snell and Dean 1992; Whitener 2001). In this literature stream it is argued that certain bundles of HR practices positively affect organizational performance; however, the link between these HR practices and output measures is often taken for granted, while very few researchers have explicitly addressed the question how HR practices impact organizational members' trust levels (cf. Gould-Williams 2003).

Since HRM literature distinguishes among different HRM "domains" or "bundles" that together form a consistent system of HRM practices (e.g. Evans and Davis 2005; Searle et al. 2011), an HRM perspective could contribute to developing a systematic theoretical framework in a franchising context. Although there is still a lack of consensus regarding which practices constitute a "high involvement" or "high performance" work system, certain practices are frequently included, such as training, information sharing, employee participation, recruitment and selection, and performance management (cf. Becker and Gerhart 1996; Snell and Dean 1992; Searle et al. 2011; Whitener 2001). In this paper we have built on several literature sources on "high performance" and "high involvement" HRM practices to develop our own list of relevant determinants used by franchisees to assess the trustworthiness of their franchisors and the franchise systems. These determinants are also related to franchise relationship management. We used the three HRM practices of Castrogiovanni and Kidwell as a starting point and added other relevant determinants as proposed in the franchising literature.

The first determinant is franchisee recruitment and selection (cf. Evans and Davis 2005; Snell and Dean 1992; Whitener 2001). This determinant refers to the extensiveness of the franchisor's franchisee selection process; are the procedures used by the franchisor for evaluating the prospective franchisee' skills, knowledge and abilities thorough enough to see whether the franchisee can successfully run a unit and sufficiently fits in with the franchise system? We propose that an extensive franchisee recruitment and selection process positively contributes to the franchisees' perceptions of the trustworthiness of the franchisor and its franchise system. This is because this approach reflects the value attached by the franchisor to the quality of the franchisees as opposed to the one-sided goal of attracting as many franchisees (including their entry fees and royalties) as possible.

The second determinant is franchisee training and development (cf. Becker and Gerhart 1996; Evans and Davis 2005; Snell and Dean 1992; Whitener 2001) which Castrogiovanni and Kidwell call "enhancement". This determinant refers to the extensiveness of the franchisees' "initial training" (i.e. when entering the franchise system) and "ongoing training" or development. A franchisor's investment in its franchisees' training and development can be considered by franchisees as a manifestation of its intentions and abilities. This is particularly the case because training and development are primarily aimed at improving the franchisees' skills in successfully running their businesses.

The third determinant entails franchisee rewards (cf. Snell and Dean 1992; Whitener 2001). Castrogiovanni and Kidwell (2010) refer to this concept as compensation. Franchisee rewards can include financial and non-financial incentives. Financial franchisee rewards are closely related to the profitability of the franchised unit(s): the revenue of the unit(s) minus all the costs. These costs include all the royalties or fees that the franchisee has to pay. The important role of the fee structure in understanding the concept of franchisee trust in franchise systems has already been pointed out by Croonen (2010). Non-financial rewards from the franchisor include for example the possibility of opening new units (cf. tournament theories of Gillis et al. 2011), or the opportunity to join some special social activities exclusively available to some franchisees, such as trips to suppliers or music concerts (mostly in a very attractive setting, cf. Croonen 2006). As regards both types of franchisee rewards the reward system has to be equitable; it has to be fair and reasonable and treat all franchisees in an equal way (cf. Snell and Dean 1992; Whitener 2001). An equitable franchisee reward system is a reflection of the franchisor's care for its franchisees.

The fourth determinant is franchisee performance management (cf. Searle et al. 2011). This item refers to how the franchisor sets its expectations and how it measures, reviews and manages the franchisee's performance. It can be argued that an accurate and transparent performance management system demonstrates that the franchisor is well capable of managing its franchisees (cf. Mayer and

Davis 1999; Searle et al. 2011). Additionally, it can be argued that a franchisor's use of such systems shows the franchisor's care for its franchisees' interests in the sense that the franchisor pays attention to recognizing well-performing franchisees and helping under-performing franchisees.

The fifth determinant concerns communication and franchisee participation (cf. Evans and Davis 2005; Searle et al. 2011). These concepts relate to the franchisor's frequency of and openness in information provision and communication and the opportunities of franchisees to participate in decision making. Franchisee participation in decision making can take the form of a Franchise Advisory Council (Cochet and Ehrmann 2007; Croonen 2010). These authors have already recognized the potentially important role of FACs in preventing franchisor misbehavior and creating franchise system trust. A well-organized FAC tells franchisees that the franchisor is willing to take their interests into account in a transparent manner. Such a context positively influences the franchisees' perceptions of their franchisors' benevolence and integrity.

The sixth determinant is associated with conflict management (cf. Becker and Gerhart 1996; Blum and Wall 1997; Cutcher-Gershenfeld 1991). It entails the procedures followed in conflict situations and the speed with which steps are taken. Clear and fair procedures for conflict management demonstrate the ability and willingness of the franchisor to manage and solve conflicts. In this capacity these rules are also an indication of the franchisors' benevolence and integrity.

The seventh determinant includes franchisor restrictions (cf. Croonen 2010; Morrison 1996) and pertains to the franchisees' assessment of the fairness of the conditions of the franchise contract. These conditions involve restrictions regarding the actual operation of the business (e.g. purchasing, methods of operation, working hours, the level of investment required) and the conditions of the franchise (e.g. the size of the exclusive territory, termination/renewal terms, the use of "implicit charges", cf. Croonen 2010).

The final determinant is related to the proportion of company-owned units in the franchise system. Researchers differ in their views on this issue. On the one hand, some researchers, for example Cliquet (2000) and Storholm and Scheuing (1994) on "dual distribution implications", argue that a high proportion of company-owned units may lead to franchisee anxiety, which negatively impacts franchisee trust. On the other hand, other researchers (e.g. Croonen 2010; Gallini and Lutz 1992) have pointed out that franchisors can use company-owned units as an instrument to signal the quality of their business formats (cf. ability) and to demonstrate that their interests are aligned with those of their franchisees (cf. benevolence). It thus seems that franchisors need to find a proper balance in their proportion of company-owned units in order to influence their franchisees' trust levels positively.

The above considerations have led to the following general proposition:

Proposition 5 The franchisee's assessment of the franchisee management in the franchise system positively influences this party's evaluation of the trustworthiness of its franchisor and the franchise system.

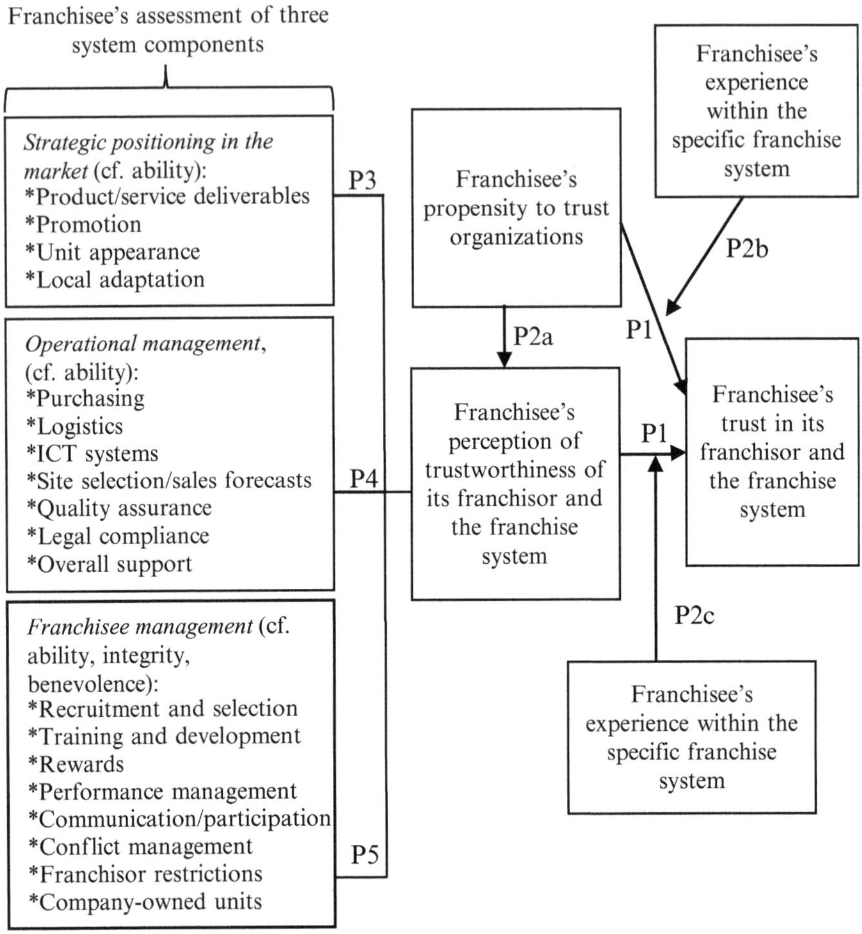

Fig. 2 Antecedents of franchisee trust in the franchisor and franchise system

3.3 Summary: Theoretical Framework

The above discussion has led to a theoretical framework based on two main antecedents of franchisee trust in the franchisor and the franchise system: (1) the franchisee's propensity to trust organizations, and (2) the franchisee's assessment of the trustworthiness of the franchisor and the franchise system. We have argued that the relationships between these antecedents and the franchisee's trust in its franchisor and the franchise system are moderated by the franchisee's experience within this system. Additionally, we have claimed that franchisees will assess the trustworthiness of their franchisors and the franchise systems via three components each comprising a group of determinants. These arguments have resulted in a comprehensive framework (Fig. 2) which explains the concept of franchisee trust.

4 Conclusions and Implications

Franchising is based on a mutual relationship between the franchisor and its franchisees, the effectiveness of which depends on the level of trust between the partners in this relationship. Franchisors have to manage a whole set of these individual relationships simultaneously, which is a very complex task. When searching the literature for extant conceptual or empirical works that provide insight into the antecedents of franchisee trust which could help franchisors in managing their franchise relationships, we had to conclude that there is a significant gap in this body of knowledge. In developing our theoretical framework we aimed at combining earlier work on trust in other organizational contexts with the scarce work on trust in franchise relationships. This framework presents a comprehensive overview of the main antecedents of franchisee trust in the franchisor and franchise system (i.e. propensity to trust and trustworthiness), and the determinants of these antecedents in a franchise context.

The contributions of this paper can be summarized as follows. First, we have added to the franchising literature by taking the franchisee's perspective as our point of departure rather than that of the franchisor, which attracted most research attention in the past (cf. Dant 2008; Davies et al. 2011; Michael and Combs 2008). Adding the franchisee perspective is important for both academics and policy makers who are trying to grasp the functioning of franchise systems, which have become so important in today's economy. Second, given the relevance of franchisee trust to franchise system performance, the lack of research on antecedents of franchisee trust represents an important knowledge gap in the franchising literature. This paper has made a start in filling this gap by developing a comprehensive framework based on a multidimensional approach to defining franchisee trust. To this end, we combined franchising literature with studies on trust in other organizational contexts. Additionally, we made a case for conceptually distinguishing between propensity to trust individuals and propensity to trust organizations. Such a distinction could also be useful in other contexts.

This paper also has some limitations. First, we ignored a potentially relevant research stream in our theoretical discussion, namely the literature on fairness and justice (see Cohen-Charash and Spector 2001 for different forms of fairness/justice and a meta-analysis). Several researchers have claimed that there is a clear link between different forms of fairness or justice perceptions on the one hand and trust on the other hand. The fairness/justice literature also provides a framework that can be used to obtain a more systematic understanding of antecedents of trust and determinants of trustworthiness. However, it can be argued that the way in which organizations implement HRM practices has a large influence on the organizational members' fairness/justice perceptions. Therefore, we included only some determinants of organizational trustworthiness related to fairness/justice. According to Searle et al. (2011), the HRM and justice/fairness research domains have progressed quite independently of one another. So a fruitful area of future research would be to integrate these two perspectives into a more comprehensive model. A second limitation of our study is that we left out some potentially relevant antecedents of trust as distinguished in other literature streams, such as the environmental context (e.g. Dahlstrom and Nygaard 1995). Besides two franchisee characteristics (i.e. propensity to trust and experience), we have largely focused on the franchisee's assessment of its franchisor's trustworthiness, which is based upon the franchisee's subjective perception of its franchisor's behaviors (following Schoorman et al. 2007; Gullett et al. 2009). Future research, however, could include the franchisee's perception of environmental characteristics in explaining its trust level toward its franchisor. This suggestion leads us to the third limitation of our paper; our paper has neither theorized nor presented any empirical data yet on the relative importance of the different determinants of a franchisee's assessment of its franchisor's trustworthiness and that of the franchise system. More insight into this issue would lessen the complexity of the theoretical model and increase the franchisors' understanding of the most important instruments that they can use to create and maintain the trust of their franchisees. The fourth and final limitation is that we focused on the antecedents of only one level of franchisee trust: the organizational level. However, since personal and organizational trust are related (cf. Zaheer et al. 1998) it would have been useful to also distinguish antecedents of a franchisee's personal trust and to theorize on how these two levels of franchisee trust (i.e. personal trust and organizational trust) affect either desired or undesired franchisee behaviors and ultimately the franchise system's performance.

This paper's limitations lead to several implications for future research. First of all, the theory presented needs to be tested in an empirical setting. Considering the volume and quality of the extant empirical work on the antecedents of trust in non-franchise contexts, it should be very well possible to develop a quantitative study to test the propositions formulated in this paper. We would propose to conduct such a first test in a relatively controlled environment, for example one large franchise system (cf. Davies et al. 2011). In this system all franchisees would operate within a chain and deal with the same franchisor. This approach would diminish the risk of disturbances by other possible factors that could influence trust (e.g. environmental factors). The study of one specific system would also provide the opportunity to work closely with the franchisor whereby additional data could be obtained about the respondents, such as objective performance data, unit and franchisee demographics, et cetera. Such data could be relevant to include as controls in the quantitative analysis. A second implication for future research is associated with our approach to franchisee trust as a multidimensional and multilevel concept. As mentioned, previous research on the consequences of franchisee trust has generally considered this concept as uni-dimensional and single level (exceptions are Croonen 2010, and Davies et al. 2011). This paper has suggested that in studying the consequences of franchisee trust, researchers should also take the different dimensions and levels of this construct into account. The focus should then be on how these elements together impact the outcomes with respect to franchisee compliance, franchisee commitment or franchisee retaliation behaviors.

This paper also has implications for practice. Franchisors have to realize that although franchisee trust is an important determinant of franchise system performance, it is a complex concept influenced by an array of other factors. The framework presented in this paper offers franchisors a checklist of possible instruments that could be used to enhance trust. In combination with some form of measuring and monitoring franchisee trust, these instruments could be used to increase the effectiveness of the franchise system. Considering the complex task of managing a network of legally independent franchisees within a franchise system, many organizations may welcome the instruments introduced in this paper.

References

Anderson JC, Narus JA (1990) A model of distributor firm and manufacturer firm working partnerships. J Mark 54(1):42–58

Bachmann R (2001) Trust, power and control in trans-organizational relations. Organ Stud 22 (2):337–365

Becker B, Gerhart B (1996) The impact of human resource management on organizational performance: progress and prospects. Acad Manage J 39(4):779–801

Bigley GA, Pearce JL (1998) Straining for shared meaning in organization science; problems of trust and distrust. Acad Manage Rev 23(3):405–421

Blum MW, Jr. Wall JA (1997) HRM: managing conflicts in the firm. Bus Horiz 40(3):84–87

Blut M, Backhaus C, Heussler T, Woisetschläger DM, Evanschitsky H, Ahlert D (2011) What to expect after the honeymoon: testing a lifecycle theory of franchise relationships. J Retailing 87 (3):306–319

Bordonaba-Juste MV, Polo-Redondo Y (2004) Relationships in franchised distribution systems: the case of the Spanish market. Int Rev Retail Distribution Consum Res 14(1):101–127

Bordonaba-Juste MV, Polo-Redondo Y (2008) Differences between short and long-term relationships: an empirical analysis in franchise systems. J Strateg Mark 16(4):327–354

Bradach JL (1997) Using the plural form in the management of restaurant chains. Adm Sci Q 42 (2):276–303

Bradach JL, Eccles RG (1989) Price, authority, and trust: from ideal types to plural forms. Annu Rev Sociol 15:97–118

Caldwell C, Clapham SE (2003) Organizational trustworthiness: an international perspective. J Bus Ethics 47:349–364

Castrogiovanni GJ, Kidwell RE (2010) Human resource management practices affecting unit managers in franchise networks. Hum Resour Manage 49(2):225–239

Chiou JS, Hsieh CH, Yang CH (2004) The effect of franchisor's communication, service assistance, and competitive advantage on franchisees' intentions to remain in the franchise system. J Small Bus Manage 42(1):19–36

Cliquet G (2000) Plural forms in store networks; a model for store network evolution. Int Rev Retail Distribution Consum Res 10(4):369–387

Cochet O, Ehrmann T (2007) Preliminary evidence on the appointment of institutional solutions to franchisor moral hazard – the case of franchise councils. Manage Decis Econ 28(1):41–55

Cochet O, Garg O (2008) How do franchise contracts evolve? A study of three German SMEs. J Small Bus Manage 46(1):134–151

Cohen-Charash Y, Spector PE (2001) The role of justice in organizations: a meta-analysis. Organ Behav Hum Decis Process 86(2):278–321

Colquitt JA, Scott BA, Le Pine JA (2007) Trust, trustworthiness, and trust propensity: a meta-analytic test of their unique relationships with risk taking and job performance. J Appl Psychol 92(4):909–927

Combs JG, Michael SC, Castrogiovanni GJ (2009) Institutional influences on the choice of organizational form: the case of franchising. J Manage 35(5):1268–1290
Combs JG, Ketchen DJ, Shook CL, Short JC (2011) Antecedents and consequences of franchising; past accomplishments and future challenges. J Manage 37(1):99–126
Croonen EPM (2006) Strategic interactions in franchise Relationships. PhD Thesis, Faculty of Management and Organization, University of Groningen
Croonen EPM (2010) Trust and fairness during strategic change processes in franchise systems. J Bus Ethics 95:191–209
Currall SC, Inkpen AC (2002) A multilevel approach to trust in joint ventures. J Int Bus Stud 33 (3):479–495
Cutcher-Gershenfeld J (1991) The impact on economic performance of a transformation in workplace relations. Ind Labor Relat Rev 44(2):241–260
Dahlstrom R, Nygaard A (1995) An exploratory investigation of interpersonal trust in new and mature economies. J Retailing 71(4):339–361
Dant RP (2008) A futuristic research agenda for the field of franchising. J Small Bus Manage 46(1):91–98
Davies MAP, Lassar W, Manolis C, Prince M, Winsor RD (2011) A model of trust and compliance in franchise relationships. J Bus Venturing 26(3):321–340
Dirks KT, Ferrin DL (2002) Trust in leadership: meta-analytic findings and implications for research and practice. J Appl Psychol 87(4):611–628
Dunn J, Ruedy NE, Schweitzer ME (2012) It hurts both ways: how social comparison harm affective and cognitive trust. Organ Behav Hum Decis Process 117:2–14
Essen van C, Pleijster F (2009) Franchiseketens; Marktaandeel en Continuïteit. Economisch Instituut voor het Midden- en Kleinbedrijf (EIM), Zoetermeer
Evans WR, Davis WD (2005) High-performance work systems and organizational performance; the mediating role of internal social structure. J Manage 31(5):758–775
Ferrin DL, Dirks KT (2003) The use of rewards to increase and decrease trust: mediating processes and differential effects. Organ Sci 14(1):18–31
Gallini NT, Lutz NA (1992) Dual distribution and royalty fees in franchising. J Law Econ Organ 8(3):471–501
Gillespie N, Dietz G (2009) Trust repair after and organization-level failure. Acad Manage Rev 34(1):127–145
Gillis WE, Combs JG (2009) Franchisor strategy and firm performance: making the most of strategic resource investments. Bus Horiz 52(6):553–561
Gillis WE, McEwan E, Crook RT, Michael SC (2011) Using tournaments to reduce agency problems: the case of franchising. Entrepren Theory Pract 35(3):427–447
Gould-Williams J (2003) The importance of hr practices and workplace trust in achieving superior performance: a study of public-sector organizations. Int J Hum Resource Manage 14(1):28–54
Gullett J, Do L, Canuto-Carranco M, Brister M, Turner S, Caldwell C (2009) The buyer–supplier relationship: an integrative model of ethics and trust. J Bus Ethics 90:329–341
Hodson R (2004) Organizational trustworthiness: findings from the population of organizational ethnographies. Organ Sci 15(4):432–445
Huselid MA (1995) The impact of human resource management practices on turnover, productivity and corporate financial performance. Acad Manage J 38(3):635–672
Ingenhoff D, Sommer K (2010) Trust in companies and in CEOs: a comparative study of the main influences. J Bus Ethics 95:339–355
Kaufmann PJ, Eroglu S (1998) Standardization and adaptation in business format franchising. J Bus Venturing 14(1):69–85
Ketchen DJ, Short JC, Combs JG (2011) Is franchising entrepreneurship? Yes, no, and maybe so. Entrepren Theory Prac 35(3):583–593
Kidwell RE, Nygaard A, Silkoset R (2007) Antecedents and effects of free riding in the franchisor–franchisee relationship. J Bus Venturing 22(4):522–544
Lado AA, Wilson MC (1994) Human resource systems and sustained competitive advantage: a competency-based perspective. Acad Manage Rev 19(4):699–727

Lewicki RJ, Bunker BB (1996) Developing and maintaining trust in work relationships. In: Kramer RM, Tyler TR (eds) Trust in organizations; frontiers of theory and research. Sage, Thousand Oaks
Lusch RF, O'Brien M, Sindhav B (2003) The critical role of trust in obtaining retailer support for a suppliers strategic organizational change. J Retailing 79(4):249–258
Mayer RC, Davis JH (1999) The effect of the performance appraisal system on trust for management: a field quasi-experiment. J Appl Psychol 84(1):123–136
Mayer RC, Davis JH, Schoorman FD (1995) An integrative model of organizational trust. Acad Manage Rev 20(3):709–734
Michael SC, Combs JG (2008) Entrepreneurial failure; the case of franchisees. J Small Bus Manage 46(1):73–90
Morgan RM, Hunt SD (1994) The commitment-trust theory of relationship marketing. J Mark 58:20–38
Morrison KA (1996) An empirical test of a model of franchisee job satisfaction. J Small Bus Manage 34(3):27–41
Morrison KA (1997) How franchise job satisfaction and personality affects performance, organizational commitment, franchisor relations, and intention to remain. J Small Bus Manage 35(3):39–67
Nooteboom B (1999) Inter-firm alliances: analysis and design. Routledge, London
Roh EY, Yoon JH (2009) Franchisor's ongoing support and franchisee's satisfaction; a case of ice cream franchising in Korea. Int J Contemp Hospitality Manage 21(1):85–99
Rotter JB (1971) Generalized expectancies for interpersonal trust. Am Psychol 26:443–452
Rousseau DM, Sitkin SB, Burt RS, Camerer C (1998) Not so different after all: a cross-discipline view of trust. Acad Manage Rev 23(3):393–404
Saparito PA, Chen CC, Sapienza HJ (2004) The role of relational trust in bank-small firm relationships. Acad Manage J 47(3):400–410
Schoorman FD, Mayer RC, Davis JH (2007) An intergrative model of organizational trust; past, present and future. Acad Manage Rev 32(2):344–354
Searle R, Den Hartog DN, Weibel A, Gillespie N, Six F, Hatzakis T, Skinner D (2011) Trust in the employer; the role of high-involvement work practices and procedural justice in European organizations. Int J Hum Resource Manage 22(5):1069–1092
Six F, Sorge A (2008) Creating a high-trust organization: an exploration into organizational policies that stimulate interpersonal trust building. J Manage Stud 45(5):857–884
Snell SA, Jr. Dean JW (1992) Integrated manufacturing and human resource management; a human capital perspective. Acad Manage J 35(3):467–504
Sorenson O, Sørensen JB (2001) Research note: finding the right mix: franchising, organizational learning, and chain performance. Strateg Manage J 22(6/7):713–724
Storholm G, Scheuing EE (1994) Ethical implications of business format franchising. J Bus Ethics 13(3):181–188
Sullivan M, Adcock D (2002) Retail marketing. Thomson, London
Szulanski G, Jensen RJ (2008) Growth through copying: the negative consequences of innovation on franchise network growth. Res Policy 37(10):1732–1741
Whitener EM (2001) Do 'High Commitment' human resource practices affect employee commitment? A cross-level analysis using hierarchical linear modeling. J Manage 27:515–535
Windsperger J (2004) Centralization of franchising networks: evidence from the Austrian franchise sector. J Bus Res 57(12):1361–1369
Yin X, Zajac EJ (2004) The strategy/governance structure fit relationship: theory and evidence in franchising arrangements. Strateg Manage J 25:365–383
Zaheer A, McEvily B, Perrone V (1998) Does trust matter? Exploring the effects of interorganizational trust and interpersonal trust on performance. Organ Sci 9(2):141–159

Franchisees' Websites and Concept Uniformity: A New Challenge for Franchisors

Rozenn Perrigot, Guy Basset, Danièle Briand-Meledo, and Gérard Cliquet

Abstract This paper aims to highlight the challenges associated with network uniformity and brand image for franchisors, more specifically when their franchisees set up and manage their own website. This practice has some impact on network uniformity which is a key concept in franchising. We analyze the presence of franchisees on the Internet of 471 networks, both in retailing and services, described in the 2011 franchise directory. We find that only 38 franchise networks are concerned about this practice. We use a qualitative approach based on multiple cases studies of these 38 franchise networks. It points out the different aspects of franchisees' websites that can damage concept uniformity. Maintaining network uniformity when there are various websites set up and run by franchisees entails challenges to franchisors that are presented in this paper within a managerial

This research was supported by France's National Research Agency (ANR "Programme Blanc" – FRANBLE – ANR-12-BSH1-0011-01) and Human Sciences Institute in Brittany (MSHB – RECOMAD). The authors sincerely thank the ANR and the MSHB for their support and Kelly PRIOUX and Thomas SIMON for their valuable assistance regarding the websites' observation.

R. Perrigot (✉)
Graduate School of Management – University of Rennes 1 & ESC Rennes, Center for Research in Economics and Management (CREM UMR CNRS 6211) & Institute for Business Law, Estate and Torts (CDA-PR UPRES EA 3195), 11 rue Jean Macé, CS 70803, 35708 Rennes Cedex 7, France
e-mail: rozenn.perrigot@univ-rennes1.fr

G. Basset • G. Cliquet
Graduate School of Management – University of Rennes 1, Center for Research in Economics and Management (CREM UMR CNRS 6211), 11 rue Jean Macé, CS 70803, 35708 Rennes Cedex 7, France
e-mail: guy.basset@univ-rennes1.fr; gerard.cliquet@univ-rennes1.fr

D. Briand-Meledo
Faculty of Law and Political Science – University of Rennes 1, Institute for Business Law, Estate and Torts (CDA-PR UPRES EA 3195), and lawyer, 9 rue Jean Macé, CS 54203, 35042 Rennes Cedex, France
e-mail: daniele.meledo@univ-rennes1.fr; brianddaniele@gmail.com

T. Ehrmann et al. (eds.), *Network Governance*, Contributions to Management Science, DOI 10.1007/978-3-7908-2867-2_13, © Springer-Verlag Berlin Heidelberg 2013

perspective linked to technical and organizational know-how. Some insights from the legal perspective are also provided.

Keywords Franchising • Franchisees' websites • Internet • Uniformity

1 Introduction

The trade press often highlights the importance of the use of the Internet in the franchising sector. The Internet is considered "a good vehicle for advertising and promotion" (Trice 2001). The Internet is also viewed as "an effective tool for promoting [the] systems, communicating efficiently with [the] franchisees and suppliers, and capitalizing on the opportunities presented by 'e-commerce'" (Plave and Almosch 2000). Franson and DeSmith (2005) asserted that "[v]ery few franchise concepts have no need of the Internet as a means of increasing communication and contact with the customer in a way that the customer finds convenient. As a customer, it is a bit shocking when a company has no Web site or online inquiry capability (almost as shocking as going to a retail store and finding they only take cash)." Other benefits associated with the use of the Internet in the franchising sector are brand recognition, lead generation, the possibility of reaching out to new customers virtually anywhere, higher-market saturation and new sources of revenues (Franson and DeSmith 2005; Rogers et al. 2007).

The impact of Internet on franchising is of great interest. First, it is essential for franchisors to establish a consistent presence on the Internet in order to maintain network uniformity and reinforce their brand image, given that uniformity and brand name are key elements of franchising (Caves and Murphy 1976; Klein 1995). From this perspective, Flosdorf (2002) asserted that "[m]aintaining a consistent brand image can help build business in any industry, but for franchises consistency is imperative for success and survival." Second, in the field of European competition Law, the Guidelines (2010/C 130/01) associated to the EC Regulation on vertical restraints n° 330/2010 of April 20, 2010, opened a new conceptual field of reflection which is now calling more extensive research. These guidelines recognize that the Internet, which should be free to be used not only by franchisors but also franchisees, is a very powerful tool for selling products. They define the set of rules to be respected when being present on the Internet and stipulate that a supplier—here, the franchisor—may control the quality of the websites used by its distributors—here, its franchisees. Third, the impact of Internet on franchising is very important since franchising has become a key sector in the economy. Franchising deserves specific attention due to its continuous growth in many developed and emerging countries. For instance, in the US, there are 2,200 franchise networks including 784,802 franchised units, generating 7.8 million jobs and 739.9 billion dollars of turnover (PricewaterhouseCoopers 2011). In Europe, as well, franchising is particularly developed with about 10,183 different franchisors and 427,000 franchised stores (European Franchise Federation 2012).

Most previous research has dealt with the use of the Internet by franchisors (Cedrola and Memmo 2009; Dixon and Quinn 2004; Kaufmann et al. 2010; Perrigot and Pénard 2012; Rao and Frazer 2006). Nevertheless, the literature on the use of the Internet by franchisees, i.e., when they set up and manage their own website, is very scarce. The purpose of this paper is to highlight the risks of such franchisees' practices which can jeopardize both network uniformity and brand image. Indeed, "[b]rand identity, or image, is one of the most prized assets of successful franchise organizations. Yet, while most franchisors today build excellent brand recognition through national advertising and marketing programs, many fail to extend brand identity to every marketing piece used by their franchisees" (Oseland 1995). In this paper, uniformity is considered on two levels: between the franchisor's website and its franchisees' websites and between franchisees' websites. We explore the following research questions: (1) What is the extent of franchisees' presence on the Internet (via their own website)?, (2) What are the main characteristics of the franchise networks within which some of the franchisees have their own website?, and (3) what are the main elements for which we find similarities or, contrarily, differences that break network uniformity and then damage brand image?

The empirical study deals with the French market. From the population of 471 franchise networks described in the 2011 franchising directory published by the *French Franchise Federation*, we build a sub-sample of franchisors whose franchisees have their own website. Using these networks, we analyze in depth the content of the franchisor's websites along with the content of the franchisees' websites in order to point out the similarities and differences in terms of network uniformity.

This research offers three main contributions. First, it builds on the franchising literature for which more franchisee-based perspectives are encouraged (Dant 2008). Secondly, this paper contributes to the literature on the use of the Internet in the franchising sector for which Cedrola and Memmo (2009) and Rao and Frazer (2006) called for more research. Thirdly, it draws on the literature on uniformity and brand name in the franchising sector (Kaufmann and Eroglu 1998) by focusing on uniformity between franchisor's and franchisees' websites, as well as among franchisees' websites themselves and not between physical stores as usually studied in the literature (Streed 2007; Streed and Cliquet 2008).

The paper is organized as follows. In the second section, we briefly review the literature on the use of the Internet in the franchising sector to highlight the need for more research on the use of the Internet by franchisees and its impact on network uniformity. The third section describes the research methodology. We then successively present and discuss the findings of our empirical study in the fourth and fifth sections. The last section is the conclusion.

2 Literature Review

2.1 The Use of the Internet by Franchisors

Some scholars have already investigated the use of the Internet by franchisors (e.g., Cedrola and Memmo 2009; Dixon and Quinn 2004; Kaufmann et al. 2010; Perrigot and Pénard (forthcoming); Rao and Frazer 2006). For instance, Dixon and Quinn (2004) analyzed the use of the Internet by 364 franchisors in the U.K. Their main research questions consisted of determining the percentage of franchisors who have websites and seeing if this percentage depended on the industry; highlighting the percentage of franchisors' websites allowing customers to order online; exploring if franchisors had pages dedicated to individual franchisees; and exploring other uses for the franchisors' websites. They found that about two thirds of the sampled franchisors were present on the Internet. This figure depended on the industry in which they run their business. They also pointed out that 15 % of the franchisors offered the ability to order and pay via the website. Moreover, they confirmed that many franchisors used their websites to provide Internet users with products and services information, company information, franchisee solicitation and franchisee location. Finally, they pointed out three main categories of franchisors' online activities: information, sales and franchising.

Rao and Frazer (2006) studied a random sample of 202 Australian franchisors. They distinguished two main activities on the franchisors' websites: franchisee solicitation and coordination activity (company history, franchising background, franchisor services, testimonials, etc.) and promotion activity (store location, product information, discounts, online sales, etc.). They found that only 11.4 % of franchisors websites allowed customers to buy online. They also examined differences in website activities according to network size and age. They found no systematic patterns and concluded that the use of the Internet by Australian franchisors was still in its infancy.

Kaufmann et al. (2010) focused their examination on a sample of 166 US networks present in industries that were "the most susceptible to direct on-line sales to end users" (p. 6) and completed this approach by using two case studies. They observed that 34 % of the websites in their sample were transactional. This high figure, in comparison with previous findings, is probably due to the types of industries selected, as well as the exploratory nature of their study. The results of their logistic regression model revealed that, beyond the significant and negative influence of the percentage of franchised stores, a significant and positive influence of network internationalization on the adoption of an E-commerce strategy by franchisors was also being exerted.

Cedrola and Memmo (2009) explored a sample of 305 franchisor websites selected from eight different industries and various markets (Australia, Brazil, Canada, Mexico, Spain, the U.K. and the US). They analyzed the way franchisors used their websites. They focused on several elements displayed on franchisor websites: franchisor presentation and website features, product, communication, pricing, distribution and customers relations. They found that 80 % of the franchisors under investigation were mainly using the Internet to provide information on their products and services, that only 44.3 % advertised on their websites using banners, pop ups, or specific online ads, that 41 % provided clear information on product and service prices and that only 19 % carried promotional campaigns on their websites. They also found that only 10.2 % of websites offered functionalities related to online purchases, again depending on the industry in which they run their business.

Perrigot and Pénard (forthcoming) focused on the E-commerce strategies of 486 franchise networks in the US market. Using the resource-based view, they formulated various hypotheses on the factors that influence the adoption of an E-commerce strategy by franchisors, namely the percentage of company-owned stores in the network, network size and age, franchisor resources (franchising fees and franchising royalties), and the allocation of exclusive territories to franchisees. Their findings suggest that the percentage of company-owned stores and the brand image, as represented by network size, both exert a significant and positive impact on the adoption of an E-commerce strategy, whereas network age and franchising royalties exert a significant and negative impact on the adoption of such a strategy.

2.2 The Use of the Internet by Franchisees and the Consequent Need for Uniformity

The papers on the use of the Internet in the franchising sector—mentioned above—have all dealt with the franchisor perspective, as does most of the research in the franchising field. Moreover, their authors have mainly focused on the transactional capabilities of the websites, even if some of them have also analyzed the content of the franchisor websites. In this paper, we adopt another perspective. More specifically, we explore the presence of the franchisees on the Internet and the associated challenges for their franchisors in terms of network uniformity and brand image. "On the surface, branding is the public perception of [the] company, but on a deeper level it is every single point of contact that a consumer makes with [the] company. It is the products and the services provided. It's the Web site, advertising and special events. It is word-of-mouth and any press coverage the franchisee received. Perhaps most importantly, branding is the message every one of the company's franchisees communicates to consumers" (Gould 2005). Franchisees' websites thus contribute to network brand image. In terms of network uniformity, as Findley (2007) wrote, "maintaining and strengthening brand identity, though sometimes difficult, is vital to franchise success. If the brand message is not consistent between units, the message can become confusing and unclear."

From this perspective, Oseland (1995) asserted that "[f]ranchisees, as well as franchisors, can play a role in managing brand identity by ensuring that all components of the brand have a consistent look and message. The logo, copy points and slogan that consumers see in a local direct mail piece should be the same as those they see in a nationally televised commercial. The graphic look should be the same, the colors the same, the feel, the signature, the tonality, all the same." Findley (2007) talked about consistency as critical to maintaining uniformity and specified that "[t]here are key areas in marketing where consistency should be expected and enforced by the franchise company. The areas include customer service, operations, logo usage, advertising campaigns and quality control systems." We can also mention marketing-mix elements, advertising, logos, signs, URLs, etc. We detail now these elements that have to be consistent on the Internet, whatever the considered website.

First, uniformity in terms of products and services deals with products and services that must be displayed on the franchisees' websites. A customer will not appreciate seeing a product or service available in a specific store of brand X, or on a franchisee X's website, that is not available on the website of another franchisee of the same brand X. So, the question of product and service assortment and overlap is relevant. Second, uniformity in terms of price is very important as well. The price represents one of the essential elements of brand positioning and can have an impact on customer satisfaction and loyalty. Drastic differences in pricing would not be understood by customers even though imposing prices is contradictory to anti-trust laws (Lafontaine 1998). Thirdly, uniformity in terms of communication deals with promotion, slogans and mascots that refer to the brand. The management of promotional activities in a uniform way across websites is essential. Fourth, there are some other elements that can be associated with the "place" of Marketing's 4P in the Internet arena. It deals with the URL (respect of the brand name use, use of the city name, etc.). As Plave and Almosch (2000) reminded, "[c]oordination of domain names is [. . .] critical, as this process protects the entire network." There is also the question of graphic chart and use of logos (respect of colors, fonts, etc.). Graphic charts can correspond to the appearance of the physical store. Customers have to have the same impression when they visit a store as when they visit any of the brand's websites. Concerning logos, Findley (2007) specified that "[l]ogo usage is another huge aspect of marketing consistency. A logo is a representation of brand identity. It means much more than just a name or symbol, which is why it's so important to remain consistent in all usage of it." Finally, there is the website design because the challenge is to create and maintain a uniform "look and feel" for all Websites. "Inconsistencies in the "look and feel" of a network's Websites may damage the public's general perception of the network's uniformity, which is the hallmark of any franchise network" (Plave and Almosch 2000).

All the above elements have to be consistent across all franchisor's and franchisees' websites, even if differences do exist among the franchisors and the franchisees in terms of technical, human and financial resources and entrepreneurship orientation. Franchisees can set up and run their own website. However, according to Paragraph 54 of the 2010 guidelines on vertical restraints, "[the] supplier may require quality standards for the use of the internet site to resell its goods, just as the supplier may require quality standards for a shop or for selling by catalogue or for advertising and promotion in general." Thus, from a practical perspective, maintaining uniformity across the network of stores and maintaining uniformity across the network of franchisee websites are almost similar tasks in terms of managerial and marketing aspects.

3 Methodology

3.1 Data

3.1.1 The Franchising Sector in France

Our empirical study deals with the Internet-related practices of franchisors and franchisees and their associated challenges in terms of network uniformity in a European market: France. In Europe, franchising is well developed with about 10,183 different franchisors and 427,000 franchised stores. As in other European countries, franchising in France has experienced a continuous growth since the 1970s. In late 2010, the number of franchisors equaled 1,477 (+ 5.8 % compared to 2009) and the number of franchised stores equaled 58,351 (+ 10 % compared to 2009), generating 335,000 jobs and more than € 47.88 billion of turnover (French Franchise Federation 2012). Franchising is present in all industries in France, including retailing and services. Many French franchisors have succeeded at the worldwide level with famous brands such as *Brioche Dorée*, *Cache-Cache*, *Ibis*, *Jacques Dessange*, *Morgan*, *Novotel* and *Yves Rocher*. France is also an attractive market for foreign franchisors. This is particularly the case for US franchisors in the fast-food industry with *Domino's Pizza*, *KFC*, *McDonald's*, *Pizza Hut*, *Subway*, etc. currently expanding in the French market.

3.1.2 The Population of Franchisors Under Investigation

The observation process—detailed below—was conducted on the 471 networks described in the franchise directory published by the *French Franchise Federation* in 2011. Figures on the characteristics of the franchise networks allow for a global overview of franchising in France. From these 471 franchise networks, we see that the average network size is 90.46 stores, including both franchised stores and company-owned stores within the French market (min: 1; max: 928; st. dev.: 116.78). The average percentage of company-owned stores within the network in the French market is equal to 33.10 % (min: 0; max: 100; st. dev.: 28.88). Mean network age is 15.03 years (min: 1; max: 98; st. dev.: 13.19). The entry fees are equal to € 16,596.62 in average (min: 0; max: 80,000; st. dev: 12,458.75). The mean franchising royalties are equal to 3.67 % of store turnover (min: 0; max: 38; st. dev. 3.13). Franchising contract duration is 6.07 years on average (min: 0; max:

20 years: st. dev. 2.10). Finally, 57 % of the sampled franchisors are in the service industry (versus the retail industry).

3.2 Methods

3.2.1 The Multiple Cases Approach

We chose to illustrate the uniformity-related issues faced by franchisors, when their franchisees set up and ran their own website, using a qualitative approach and, more specifically, the multiple cases approach. Qualitative research offers several advantages, among which is richness of data (Hair et al. 2008). The multiple cases approach allows scholars to explore phenomenon as multi-unit franchising (Weaven and Frazer 2007a, b) or plural form (Brookes and Roper 2012; Perrigot and Herrbach 2012) by using a general perspective. In this paper, the main form of data collection relies on the observation of franchisors' and franchisees' websites. This observation was conducted in a short period of time (from April 28, 2011 to May 20, 2011) in order to limit all kinds of biases associated with website observation (modification/update of the content of the websites, creation/suppression of the websites, etc.). The URLs of the franchisors' websites were displayed in the 2011 franchise directory published by the *French Franchise Federation*. Regarding the franchisees' websites, we searched their URL using the *Google* search engine, with the name of the franchise network as the keyword. We looked at the first 20 pages of the *Google* results and if a franchisee's website appeared on one of the last five pages of the *Google* results, i.e., on pages 15–20, we pursued the search process to include five additional pages of the *Google* results.

One of our first findings was that only 38 networks out of the 471 under investigation, i.e. 8.07 % of the sampled franchisors, have franchisees (at least one) running their own website. Table 1 describes the characteristics of these 38 franchise networks.

Results of t-tests displayed in Table 2 show that there are some significant differences, in terms of network characteristics, between franchisors who have franchisees running their own website and those who do not have. More specifically, franchisors who have franchisees running their own website are significantly older, with a lower percentage of company-owned stores and requiring less franchising royalties than those who do not have franchisees running a website.

3.2.2 The Observation of Franchisors' and Franchisees' Websites

We examined in detail all the franchisees' websites of the 38 franchise networks under investigation, along with those of 38 corresponding franchisors. We made screenshots of all the pages of each website. This led to 38 Word documents of 121 pages on average, with two screenshots per page. We then had a total corpus

Table 1 Characteristics of networks where at least one franchisee runs its own website

Network name	Industry	Network age	Network size	Percentage of company-owned stores	Franchising royalties	Franchising fees	Contract duration
Abithea	Real estate	6	26	7.69		7,500	5
Akena	Hotel	19					10
Anne Carole	Real estate	11	26	3.85		12,000	5
Archea	Household equipment	11			2.25	19,000	5
Atout Ménage	Services for people	4	22	9.09	4	13,500	5
Axeo Services	Services for people	4	126	6.35	5	25,000	5
Balladins	Hotel	26	140	45.71	4		5
Bistrot du Boucher	Restaurant	23	36	2.78	3	45,000	7
Brasseries Flo	Restaurant	42	22	63.64	3	30,000	10
Brit Hotel	Hotel	11	100	15		2,000	5
Café Leffe	Restaurant	24	17	0	3	47,000	5
Camille Albane	Aesthetics	17	226	3.1	3	8,500	5
Cavavin	Specialized food	26	83	3.61	1	14,000	7
Dafy Moto	Services for car	37	100	30	5	9,000	5
De Neuville	Specialized food	25	131	11.45	4	15,000	7
DistriClub Medical	Services for people	36	37	2.7	2.45	15,000	5
Easy Cash	Other businesses	10	54	0		24,750	9
Ecotel	Services for companies	35	35	2.86		23,000	3
Eric Stipa	Aesthetics	20	109	26.61		2,750	5
Guy Hoquet	Real estate	17	500	0	2	28,000	5
In & Fi	Services for people	7			5	27,650	5
Inter Caves	Specialized food	28			3	0	5
Jack Holt	Aesthetics	26	61	6.56	4	7,000	7
Jardiland	Household equipment	29	149	54.36	1.1	25,000	5
La Maison des Travaux	Real estate	3	32	0	5	12,000	5
La Mangoune	Restaurant	6	4	50	4	40,000	7

(continued)

Table 1 (continued)

Network name	Industry	Network age	Network size	Percentage of company-owned stores	Franchising royalties	Franchising fees	Contract duration
La Pataterie	Restaurant	8	79	3.8	4	45,000	7
Les Domaines qui montent	Specialized food	5	17	17.65	1.5	17,000	7
Maison de la Literie	Household equipment	31	287	34.15	3	20,000	5
Maxauto	Services for car	30	78	8.97	2	20,000	5
Novotel	Hotel	44	115	65.22			12
Saint Algue	Aesthetics	26	268	4.1		10,000	5
Simply Market	Specialized food	29	324	84.88		0	7
Solvimo	Real estate	8	160	0	2	20,000	5
Speed Rabbit	Restaurant	18	119	6.72	5	15,000	10
La Taverne de Maître Kanter	Restaurant		40	12.5	3	30,000	10
Tonic Hôtel	Hotel	17			1.5		5
Villaverde	Household equipment	25	86	16.28	0.6	6,100	5
		20.11	109.36	18.17	3.09	18,698	6.18

Table 2 Results of *t*-tests

Variables	Franchisors without any franchisees running their own website	Franchisors with franchisees running their own website
Network age**	14.58 (13.25)	20.11 (11.48)
	$n = 429$	$n = 37$
Network size	88.71 (117.61)	109.36 (107.23)
	$n = 357$	$n = 33$
Percentage of company-owned stores***	34.48 (29.03)	18.17 (22.68)
	$n = 357$	$n = 33$
Franchising royalties**	3.72 (3.24)	3.09 (1.33)
	$n = 300$	$N = 28$
Franchising fees	16,418.40 (12,439.13)	18,698.53 (12,685.02)
	$n = 401$	$N = 34$
Contract duration	6.06 (2.12)	6.18 (2.01)
	$n = 417$	$N = 38$

Legend: standard deviation values are indicated in brackets
*Significant at the 0.10 level; **0.05 level; ***0.01 level

of 4,641 pages, i.e. 9,282 screenshots. The information provided in Table 3 shows the significant differences across the networks, and also within each network, in terms of number of franchisees that have their own website and the minimum and maximum number of pages on the franchisees' websites. For instance, networks such as *Atout Ménage*, *Simply Market*, and *Solvimo* each have only one franchisee with their own website, whereas *Brit Hotel* has 28 franchisees running their own website. Table 3 also indicates that only 334 franchisees out of several thousands have their own website.

4 Findings

4.1 Uniformity and Product Policy (Product and Service Assortment)

Uniformity-related issues are highlighted in the following four cases, as far as product and service assortment is concerned. The first case deals with *Jardiland*. On the franchisor's website, the assortment includes six categories of products: "houseplants," "pet shop," "seeds & bulbs," "breeding-ground," "garden fittings" and "inspiration." The number of product categories displayed on the franchisees' websites varies from one to five. Regarding the categories similar (or almost similar) to these displayed on the franchisor's website, we have "nurseryman" and "garden" appearing on three of the four franchisees' websites, "pet shop" and "other activities" appearing on two websites, "garden furniture" is close to "garden

Table 3 Information on franchisors and franchisees websites

Network name	Number of pages on the franchisor website	Number of franchisees having their own website	Average number of pages on the franchisee website	Minimum number of pages on the franchisee website	Maximum number of pages on the franchisee website	Total number of analyzed pages for the network
Abithea	23	7	12	6	16	107
Akena	29	15	10	5	19	175
Anne Carole	45	3	9	13	11	77
Archea	30	4	13	10	21	81
Atout Ménage	23	1	5	5	5	28
Axeo Services	46	6	11	4	23	114
Balladins	33	24	12	1	39	312
Bistrot du Boucher	31	5	8	5	18	73
Brasseries Flo	8	14	5	1	16	71
Brit Hotel	57	28	15	2	34	465
Café Leffe	6	5	8	1	14	46
Camille Albane	25	11	9	4	33	127
Cavavin	14	13	8	1	29	124
Dafy Moto	36	15	12	7	37	214
De Neuville	29	3	9	6	15	56
DistriClub Medical	31	8	8	4	13	94
Easy Cash	28	2	9	9	9	46
Ecotel	16	5	9	6	10	59
Eric Stipa	36	8	7	4	9	92
Guy Hoquet	18	27	8	1	19	244
In & Fi	58	12	15	7	28	239
Inter Caves	23	13	7	3	21	117
Jack Holt	30	3	8	7	11	55
Jardiland	73	8	16	7	28	199
La Maison des Travaux	17	14	11	7	20	166
La Mangoune	5	1	12	12	12	17
La Pataterie	20	6	10	6	19	80
Les Domaines qui Montent	46	8	14	5	28	160
Maison de la Literie	28	16	10	6	15	180
Maxauto	22	6	9	2	14	73
Novotel	21	9	12	2	19	126
Saint Algue	30	6	8	5	12	75
Simply Market	29	1	2	2	2	31
Solvimo	3	1	7	7	7	10
Speed Rabbit	30	6	14	6	28	115

(continued)

Table 3 (continued)

Network name	Number of pages on the franchisor website	Number of franchisees having their own website	Average number of pages on the franchisee website	Minimum number of pages on the franchisee website	Maximum number of pages on the franchisee website	Total number of analyzed pages for the network
Taverne de Maître Kanter	8	14	13	1	20	183
Tonic Hôtel	11	3	22	6	35	77
Villaverde	–	3	44	34	55	133
TOTAL	**1,018**	**334**	**421**	**220**	**764**	**4,641**

fitting" and "inspiration" that appear on the franchisor's website. "Florist" is close to "houseplants" and "seeds & bulbs." One category is particularly threatening in terms of network uniformity: that is "swimming-pool" because it is not at all consistent with the franchisor's assortment.

For the second case, *Ecotel*, the franchisor's website displays eight categories ("art of entertaining," "disposable," "cooking," "meal transport and caterer," "hygiene," "work clothes," "hotel business" and "furniture") while the franchisees' websites only display two to seven categories. One franchisee's website displays six categories that are similar to the franchisor's website. One is quite similar: "upkeep" instead of "hygiene," but doesn't include "meal transport and caterer." Two franchisees' websites differ from the franchisor's, but both look alike with categories such as "bar and pub supplies," "restaurant supplies," "hotel furniture," "setting out," and one of the two websites has another category: "hotel supplies." Finally, two franchisees split their assortment into only two categories: "restaurant equipment" and "public sector equipment."

Concerning the third case, *Maxauto*, on the franchisor's website the assortment includes seven categories of products: "workshop services," "tires," "technical parts," "my check-up," "maintenance products," "equipment & comfort," "discounts and selected deals." Assortments displayed on the five franchisees' websites include three to four categories. The five of them include "car garage," four include "tires," and "car maintenance centre," two include "industrial vehicles." All those categories, though they are different from those of the franchisor's website, still deal with the franchisor's main products and services. The most worrying part regarding this case is that one franchisee displays two categories that are not included in the franchisor's assortment: "car rental" and "windscreen and sunroof."

The last case deals with *La Maison des Travaux*. We studied 12 franchisee's websites that displayed different assortments. On average, the assortments included 4.33 categories (min: 3; max: 5). "Building" is the most quoted category and appears in seven websites. It is followed by "insulation" and "extension" quoted six times. "Roofspace fitting" and "renovation" appear five times while "roofing," “refacing” and “others” appear three times. “Bathroom” is quoted twice and “fitting” and “windows” appear only once.

In highlighting uniformity-related issues as far as products and services assortment is concerned, we notice that most of the franchisees, in these examples, do not offer Internet users the entire assortment advertised by the franchisors, i.e. what is available on the franchisors’ websites. Variations in terms of assortments also exist between franchisees’ websites. Even if selecting the product and service assortment to be sold is one of the franchisees’ rights, these differences in terms of product assortment—between franchisor’s and franchisees’ websites and simply franchisees’ websites—break the uniformity and can have a negative impact on the brand image of the franchise network as a whole. Customers perceiving this lack of uniformity related to products and services assortment may be confused, above all when information is so close—at one click—on the Internet.

4.2 Uniformity and Price Policy (Price Indication and Level)

Four cases highlight uniformity-related issues as far as price indication and price level are concerned. The first case that shows price consistency on the Internet deals with *Saint Algue*. The franchisor explicitly displays the prices of the services offered on its website, and three franchisees out of six do the same on their own website. We notice that, in general, prices on franchisees’ websites are one or two euros higher than those displayed on the franchisor’ website.

As far as the second case is concerned, *Café Leffe*, the franchisor does not display product prices on its website, whereas all the franchisees explicitly do so on their websites. We can observe price differences across franchisees’ websites. For instance, a rib steak is priced at € 12.00, € 12.90, € 14.50 or € 18.20, and mussels are priced at € 11.80, € 14.70 or € 14.95, depending on the franchised restaurant.

In the third case, *La Pataterie*, the franchisor makes its summer menu and all prices available on its website. One franchisee displays its menu without prices. Three franchisees out of six show neither the menu, nor the prices. Another franchisee shows its winter menu, thus indicating that the website has not been updated.

The same comments apply in regards to the fourth case, *Bistrot du Boucher*. The franchisor makes its menu and prices available on its website as does one franchisee out of five who have their own website. We also notice a difference between the prices displayed on the franchisee’s website and those displayed on the franchisor’s website, with some items more expensive and others cheaper.

In highlighting uniformity-related issues as far as price indication and price levels are concerned, we notice that even if these franchisees are free to display prices on their websites—prices that they set themselves as independent business owners—the differences in terms of price indication and level, can raise some issues and damage the franchise network's brand image as a whole. Customers may also see this lack of uniformity related to the franchisor's the franchisees' price policy.

4.3 Uniformity and Promotion Policy (Communication and Promotions)

Four cases highlight uniformity-related issues as far as promotions are concerned. The first case deals with *Saint Algue*. The franchisor displays on its website a 20 % discount on the fixed price of haircolor and highlights, this discount being available in participating stores only. But none of the franchisees mention this discount on their websites. One franchisee offers a 20 % discount for customers under 20 years old and a *Saint Algue* loyalty card. Another franchisee displays offers specific to its store: a "discovery offer" gives a 10 % discount for a first time visit along with a "birthday offer" (on their birthdays, customers get a hair-product worth € 7).

In the second case, *Intercaves*, at the time of observation, the franchisor neither mentioned a specific promotion, nor any loyalty program, on its website. However, many franchisees display information about promotions on their websites, e.g., "Special offer – beautiful days," "Special offer – Club and association," "Offer – Beers," "For 45€ of purchase... a barbecue set offered," "Week-end package," etc. Consequently, all these promotions are instigated by the franchisees, underlining a lack of uniformity in terms of communication strategies. Regarding the loyalty program, only one franchisee highlights its loyalty program; the loyalty card is usable in the specific store only and not in all network stores. Finally, one of the franchisees, in May, was still advertising promotions that were available for St. Valentine's Day, three months before. This lack of information updating can damage the brand image of the whole network, not only the brand image of the specific store associated with this website.

The third case we can mention is *Les Domaines qui Montent*. There are two types of promotions: discounts offered for large quantities and discounts following the seasons and/or limited to a specific period of time. Concerning the franchisor's website, one promotion is available only in company-owned stores: "currently, in the company-owned stores, € 5.75 per bottle by box of six bottles, instead of € 6.90." One franchisee, in addition to proposing the same discount, offers a similar discount on another product with a different price reduction. Another franchisee shows the specific product but does not mention any associated discount. One discount is announced on the franchisor's website as being available in the whole network, "the winning 15," even though it is only mentioned on one franchisee's website. Finally, two franchisees mention their own promotions (a 15 % discount and gift cards of € 10, € 20 or € 50).

Regarding the fourth case, *Dafy Moto*, the franchisor offers a loyalty card called "Dafydélité," (fidélité meaning loyalty in English). It also proposes a game and a discount on *Michelin* tires. The loyalty card is mentioned on one franchisee's website only. Concerning promotions set up by franchisees, 3 out of 16 offer discounts with dedicated tabs ("good deals") on their website.

In highlighting uniformity-related issues as far as promotions are concerned, we notice that there is a clear lack of uniformity in terms of communicating about the promotions available in the network. When the network is a plural form network, composed of franchised stores and company-owned stores, it can be difficult for customers to understand that promotions are only available in part of the network. The same comment applies to the loyalty card that is specific to some stores and sometimes not usable in the whole network.

Some comments should be made as well regarding slogans and mascots because they are other means of brand promotion and communication. For instance, concerning *Axeo Service*, the franchisor's slogan is "the commitment in a provided service," and it has a mascot on its website. This slogan is used by four franchisees out of six, while the mascot is used by only two of them. The two franchisees that do not display the franchisor's slogan have their own slogans: "a personalized and fitted assistance!" and "a daily helping hand!"

Another case is *Maxauto* whose franchisor's slogan is "Our job is your car." This slogan is taken up by only one franchisee out of the five who have a website. The other franchisees use different slogans on their websites, e.g., "Maxauto, a MAX of experience," "Maxauto, everything for your car," "A team at your service," "At Maxauto, we take care of your vehicle."

A last case is *Abithea*. The franchisor's slogan is "For us, proximity has a signification," which is taken up by one franchisee out of three. But even though this franchisee uses this slogan, it also uses another one: "Proximity to help you better." The two other slogans on franchisees' websites are "A prime example in real estate" and "Your partner in real estate projects."

To conclude, as slogans and mascots are significant elements related to brand image, a lack of uniformity in their use on franchisees' websites has some negative impact.

4.4 Uniformity and Place Policy

4.4.1 URL

Four cases highlight uniformity-related issues as far as URLs are concerned. The first case is *Speed Rabbit Pizza*. The URL for the franchisor's website is www.speedrabbitpizza.com. The franchisees have URLs ending with ".fr" (two), ".com" (three), and ".net" (one), and all of them mention the name of the network in their URLs. All of them indicate a geographic indication except one. We can also highlight two surprising cases. The first mentions the word "pizza" three times and the name of the network twice in the same URL. Last, but not least, a franchisee uses the name of a competitor's network, "dominos," in its URL, in addition to the network's name "speedrabbitpizza" as http://pizza-livraison-emporter-pizzeria-dominos.speedrabbitpizza-bobigny.com/.

In the second case, *La Taverne de Maître Kanter*, the URL for the franchisor's website is www.taverne-maitre-kanter.fr. Three franchisees do not really mention the name of the network in their URLs. Most franchisees indicate a geographic zone. One franchisee mentions the name of another network in its website URL, *Ibis*, which is a hotel network. Finally, three URLs can be considered challenging for the network, because they are very close to the franchisor's website URL (www.taverne-de-maitre-kanter.com, www.la-taverne-de-maitre-kanter.fr and www.tavernemaitrekanter.fr).

Concerning the third, *Distri Club Medical*, the URL of the franchisor's website is www.distri-club-medical.fr. Only three franchisees mention the name of the network in their URLs. Some use a geographic indication in their websites' URLs, while others mention their own name. One URL appears quite strange; it includes six key words (e.g., "incontinence") in addition to the classic URL.

The last case is *In & Fi*. Here, four franchisees do not mention the network name in their website URL but provide a geographical situation. Two other franchisees mention neither the network name nor a geographical indication in their URL (www.courtage-express.fr and www.destination-credit.fr). Without any clear reference to the brand name, this totally breaks the link with their franchisor.

While these four cases highlight uniformity-related issues as far as URLs are concerned, the dissimilarities in the URLs of the franchisees' websites indicate that franchisees do not follow any guidelines when they set up their websites. This can complexify the search process on the web, on the one hand, and raise some issues and damage the brand image of the franchise network as a whole, on the other hand. Moreover, customers may perceive a lack of uniformity related to the visibility of the franchisor and of the franchisees as well.

4.4.2 Graphic Chart and Logo

Four cases highlight uniformity-related issues as far as graphic charts and logos are concerned. The first case is *Intercaves*. In November 2009, one of the specialized websites dedicated to franchising displayed *Intercaves* new logo. Two years later, in May 2011, our observation of the franchisees' websites pointed out that ten out of the 14 franchisee websites still displayed the former logo and not the updated one. This also applied to their display of the former graphic chart with warm colors (i.e., red and chestnut) rather than the current trendy colors (i.e., purple and apple green).

Concerning the second case, *Guy Hoquet*, three years after the logo changed (in July 2008), our observation of the franchisees' websites in May 2011 pointed out that five out of the 28 franchisees' websites displayed the former logo and not the updated one, which is more modern and dynamic than the former one. Moreover, two franchisees used their own logo in addition to the franchisor's logo. Regarding the graphic chart, seven franchisees' websites were designed by the franchisors with links between franchisor and franchisee's websites, so the graphic chart was fully respected. One multi-unit franchisee had four websites very similar to one another. The most challenging case relates to nine franchisee's websites that had completely different graphic charts than the franchisor's website, as well as those of the other franchisees' websites (e.g., use of different colors: black, grey or yellow background instead of the blue recommended by the network).

As far as the third case is concerned, *Novotel*, three franchisees out of nine respect the franchisor's graphic chart with similar tabs and colors. A multi-unit franchisee uses, on its two distinct websites not only the blue color but also other colors such as yellow, purple and orange, and uses a slogan different from this of the franchisor.

Finally, regarding the fourth case, *Cavavin*, the franchisor uses a light green background on its website but none of its franchisees use this color on their websites. Instead of this specific color associated with the brand, they use dark green, yellow, black, brown, red, white or grey. Regarding the logo, nine franchisees display the franchisor's logo on their websites, but the three remaining franchisees do not display it.

This lack of uniformity in terms of use of the logo and graphic chart raises an issue related to the identity of the franchise network. Internet users may be lost when visiting several franchisee websites associated with the brand, and by extension when visiting the physical stores.

4.4.3 Website Design

Four cases highlight uniformity-related issues as far as website design is concerned. The first case is *Akena*. Among the 17 franchisee's websites we studied, seven were built through the "Yellow pages" service, the ten remaining are traditional websites but display very heterogeneous designs. Three franchisee's websites appear modern and professionally-built. Four websites appear to be either house-made or professionally-built but already several years old with some signs of obsolescence. And three franchisee's websites display a very obsolete and unattractive arrangement and set of colors.

In the second case, *Intercaves*, whereas the franchisor's website seems to have been "in-house" created, due to a clear lack of professionalism in terms of design, ergonomics, etc. and the frequent unavailable pages, some of the franchisees have recruited web experts to design their websites. On the fourteen franchisees' websites, we noticed that at least six web experts have worked on an *Intercaves* website. One of these six web experts has designed the websites of four different franchisees. This can contribute to maintaining brand uniformity in terms of website design. However, some franchisees created a blog instead of a website, not in accordance with the other websites, and two franchisees created their websites via the "Yellow Pages" services.

Concerning the third case, *Camille Albane*, among the ten franchisee's websites under investigation, seven were built using the "Yellow pages" service, and one uses a blogging platform that allows very few ways to personalize the pages. The two remaining websites appear professionally-built and display a modern arrangement and set of colors.

It is important to mention the last case, *Brasseries Flo*. All the franchisees' websites have been built upon the franchisor's initiative, and we can find links to franchisees' websites on the franchisor's websites. Thus, the graphic chart is strictly followed, but to underline franchisees' independence, a different color has been allocated to each franchisee.

In conclusion, these cases highlight significant differences across websites in terms of website creation, design and updating that have a negative impact on brand image.

5 Discussion

5.1 Research Contributions

This paper contributes to the literature on franchising providing a different perspective—the franchisee one—than those usually adopted in the previous literature, i.e., the franchisor one. We followed the recommendation of Dant (2008) who insisted on the need for further investigation of franchising issues from a franchisee perspective. More specifically, we analyzed franchisees' websites along with their respective franchisor's websites, as well as the consequences of the content of these franchisees' websites on network uniformity and franchisor's brand image. We thus observed that franchisees' practices have some impact on the whole network.

Secondly, we built on the literature on the use of the Internet in the franchising sector in several ways. Contrary to previous literature dedicated to the E-commerce strategy of franchisors (Cedrola and Memmo 2009; Dixon and Quinn 2004; Kaufmann et al. 2010; Perrigot and Pénard (forthcoming); Rao and Frazer 2006), we focused on franchisees' presence on the Internet. Moreover, regarding the extent of their presence, we found that only 8.07 % of the franchisors listed in the 2011 franchising directory, published by the *French Franchise Federation*, have franchisees managing their own website. This figure is low and suggests that Internet-related challenges remain specific to some individual franchisors for the moment and has not yet become a generalized issue. Nevertheless, due to the continuous growth of the Internet and the European legal environment, this figure should increase in the future, and franchisors have to anticipate this trend.

Finally, this paper draws on the literature on franchising, uniformity and brand image. More specifically, we focused on the uniformity issue in the Internet arena, contrary to previous researchers who analyzed uniformity across the physical stores

(Kaufmann and Eroglu 1998). We pointed out that the issues are the same, whether in physical stores or the Internet, and that all the dimensions have now to be considered. Franchisors do not only have to focus on the compliance of franchisees to respect the uniformity basics in their stores, but uniformity has to be also considered between physical stores and Internet websites in order to maintain and strengthen their brand image. The issue then becomes more complex.

5.2 *Managerial Implications*

In this paper, we highlighted the consequences of franchisees who set up and run their websites on network uniformity and brand image. In fact, franchisors have two main ways to face this challenge and maintain network uniformity. Both are linked to know-how which is a core element of franchising; one deals with technical know-how, the other concerns organizational know-how.

5.2.1 Technical Know-How

The first way is to consider Internet use and activities as part of technical know-how that the franchisor transfers to its franchisees. According to the European Franchise Federation, "[technical] know-how means a body of non-patented practical information, resulting from experience and testing by the Franchisor, which is secret [ ], substantial [ ] and identified [ ]". The definition of know-how provided by the Commission Regulation (EU) (No 330/2010) is almost similar: "'know-how' means a package of non-patented practical information, resulting from experience and testing by the supplier, which is secret [], substantial and identified." This technical know-how is usually described in an operational manual, also called the "Bible." It provides franchisees with all the standards and rules to be applied in the physical stores. Why not inserting standards and rules regarding online activities in this operations manual? Or why not creating an E-Bible, i.e., an operations manual dedicated to technical know-how related to Internet activities, in particular the rules to be followed when setting up and managing a website (transactional or not)? The E-chapters of the already existing Bible or a separate E-Bible would complement the already existing chapters/Bible as far as online activities of the franchisees are concerned. This could for instance include templates for setting up the website, rules to include mandatory information and categories of products and services, advice for promotions and for price settings, etc. From this perspective, Plave and Almosch (2000) for instance mentioned that "any network will benefit from using a model that allows for easy updating of the information circulated to the public via the Internet (such as seasonal promotions, products changes or franchisee information)." This E-Bible could significantly contribute to keeping the uniformity of the network.

5.2.2 Organizational Know-How

The second way is to consider Internet use and activities as part of an organizational know-how that the franchisor internally masters. It deals with know-how relative to network engineering and management without being necessarily transferred to the franchisees (El Akremi et al. 2009; Perrigot et al. 2011). Such organizational know-how reflects the systemic and cross-disciplinary capacities that allow a franchisor to coordinate on a sustainable basis the generation and use of its strategic assets, along with its professional skills, in pursuit of achieving objectives. Based on a study conducted among 211 franchisors, several categories of organizational know-how have been pointed out (El Akremi et al. 2009). They deal with codification/transmission/replication, as well as support for human resources management, monitoring/oversight of store operations, external communication, internal cohesion/uniformity building, organizational flexibility, purchasing/logistics and access to financing sources. According to some franchising experts, the benefits for a franchisee in setting up and running its own website are greatly diminished if the franchisor has been able to previously develop a website offering an effective communication platform, eventually with an E-commerce functionality, provided the franchisor has successfully integrated its franchisees into a multi-channel strategy. It may be considered that over time, an online activity proves to be a standalone organizational know-how or a component of one of the previously identified organizational know-how, particularly as regards organizational flexibility, external communication and logistics.

In this specific case, the franchisor has to do everything in its power to ensure a mastery of the know-how components tied to the use of the Internet by investing sufficient amounts of financial, technical and human resources. The franchisor's website has to derive its full legitimacy and meet franchisees' expectations in terms of establishing the brand's web presence. Under such a scenario, franchisees will find no great benefit in setting up and managing their own website and facing the associated challenges, whether logistical, financial, etc. They will not ""redesigns the wheel" [by] developing individual sites" (Plave and Almosch 2000). And as said in the trade press, "[t]he most effective way for franchise systems to manage the Web site issue is for the franchisor to maintain one site for the system" (Trice 2001).

5.3 Legal Implications

This research opens the field for further collaborations between experts of different fields. In a legal perspective, franchising is "schizophrenic". On the one hand, franchising management, based on the concept of uniformity, is considered as promoting an economic and social progress: equal level in the quality of products and services, creation of a physical network which implies local employment, improvement of the access to consumption, etc. On the other hand, as franchised stores are independent firms, according to competition Law, they must preserve their own management: price policy, choice of suppliers, products' assortment, etc. The legal significance of these risks presents an additional issue related to franchisee's websites. The findings mentioned in this study highlight the ways franchisors strike to find a balance between the distinct aims in the respect of competition law. This "schizophrenic" character of franchising raises a significant challenge to be overcome.

5.4 Limitations and Tracks for Future Research

This research has some limitations that future research could address. First, the empirical study is based on observations of franchisors' and franchisees' websites. It could still be argued that subjectivity has been introduced into this study via the interpretation of promotional aspects, web designs, etc. In order to minimize research bias, we initiated the observation together by comparing each individual observation to that of the other team members for the purpose of harmonizing our website observation process. Screen captures of all websites were also produced. Moreover, this approach only provides a snapshot in time of the content of franchisors' and franchisees' websites. This content likely evolved over the course of the observation period. A longitudinal approach spanning several months or years would be instructive in an effort to better understand the evolution of the content of franchisors' and franchisees' websites and the associated uniformity-related challenges for the franchisors.

Secondly, our research has been limited to the French market. Though Dant (2008) and Dant et al. (2008) mentioned the importance of studying franchising issues in other markets than the US, in pointing out the current predominant mono-cultural view towards franchising research, the exploration of uniformity-related issues raised by the franchisees' presence on the Internet with different legal context could be of interest. Some researchers have begun adopting multi-country perspectives in their studies of franchising issues (Dant et al. 2008; Dos Santos Silva and de Azevedo 2007; Dunning et al. 2007; Perrigot et al. (forthcoming). For instance, such comparisons have highlighted significant differences in the strategies developed by US and French franchisors. It would thus be interesting to examine their respective strategies in terms of presence on the Internet.

Thirdly, in addition to the issues related to network uniformity and brand image, it would be of interest to focus on internal conflicts (franchisor/franchisee and also franchisee/franchisee) that are expected to increase with the growing presence of franchisors and franchisees on Internet. For instance, franchisees, who often benefit from exclusive territories, might consider the creation of a transactional website by the franchisor or by franchisees of the network as unfair competition, in extending the key concept of encroachment (Emerson 2010; Kalnins 2004).

Fourthly, this paper is managerially-oriented even if some insights from law have been introduced in the manuscript. The concept of uniformity will deserve a detailed attention from a law perspective in future research. First, uniformity may be perceived as being the main element of protection of intellectual and industrial properties. Secondly, uniformity may be perceived as being in contradiction to the free competition principle. Thirdly, uniformity may be perceived as being in contradiction with international law.

6 Conclusion

This paper is a first attempt to highlight the challenges for franchisors facing the existence of various websites set up and run by their franchisees. "Brand image and identity is the responsibility of both franchisee and franchisor. It's the part of the partnership that separates your business from the competition, keeps customers coming back, and encourages growth and expansion. By each taking a role in managing the brand, franchisors and franchisees can synergistically maximize its power, and its profit" (Oseland 1995). So, the use of the Internet by franchisors and franchisees has to be carefully examined. More research, from both business and law fields, is needed in the future. Other practices leading to the same kinds of issues in terms of the use of the Internet and network uniformity could be explored in further research. These practices deal with the presence of brands on social networks such as *Facebook*, *LinkedIn*, *Viadeo*, etc. Recently, the US franchisor, *Applebees*, worked with a media company to propose to its franchisees specific and regular contents to be included on their own *Facebook* page. This move highlights the know-how of the franchisor and its involvement. It is also a way to maintain the uniformity of *Applebees* on this social network.

References

Brookes M, Roper A (2012) Realising plural-form benefits in international hotel chains. Tour Manage 33(3):580–591

Caves RE, Murphy WF (1976) Franchising: firms, markets and intangible assets. South Econ J 42:572–586

Cedrola E, Memmo S (2009) Internet for franchising: current use and areas for improvement – results from an empirical research. J Euromark 18:5–21

Dant RP (2008) A futuristic research agenda for the field of franchising. J Small Bus Manage 46 (1):91–98

Dant RP, Perrigot R, Cliquet G (2008) A cross-cultural comparison of the plural forms in franchise networks: USA, France, and Brazil. J Small Bus Manage 46(2):286–311

Dixon H, Quinn B (2004) Franchising and the Internet: an exploratory study of franchisor Web sites. Internet Res 14(4):311–322

Dos Santos Silva VL, de Azevedo PF (2007) Interfirm arrangements in different institutional environments: McDonald's France and Brazil case study. J Mark Channels 14(3):103–125

Dunning JH, Pak YS, Beldona S (2007) Foreign ownership strategies of UK and US international franchisors: an exploratory application of Dunning's envelope paradigm. Int Bus Rev 16 (5):531–548

El Akremi A, Perrigot R, Herrbach O, Mignonac K (2009) What are the know-how that are sources of a competitive advantage in franchising?, Report for the French Franchise Federation (FFF)
Emerson RW (2010) Franchise encroachment. Am Bus Law J 47(2):191–290
European Franchise Federation (2012) European Franchise Website, www.eff-franchise.com
Findley G (2007) Consistency: the key to branding. Franchising World, April: 30–31
Flosdorf G (2002) When it comes to uniforms, franchisees need a perfect fit. Franchising World, February-March: 49–50
Franson B, DeSmith D (2005) Does the Internet solve all of your franchisees' problems? Franchising World, August: 14–16
French Franchise Federation (2012) Toute la Franchise, les Textes, les Chiffres, les Réseaux, Paris
Gould D (2005) Franchisees must be on board to strengthen brand message. Franchising World, December: 15–18
Guidelines on Vertical Restraints (2010) 2010/C 130/01 http://eur-lex.europa.eu/LexUriServ/LexUriServ.do?uri=OJ:C:2010:130:0001:0046:EN:PDF
Hair JF, Wolfinbarger M, Bush RP, Ortinau DJ (2008) Essentials of marketing research. McGraw-Hill/Irwin, New York
Kalnins A (2004) An empirical analysis of territorial encroachment within franchised and company-owned branded chains. Mark Sci 23(4):476–489
Kaufmann PJ, Cliquet G, Achabal DD (2010) Implementing multi-channel strategies: issues and opportunities for franchise system websites. Proceedings of the 24th ISoF Conference, Sydney
Kaufmann PJ, Eroglu S (1998) Standardization and adaptation in business format franchising. J Bus Venturing 14:69–85
Klein B (1995) The economics of franchise contracts. J Corp Finance 2:9–39
Lafontaine F (1998) Franchising versus corporate ownership: the effect on price dispersion. J Bus Venturing 14(1):17–34
Oseland BJ (1995) Managing brand identity. Franchising World, January/February: 27–28
Perrigot R, Herrbach O (2012) The plural form from the inside: a study of franchisee perception of company-owned outlets within their network. Int J Retail Distrib Manage 40(7):544–563
Perrigot R, Herrbach O, El Akremi A (2011) L'impact de la mixité sur la maîtrise des savoir-faire organisationnels des franchiseurs. Economies et Sociétés 13(6):1007–1024
Perrigot R, Pénard T (forthcoming – 2013) Determinants of E-commerce Strategy in Franchising: A Resource-Based View, International Journal of Electronic Commerce
Perrigot R, López-Fernández B, Eroglu S (forthcoming) Intangible resources and plural form as drivers of franchise internationalization: examination within a two-country perspective. J Small Bus Manage (forthcoming)
Plave LJ, Almosch RB (2000) How to implement dot.com franchising solutions: two aspects of a multi-faceted issue. Franchising World, September/October, 17–19
PricewaterhouseCoopers (2011) Highlights of the 2011 Franchise Business Economic Outlook
Rao S, Frazer L (2006) Australian franchisor websites; moving towards network behaviour. J Theor Appl Electron Commer Res 1(1):56–64
Rogers J, Bennett C, Grewal S (2007) E-commerce for successful franchisors. Franchising World 39(8):23–24
Streed O (2007) The issue of brand uniformity in franchising. Proceedings of the 21st ISoF Conference, Las Vegas
Streed O, Cliquet G (2008) Concept uniformity: control versus freedom in business franchising. In: Hendrikse G, Tuunanen M, Windsperger J, Cliquet G (eds) Strategy and Governance of Networks, Cooperatives, Franchising, and Strategic Alliances Series. Palgrave Macmillan, New York, pp 205–220
Trice M (2001) Managing Internet issues within the franchise system: a legal perspective. Franchising World 33(5):37
Weaven S, Frazer L (2007a) Mature franchise systems use multiple unit franchising to leverage learning economies and sustain system wide growth. Asia Pac J Mark Logist 19(2):107–126
Weaven S, Frazer L (2007b) Expansion through multiple unit franchising: Australian franchisors reveal their motivations. Int Small Bus J 25(2):173–205

Can Service Quality Be Standardized in a Franchise Network? The Case of McDonald's in Egypt

Hanane Elzeiny and Gérard Cliquet

Abstract This study investigates service quality variation among McDonald's fast food franchise chain outlets. Findings from the data collected from 162 customers indicate that McDonald's fast food chain restaurant is able to ensure service quality standardization across its franchised outlets located in Egypt, while it fails to ensure this standardization across franchised units in Egypt on the one hand and franchised units abroad on the other hand. The study concludes that, although standardization is expected from the franchised outlets, some outlets are not able to follow the same standards, especially, when they are functioning in different conditions.

Keywords Egypt • Franchise network • McDonald's • Service quality

1 Introduction

Franchising is an organizational form chosen in order to compete in the retail and service sectors that require highly decentralized operations at a chain with multiple sites (Michael 2000). The essence of franchising is capitalizing on both the economies of scale associated with large systems and the benefits derived from small, localized operations. The franchisor, as the creator, builder, and guardian of a unique business format, is responsible for efficiently managing a complex system of independent business owners (Kaufmann and Eroglu 1998). The very strength of a

H. Elzeiny (✉)
Faculty of Commerce, University of Alexandria, Alexandria, Egypt
e-mail: htzeiny@hotmail.com

G. Cliquet
CREM UMR 6211, IGR-IAE, University of Rennes 1, 11, Jean Macé Street, CS 70803, 35708 Rennes Cedex 7, France
e-mail: Gerard.cliquet@univ-rennes1.fr

T. Ehrmann et al. (eds.), *Network Governance*, Contributions to Management Science,
DOI 10.1007/978-3-7908-2867-2_14, © Springer-Verlag Berlin Heidelberg 2013

franchise system resides in its ability to standardize its operations across heterogeneous locations. The success of some American concepts abroad was partly due to their ability to export American culture and introduce cultural change. The diffusion of U.S-based franchises abroad is often related to the foreign society desire to assimilate the American way of life (Grünhagen et al. 2010). However, franchising business is expected to provide standardization in terms of service quality.

Most of the previous research on the franchising business focused on the motivation, relationship between franchisor and franchisee (Grace et al. 2009) or standardization in franchising system from the operational point of view (Kaufmann and Eroglu 1998). But a very limited number of studies have examined the issue of service quality variation within franchise networks from the perspective of the customers (Streed and Cliquet 2008). Customers' perceptions play a significant role in determining the level of quality for the service provided. Customers' perceptions are affected by several factors that the service provider has no control over such as: culture, educational level, sex and experience. One should not neglect the dominant role of these perceptions when evaluating service quality (Parasuraman 1998).

Although standardization is expected from franchisee outlets, some outlets are not able to follow the same standards for products or services in the franchised system (Baucus et al. 1996; Abd Rahman and Si 2011). This is possibly due to the existing variations in the different environments where the system is operating (e.g. cultural, social, economic, etc.) that affect the customers' expectations. These variations may also be a result of the desire to adapt the product or service to meet these expectations. The concepts of uniformity and standardization in franchise system consequently are therefore challenged. This question is tackled in an emerging market—such as Egypt—where stakes are of great interest for the development of these countries.

Franchising in Egypt began in 1973 with President Anwar Elsadat's "open door" policy. This followed a period of strict social and economic control during which the market had been stagnated and distanced from the rest of the world. Wimpy was the first brand to take advantage of the new openness, by signing an Egyptian franchise contract, and opening its first restaurant in Cairo. The franchise was so successful that, initially, it couldn't cope with the endless lines of customers. Hamburgers quickly became fashionable, and Wimpy expanded throughout Egypt. Following the triumph of the Wimpy experiment, the word spread and franchising began to take off with other international chains rushing to follow this example.

Focusing on fast food franchise systems in Egypt, this study investigates service quality variation—from the customers' view point—among the McDonald's chain located in Egypt, on the one hand, and the variation among these outlets located in Egypt and others located abroad, on the other hand. In this study, the following research questions will be answered:

1. Does the customer perceive variation in terms of service quality among fast food franchise outlets in Egypt?
2. Does the customer perceive variation in terms of service quality between fast food franchise outlets in Egypt and fast food franchise outlets abroad?

3. If variation in terms of service quality between fast food franchise outlets in Egypt and fast food franchise outlets abroad exists, is this variation appreciated from the customers' view point?

The attempt to answer these questions will enrich the franchising literature regarding standardization and adaptation, on one hand, and the service management literature regarding the crucial role of the customers in the service context, on the other hand. The second section of this paper deals with standardization and service quality problems. In the third section, the methodology and the research context are discussed, before presenting the results. Finally, research limitations and future perspectives are developed.

2 Standardization and Service Quality

2.1 *Drivers of Standardization in Business Format Franchising*

Franchising refers to a form of "business cloning" (Hoffman and Preble 1993). It is a business format where franchisors seek to have franchisees to replicate their business in local community based on the entire business concept, including product or service, trade name and methods of operation. This kind of franchise system is usually found in many service sectors and among specifically fast-food service restaurants such as KFC, Pizza Hut and McDonald's (Bradach 1998).

One of the primary motives for standardizing across markets is the desire to reduce cost (Samiee and Roth 1992). These cost savings result from economies of scale due to purchasing (Douglas and Wind 1987), marketing (Levitt 1983), research and development (Buzzel 1968), as well as savings due to easier implementation and management of programs (Samiee and Roth 1992). As international expansion, cost minimization is also an important rationale for standardization across domestic markets (Kaufmann and Eroglu 1998).

But cost reduction is not the only objective of a standardization policy. In franchise networks, it contributes to the concept uniformity which is one of the four main challenges a chain should meet according to Bradach (1997) (see also Cliquet and Pénard 2012): (1) adding new units; (2) maintaining the uniformity of the concept; (3) responding local; (4) adapting globally the system. The key point is that maintaining the uniformity of the concept helps to diffuse and protect the image and hence the brand which is the essential asset of a franchisor. Actually, concept uniformity reinforces the brand image (Kaufmann and Eroglu 1998) which is more important in service networks because maintaining the quality of services in every unit (store, restaurant, hotel...) of the chain wherever it is located is a daily challenge (Caves and Murphy 1976). Franchisors strive to implement several methods to maintain the uniformity of their concept.

Hence, the second Bradach's challenge (1997) concerning uniformity poses the key question: adaptation vs. standardization (Kaufmann and Eroglu 1998) in franchise chains. This dilemma is very closely related to the following international marketing issue: localization vs. globalization. Globalization was defined by Levitt (1983) whereas other researchers questioned this approach by talking about the myth of globalization (Douglas and Wind 1987) and even favouring localization (Rigby and Vishwanath 2006). Douglas and Wind (1987) are not opponents to the globalization process: they just assert that it is not possible to really globalize everything and, depending on local conditions, some elements of the marketing can be globalized to some extent.

Maintaining the uniformity of the concept cannot be realized through central control. Standardization can only be implemented by persuading franchisees and persuasion appears to be the key word in managing a franchise network (Cliquet 2000). In any case, some legal considerations, about for instance pricing, and hence promotions, prevent franchisors to impose their power because prices cannot be officially standardized without breaking anti-trust laws (Lafontaine 1999). It seems that plural form networks have a better chance to meet these Bradach's challenges (1998) and hence to survive (Botti et al. 2009; Perrigot 2008; Perrigot et al. 2009).

Kaufmann and Eroglu (1998) ask the question to which degree standardization or adaptation should be realized on a continuum (if such a continuum between "wholly standardized" and "wholly adapted" could really exist). They make a distinction between core elements of the concept (product/service deliverables, benefit communicators and system identifiers) which should stay invariant throughout the chain and peripheral elements of the concept which may be adapted to local specificities. The problem is then to adapt the concept not only on a spatial basis but also on a temporal basis because every chain should adapt its concept along time. This is the fourth and last Bradach's challenge (1998): system wide adaptation. Once this adaptation has been decided, the updated concept should then be implemented in every unit and we are back to a problem of dissemination. This updating process is not so easy to develop and once again plural form networks appear to be in a better position to succeed in such a project (Bradach 1998; Cliquet 2000). Sorensen and Sorensen (2001) oppose in franchise chains the exploitation process to the exploration process. The exploitation is implemented to develop organizational routines and company-owned units are more appropriate for that purpose whereas the exploration process which can lead to new ideas for the concept is better managed by franchisees. Franchisees are in a better position to both envisage new ideas and implement these new ideas adapted to their local context (Wang and Altinay 2008). This dual organization is then of a great help in developing franchise networks.

Adaptation and standardization will remain very pregnant management concepts in franchising future. One of the most predictable developments for franchising concern emerging markets and more specifically Based-Of-the-Pyramid (BOP) markets (Kistruck et al. 2011). Adaptation is then a key problem for international companies because the environment is most of the time far from being what it is in the country where the concept has been perfected and the chain was first expanded (Jensen 2007; Matusitz J 2010).

One important area where the standardization of the format is effective in reducing cost relates to monitoring. A central concern of the operations function in franchise systems is quality control. At the most basic level, this refers to managing the overall system for optimal performance. Management of the overall system, however, implies the ability of a franchisor to identify poor performance on the part of individual franchisee. Standardization not only makes comparison of executioner ability possible, but also at the minimum cost possible (Kaufmann and Eroglu 1998). Standardization reduces the cost associated with quality control monitoring by giving the franchisor the ability to quantify an otherwise subjective property, such as quality. Hence, it provides the operations division with the ability to efficiently and objectively monitor the performance of franchisees. Cost savings through standardization are also realized in the area of standardized inputs, an issue related to both product/service deliverables and format facilitators. Franchisees gain competitive advantage over similar independent businesses via access to needed inputs at low cost.

Standardization also permits image continuity and stability across markets (Jain 1989). A franchise system's image represents the total expected reinforcement that a consumer associates with patronizing any of its outlets (Kunkel and Berry 1968). As such, all format components are, either directly or indirectly, instrumental in creating and maintaining the desired image. Franchisors strive continuously to create and maintain an image of their concept that is both desirable and uniform.

2.2 Drivers of Standardization in Business Format Franchising

Several authors have discussed the unique importance of quality of service firms (Norman 1984; Shaw 1978) and have demonstrated its positive relationship with profits, increased market share, return on investment, customer satisfaction, and future purchase intentions (Anderson et al. 1994; Boulding et al. 1993; Rust and Oliver 1994). "Service quality" is now a very important topic in marketing. A definition of service quality is given by Lewis and Booms (1983: p. 26):

> Service quality is a measure of how well the service level delivered matches customer expectations. Delivering quality service means conforming to customer expectations on a consistent basis.

Parasuraman (1998) prefers to define service quality as "... a global judgment or attitude relating to the overall excellence or superiority of the service", introducing then the notion of optimization ("excellence") and competition ("superiority") applying then the Oliver's conception (1980, 1993) and his disconfirmation model. The optimization is based on a comparison between service quality perception by the customer and customer's expectations: service quality is then measured by the gap between these two above notions.

Several streams of research have emerged according to their interpretation of the meaning of service quality. The problem stands in the definition of customer satisfaction. Following Cardozo's initial definition (1965), various definitions have been proposed (Hempel 1977; Churchill and Surprenant 1982). Parasuraman et al.

(1988) posited and operationalized service quality as a difference between customer expectations of "what they want" and their perceptions of "what they get". Based on their conceptualization and operationalization, they proposed the service quality measurement scale ServQual and they identified a set of 22 variables/items tapping five different dimensions of service quality construct, namely: tangibles, reliability, responsiveness, assurance, and empathy.

However a consensus seemed to appear with the disconfirmation approach of service quality (Parasuraman et al. 1988) which led to the elaboration of the ServQual scale. But Carman (1990) criticized it reproaching its lack of stability entailing researchers to adapt it according to the service category. One serious problem with the ServQual scale is that it entails a gigantic data collection task. When employing a lengthy questionnaire, one is required to collect data about consumers' expectations as well as perceptions of a firm's performance on each of the 22 service quality scale attributes. Other objections against the scale relate to the predictive power of the instrument and validity of the five-dimension structure (Babakus and Boller 1992; Cronin and Taylor 1992; Dabholkar et al. 2000; Teas 1993).

It is because of these criticisms, that some researchers stressed the need for developing a methodologically more precise scale (Babacus and Boller 1992; Bolton and Drew 1991; Brown et al. 1993; Carman 1990). The ServPerf scale (Cronin and Taylor 1992) is one of the important variants of ServQual. Since it is based on the perception component alone, it has been conceptually and methodologically posited as a better scale than the ServQual scale. Methodologically, the ServPerf scale represents marked improvement over the ServQual scale (Cronin and Taylor 1994). Not only is the scale more efficient in reducing the number of items to be measured by 50 %, it has also been empirically found superior to the ServQual scale for being able to explain greater variance in the overall service quality measured through the use of single item scale (Jain and Gupta 2004).

Service quality in food-industry has been recognized as individualized, intangible and subjective in nature (Johns and Howards 1998). In order to retain customers and to attain survival and growth in an increasingly competitive environment, many marketers realize that they must ensure a high quality of service that go beyond customers' expectations (Dabholkar et al. 2000).

Service marketing differs fundamentally from goods marketing in terms of intangibility. This feature makes it difficult to determine how consumers perceive service quality (Brogowicz et al. 1990; Behara and Gundersen 2001). Other service marketing characteristics—heterogeneity, perishability and the inseparability of production and consumption—further compound the issues of defining and measuring service quality. Beyond these distinguishing characteristics are differences between the measurement of service quality and tangible product quality. Service quality refers specifically to subjective, perceived quality, with the purpose of reflecting customer attitudes toward services and capturing customer perceptions regarding the service provider's excellence and superiority (Parasuraman et al. 1985, 1988). Parasuraman et al. (1988) identified five dimensions of service quality: tangibles, reliability, responsiveness, assurance, and empathy.

2.3 *Standardization in Business Format Franchising and Service Quality*

Customers are supposed to have high expectations on standardized product and service at every location with such uniform image and identity (Falbe and Dandridge 1992). Therefore, in order to attract and retain customers, system standardization and consistency are very important. Michael (2000) defines quality in fast food chain not just as product quality itself, but also the franchisor's operating instructions in order to convey a standardized product. Bradach (1997) suggests that under the chain builder strategy, franchisee of the franchised chain try to create and maintain a superior brand reputation by providing and delivering the same high quality of service in all locations. Based on these arguments, the following hypotheses are proposed:

Hypothesis 1 (H1) Fast food chain restaurant under the same franchise system will have similar level of service quality in the same country.

Hypothesis 2 (H2) Fast food chain restaurant under the same franchise system will have similar level of service quality across countries.

3 Empirical Analysis

3.1 *Data and Methodology*

In this study, McDonald's Egypt is selected in the investigation of service quality variation. A questionnaire survey was given to 259 Egyptian students, all English speaking, from the faculty of commerce, Alexandria University to examine the service quality provided by McDonald's fast food chain restaurant in Egypt and abroad. A convenience sampling which is also known as non-probabilistic sample was used in selecting the respondents. Those students were able to understand and answer the English questionnaire without translation. This way, the translation and back translation procedures were avoided (Brislin 1970). The students' age varied from 18 to 21 years and 78 % of them had experienced McDonald's out of Egypt. Respondents were asked to recall their dining experience from three separate McDonald's, the first two are located in Egypt and the third is located abroad.

The questionnaire solicits information on their perception of the quality of the service provided by McDonald's. The study uses a 7-point Likert scale to solicit the respondent's degree of agreement or disagreement with each of the statements in the questionnaire.

Consensus generally exists that service quality is a distinct construct, but there are discrepancies regarding service quality measurement. The ServQual instrument proposed by Parasuraman et al. (1988) focuses on computed disconfirmation, in which the difference between customer expectation and the actual performance (both measured after the service is performed) is calculated. The alternative approach, ServPerf, is concerned only with customer perceptions of service performance (Cronin and Taylor 1992; Gronroos 1990). The increasing body of research in this area overwhelmingly supports the "perception only" approach to measure service quality (Gronroos 1990; Zeithaml et al. 1996). Accordingly the "perception only" or ServPerf approach is adopted in this study.

At the end of the data collection period, a total of 161 questionnaires for two McDonald's located in Egypt, and 143 questionnaires for McDonald's abroad, were used for data analysis.

3.2 *Franchising in Egypt*

Franchising is one of the most important tools to develop a country and generate decent jobs to face the growing increase in population. Egypt's population currently stands at nearly 80 million people with more than 50 % under the age of 20, and Egypt is expected to reach 100 million in 8 years (CAPMAS 2009). Franchise has dramatically increased over the last 10 years from approximately 25–310 franchise systems. It directly employs more than 45,000 employees and generates more than 9 billion EGP of annual sales. More than 500,000 jobs were created through franchise supply chains. Franchise direct investments exceeded 40 billion EGP. In addition, a range of feeding industries have developed to service the sector. Retail sector represents 48.9 % of the total systems conducted as it covers: clothing and fashion, supermarkets, home products, etc. Other categories represent 51.1 %, distributed among 20 categories (including 23 % for food outlets) Egyptian Franchise Development Association (EFDA 2009).

Since 2000, franchising in Egypt has been supported by a growing number of organizations and initiatives. In 2001 the Egyptian Franchise Development Association (EFDA) was established. Two years later, EFDA was accepted as a full member of the World Franchise Council (WFC). Egypt is the second country in Africa—following South Africa—and the first in the Middle East to qualify. A law for intellectual Property rights No.82 of 2002 was passed on June 2002. This law applied the rules of the Trade Related Aspects of Intellectual Property Rights Agreement (TRIPS Agreement) and makes protection of intellectual profits under a franchise agreement more secure. Being a member of the World Intellectual Property Organization (WIPO), Egypt is a signatory to a number of major international agreements such as Madrid international convention protecting trade and industrial marks.

In 2004, McDonald's opened its first two restaurants simultaneously in Cairo. Today, over 40,000 customers are served under the Golden Arches at over 59 outlets throughout the country every day. Standardization for McDonald's is a key ingredient for success. This chain seeks to serve its customers with the same quality product and experience, whether that restaurant is located in Moscow, Idaho, or Moscow, Russia (Manrodt and Vitasek 2004).

4 Results and Discussion

4.1 Data Analysis and Findings

The consistent reliability of items for each variable in this study is measured by the value of Cronbach's alpha coefficient. The items used in each variable are reliable as the coefficient value is greater than 0.6 (between 0.69 and 0.82) as recommended by Malhotra (2010). Paired sample *t*-test is used in this research to test all hypotheses. The perception of the respondents on the service quality of McDonald's restaurants was measured on three substances: two different McDonald's restaurants located in Egypt, and one McDonald's located abroad. Tables 1, 2 and 3 summarize the results of the analysis.

Pair sample test for service quality in Table 1 showed that there is no significant difference between the service quality delivered by the two McDonald's located in Egypt ($t = 2.557, p = 0.011$). Respondents see that McDonald's delivers a somehow standardized quality in Egypt. No significant difference can be perceived in the core service delivered by McDonald's across the two McDonald's restaurants while, pair sample test in Tables 2 and 3 showed that there is a significant difference between the service quality delivered by McDonald's Egypt and McDonald's located outside Egypt ($t = 10.738, p = 0.000$) ($t = 5.826, p = 0.000$). Significant difference can be perceived in the core service delivered by McDonald's located inside and outside Egypt.

Thus this study confirms H1, which states that fast food chain restaurant under the same franchise system will have similar level of service quality in the same country, and fails to accept H2, which states that fast food chain restaurant under the same franchise system, will have similar level of service quality across countries.

4.2 Discussion

The findings of the study imply that although operating under franchising system, McDonald's fail to demonstrate the same level of standard in terms of service quality across the different countries. The concepts of uniformity and standardization of the franchising concept across different countries is shown to be less relevant in the context of Middle Eastern countries (Egypt). The Egyptian consumers perceive that a fast-food franchise network offers a standardized service quality when comparing different outlets of the same chain located in Egypt, but a different service quality compared to outlets of the same chain located outside Egypt.

Table 1 Paired samples statistics for service quality (inside Egypt)

Variable	Mean	Std. deviation	*t*	df	Sig. (2-tailed)
Service quality (restaurant 1)	5.8261	1.6108	2.557	160	0.011
Service quality (restaurant 2)	5.4783	1.1185			

Table 2 Paired samples statistics for service quality (inside and outside Egypt)

Variable	Mean	Std. deviation	*t*	df	Sig. (2-tailed)
Service quality (restaurant 1)	5.4543	1.1304	10.738	142	0.000
Service quality (restaurant 3)	6.6154	0.6271			

Table 3 Paired samples statistics for service quality (inside and outside Egypt)

Variable	Mean	Std. deviation	*t*	df	Sig. (2-tailed)
Service quality (restaurant 2)	5.7762	1.6420	5.826	142	0.000
Service quality (restaurant 3)	6.6154	0.6271			

This study provides empirical evidence that non standardized service quality is not always appreciated. Eckhardt and Mahi (2004) found only a rejection reaction to McDonald's fast-food menu in the specific context of India. In contrast, Egypt, as an Islamic country, has no prohibitions against meat consumption other than pork (Grünhagen et al. 2010). It is interesting to note that the average score for McDonald's Egypt service quality is lower than the average score for McDonald's abroad service quality. Respondents perceived foreign outlets providing better service quality than Egyptian ones. This could be due to variations in terms of management team, experience, and attitude of employees (Abd Rahman and Si 2011), or due to adaptations' effort. This result is supported by Cox and Mason (2007) and Abd Rahman and Si (2011) who found that there is no standardization among fast food restaurants. Franchisees may find themselves deviating from the standard franchise format due to cultural and geographical differentiated nature of markets and resource availability. Kaufmann and Eroglu (1998) and Sorensen and Sorensen (2001) claimed that there is deviation from standardization in franchise chain due to local adaptation. Local adaptation may not always be appreciated by the customers. In this paper, it was found that the respondents perceive the level of service quality offered by McDonald's Egypt as lower than the service quality offered by McDonald's abroad. Findings on variation in terms of service quality between fast food chain restaurants under the franchise system can provide guidelines to franchisees and franchisors in understanding the customer perception towards service quality. Franchisors should take initiative actions to ensure standardization across all outlets if customer satisfaction is to be enhanced.

Stringent control and monitoring across all chain outlets should be constantly emphasized in order to secure customer's trust and loyalty to the brand.

The findings suggest several marketing strategy implications for Egypt and other Arabic countries. Western-style fast-food restaurants appear as a service that is increasingly compatible with young population (Grünhagen et al. 2010). Almost 95 % of the respondents have experienced McDonald's services. The rapidly expanding youth segment of the Egyptian population is an opportunity for fast-food restaurant franchisors. In addition, this study stresses the importance of standardization and uniformity of the franchised outlets across different countries. Young Egyptians want to experience and assimilate American way of life by buying a standard offering.

4.3 Limitations and Future Research

Though convenience sample is easy to access, it suffers from a number of biases. It can lead to the under-representation or over-representation of a particular group. The sample is also unlikely to be representative of the population being studied. These biases undermine our ability to make generalizations from the sample to the population we are studying. In this study, students from the Faculty of Commerce, Alexandria University who responded to answering the questionnaire, were relied on. Also, college students by definition are quite educated relative to a 28 % estimated illiteracy rate in all Egypt (World Bank statistics 2010). Thus, perceptions among those students may deviate from the remainder of the Egyptian nation (Grünhagen et al. 2010). In future studies on this topic, research needs to include other age and educational categories of the respondents.

McDonald's as an example of a well-known fast-food franchised chain, was shown here to offer a non standardized service quality across its outlets. An important question remains and this is whether other service franchised networks offer a standardized service quality or not. Differences may exist in order to adapt the offering to the local environment. Is this adaptation always appreciated from the customers' point of view, or should one stick to the standards? Finally, emerging countries in general and Arabic Middle Eastern countries in particular are understudied. The franchise context seems to be promising for investigations into such cultures.

References

Abd Rahman A, Si P (2011) Measuring food and service quality variation among food franchise chain outlet in Malaysia. In: 2nd international research symposium in service management, Yogyakarta

Aboul Fath H (2008) Franchising in Egypt in 2008: country report for the World Franchise Council. Egyptian Franchise Development Association (EFDA). http://www.docstoc.com/docs/23831028/Egyptian-Franchise-Development-Association-FRANCHISING-IN-EGYPT-IN

Anderson E, Fornell C, Lehmann D (1994) Customer satisfaction, market share and profitability: findings from Sweden. J Mark 58:53–66

Babakus E, Boller G (1992) An empirical assessment of the Servqual scale. J Bus Res 24:253–268

Baucus DA, Baucus MS, Human SE (1996) Consensus in franchise organizations: a cooperative arrangement among entrepreneurs. J Bus Venturing 11:359–378

Behara RS, Gundersen DE (2001) Analysis of quality management practices in services. Int J Qual Reliability Manage 18:584–604

Bolton R, Drew J (1991) A multistage model of customer's assessment of service quality and value. J Consum Res 17:375–385

Botti L, Briec W, Cliquet G (2009) Plural forms versus franchise and company-owned systems: a DEA approach of hotel chain performance. Omega 37:566–578

Boulding W, Karla A, Staelin R, Zeithaml V (1993) A dynamic process model of service quality: from expectations to behavioral intentions. J Mark Res 30:7–27

Bradach JL (1997) Using plural form in the management of restaurant chains. Administrative Sci Q 42:276–303

Bradach JL (1998) Franchise organizations. Harvard Business School Press, Boston

Brislin R (1970) Back translation for cross-cultural research. J Cross Cultur Psychol 1:185–216

Brogowicz A, Delene L, Lyth D (1990) A synthesized service quality model with managerial implications. J Serv Manage 1:27–45

Brown T, Churchill G, Peter J (1993) Improving the measurement of service quality. J Retailing 69:127–139

Buzzel R (1968) Can you standardize multinational marketing? Harv Bus Rev 46:102–113

CAPMAS (Central Agency for Public Mobilization and Statistics) (2009) Annual report, The Egyptian ministry council

Cardozo RN (1965) An experimental study of customer effort, expectation, and satisfaction. J Mark Res 24:244–249

Carman JM (1990) Consumer perceptions of service quality: an assessment of the ServQual dimensions? J Retailing 66:33–55

Caves R, Murphy W (1976) Franchising: firms, markets, and intangible assets. South Econ J 42 (4):572–586

Churchill GA Jr, Surprenant C (1982) An investigation into the determinants of customer satisfaction. J Mark Res 19:491–504

Cliquet G (2000) Plural forms in store networks: a proposition of a model for store network evolution. Int Rev Retail Distribution Consum Res 10:369–387

Cliquet G, Pénard T (2012) Plural form franchise networks: a test of Bradach's model. J Retailing Consumer Services 19(1):159–167

Cox J, Mason C (2007) Standardization versus adaptation: geographical pressures to deviate from franchise format. Serv Ind J 27:1053–1072

Cronin JJ, Taylor SA (1992) Measuring service quality: a reexamination and extension. J Mark 56(3):55–68

Cronin JJ Jr, Taylor SA (1994) ServPerf versus ServQual: reconciling performance-based and perceptions-minus-expectations measurement of service quality. J Mark 58:125–131

Dabholkar PA, Shephered DC, Thorpe DI (2000) A comprehensive framework for service quality: an investigation of critical conceptual and measurement issues through a longitudinal study. J Retailing 76:139–173

Douglas S, Wind Y (1987) The myth of globalization. Columbia J World Bus 22:19–29

Eckhardt G, Mahi H (2004) The role of consumer agency in the globalization process in emerging markets. J Macromarketing 24:136–146

Egyptian Franchise Development Association (EFDA) (2009) Franchising in Egypt in 2008: Country Report for the World Franchise Council, http://www.docstoc.com/docs/23831028/Egyptian-Franchise-Development-Association-FRANCHISING-IN-EGYPT-IN
Falbe CM, Dandridge TC (1992) Franchising as a strategic partnership: issues of cooperation and conflict in a global market. Int Small Bus J 10:40–52
Grace D, Weaven S, Zhai S (2009) An exploration of service quality in franchising relationship. In: ANZMAC conference proceedings, Melbourne, Australia http://www.duplication.net.au/ANZMAC09/papers/ANZMAC2009-065.pdf
Gronroos C (1990) Service management and marketing. Lexington Book, Lexington
Grünhagen M, Witte C, Pryor S (2010) Effect of US-based franchising in the developing world: a middle-eastern consumer perspective. J Consum Behav 9:1–17
Hempel DJ (1977) Consumer satisfaction with the home buying process: conceptualization and measurement. In: Hunt HK (ed) The conceptualization of consumer satisfaction and dissatisfaction. Marketing Science Institute, Cambridge
Hoffman RC, Preble JF (1993) Franchising into the twenty-first century. Bus Horiz 36:35–42
Jain S (1989) Standardization of international marketing strategy: some research hypotheses. J Mark 53:70–79
Jain SK, Gupta G (2004) Measuring service quality: SERVQUAL vs. SERVPERF scales. Vikalpa 29:24–37
Jensen R (2007) Replication and adaptation: the effect of adaptation degree and timing on the performance of replicated routines. In: Academy of management annual meeting proceedings, Philadelphia, PA, pp 1–6
Johns N, Howard A (1998) Customer expectations versus perceptions of service performance in the foodservice industry. Int J Contemp Hosp Manag 9(3):248–265
Kaufmann PJ, Eroglu S (1998) Standardization and adaptation in business format franchising. J Bus Venturing 14:69–85
Kistruck GM, Webb JW, Sutter CJ, Ireland RD (2011) Microfranchising in base-of-the-pyramid markets: institutional challenges and adaptations to the franchise model. Entrepren Theory Prac 35:503–531
Kunkel JH, Berry LL (1968) A behavioral conception of retail image. J Acad Mark Sci 20:165–175
Lafontaine F (1999) Franchising versus corporate ownership: the effect on price dispersion. J Bus Venturing 14:17–34
Levitt T (1983) The globalization of markets. Harv Bus Rev 61:92–101
Lewis RC, Booms BH (1983) The marketing aspects of service quality. In: Berry L, Shostack G, Upah G (eds) Emerging perspectives on services marketing. American Marketing, Chicago, pp 99–107
Malhotra N (2010) Marketing research: an applied orientation, 6th edn. Prentice Hall, Upper Saddle River
Manrodt K, Vitasek K (2004) Global process standardization: a case study. J Bus Logistics 25:1–23
Matusitz J (2010) Disneyland Paris: a case analysis demonstrating how glocalization works. J Strateg Mark 18:223–237
Michael SC (2000) The effect of organizational form on quality: the case of franchising. J Econ Behav Organ 43:295–318
Norman R (1984) Service management. Wiley, New York
Oliver RL (1980) A cognitive model of the antecedents and consequences of satisfaction decisions. J Mark Res 17:460–469
Oliver RL (1993) Cognitive, affective, and attribute bases of the satisfaction response. J Consumer Res 20(3):418–430
Parasuraman A (1998) Customer service in business-to-business markets: an agenda for research. J Bus Ind Mark 13:309–321

Parasuraman A, Zeithaml VA, Berry LL (1985) A conceptual model of service quality and its implications for future research. J Mark 49:41–50

Parasuraman A, Zeithaml VA, Berry LL (1988) SERVQUAL: a multiple-item scale for measuring customer perceptions of service quality. J Retailing 64:12–40

Perrigot R (2008) Franchising networks survival: an approach through population ecology and survival analysis. Res Appl Mark 23:21–37

Perrigot R, Cliquet G, Piot-Lepetit I (2009) Plural form chain and efficiency: insights from the French hotel chains and the DEA methodology. Eur Manage J 27:268–280

Rigby DK, Vishwanath V (2006) Localization: the revolution in consumer markets. Harv Bus Rev 84:82–92

Rust R, Oliver R (1994) Service quality – new directions in theory and practice. Sage, New York

Samiee S, Roth K (1992) The influence of global marketing standardization on performance. J Mark 56:1–17

Shaw J (1978) The quality-productivity connection. Van Nostrand, New York

Sorensen O, Sorensen JB (2001) Finding the right mix: franchising, organizational learning and chain performance. Strateg Manage J 22:713–724

Streed O, Cliquet G (2008) Concept uniformity: control versus freedom in business format franchising. In: Hendrikse G, Tuunanen M, Windsperger J, Cliquet G (eds) Strategy and governance of networks. Physica, Heidelberg

Teas K (1993) Expectations, performance evaluation, and customers' perceptions of quality. J Mark 57:18–34

Wang CL, Altinay L (2008) International franchise partner selection and chain performance through the lens of organisational learning. Serv Ind J 28:225–238

World Bank statistics (2010) http://data.worldbank.org/indicator/SE.ADT.LITR.ZS

Zeithaml VA, Berry LL, Parasuraman A (1996) The behavioral consequences of service quality. J Mark 60:31–46

Quasi-Franchising: A New Model for Strategic Business Cooperation

Andrew Terry and Cary Di Lernia

Abstract Franchising's capacity for reinventing itself is a matter of record. Indeed its continual adaptation to accommodate changing circumstances and market conditions is a major factor in its increasing influence throughout the world. The franchising relationship is based on a prescribed business model developed by the franchisor and carried out under the franchisor's guidance and oversight by franchisees who are granted the right to trade under the franchisor's brand and system. The manner in which the franchise model is implemented is nevertheless capable of infinite variation. It is its capacity for adaptation and innovation which drives its relentless development.

This paper suggests a role for a form of franchising which incorporates only back-of-house elements—the tried, tested and proven systems and procedures which are not directly visible to the customer—and eschews brand and other visible manifestations of a standardised "one-size-fits-all" approach to service provision. It proposes a form of quasi-franchising where brand and related front-of-house features are removed or, at least, significantly reduced. The "franchisee" acquires the right, and the obligation, to use the "franchisor's" back-of-house system while retaining flexibility for entrepreneurial endeavour in building an idiosyncratic, eclectic and individualised business.

Keywords Anti-brand movement • Branding in franchising • Microfranchising • Systems and back of house functions in franchising

A. Terry (✉) • C. Di Lernia
The University of Sydney Business School, The University of Sydney, Sydney, Australia
e-mail: andrew.terry@sydney.edu.au; cary.dilernia@sydney.edu.au

T. Ehrmann et al. (eds.), *Network Governance*, Contributions to Management Science,
DOI 10.1007/978-3-7908-2867-2_15,

1 Introduction

Franchising's capacity for reinventing itself is a matter of record. Indeed its continual adaptation to accommodate changing circumstances and market conditions is a major factor in its increasing influence throughout the world. The original model—national franchisor and single unit franchisee—has been through many iterations, from the development of sub-franchising and area development arrangements to multi-unit, multi-concept and combination franchising. The franchising relationship is based on a prescribed business model developed by the franchisor and carried out under the franchisor's guidance and oversight by franchisees who are granted the right to trade under the franchisor's brand and using its system. The manner in which the franchise model is implemented is nevertheless capable of infinite variation. Franchising is not a business in itself but is a method of doing business—an innovative and dynamic method of distributing goods and services. It encompasses a wide variety of different practices that are used in different ways, and with varying degrees of sophistication, in virtually all industry sectors. It is an essentially practical strategy which, in the words of Martin Mendelsohn, 'did not derive from one moment of inventiveness by an imaginative individual [but from] the solutions developed by businessmen in response to the problems with which they were confronted in their business operations' (Mendelsohn 2004). It is franchising's capacity for adaptation and innovation which drives its relentless development.

The appeal of franchising for a franchisee lies, in the words of Australia's recent *Opportunity not Opportunism* report, in 'the potential benefits of being able to conduct the business under an established brand name using tested operational systems' (Parliamentary Joint Committee on Corporations and Financial Services 2008). The essence of franchising is, in the words of Stephen Kos, 'to convey the appearance of a single entity largely indistinguishable from a single owner chain comprising branches at separate locations' (Kos 1990). This paper nevertheless proposes a form of franchising which incorporates, to use the language of the Australian report, only the 'tested operational systems' but not the 'established brand name'. The *brand* represents the visible face of franchising—the name, the image, the "look and feel", the servicescape, the standardised customer experience. The *system* represents the unseen face of franchising—the specifications, the processes, policies and procedures, the technology, the training and support structures which are mostly imperceptible to customers but which are nevertheless indispensible drivers of the franchised business. This paper adopts the terminology 'front-of-house' to describe the brand-related aspects and 'back-of-house' to encompass the systems-related aspects underlying the external manifestations of the brand. This paper suggests a role for franchising in which the brand, and its tangible and intangible elements—the "front-of-house" elements—have greatly reduced significance. It suggests a role for a form of franchising which incorporates only, or primarily, the systems underlying the external manifestations of the chain—the "back-of-house elements". It is suggested that in certain market niches this business strategy may be more attractive to consumers through not being associated with the standardised and formulaic uniformity which is the hallmark of business format franchising. This derivation of franchising indeed presents an opportunity for a fuller expression of entrepreneurship sought by many prospective business proprietors but not offered in traditional business format franchising due to the restrictive controls imposed by the obligations associated with maintaining a consistent brand and image.

Given that every definition of franchising ever promulgated or proposed involves the element of brand as an essential ingredient, the terminology of franchising may be inappropriate to describe this concept of franchising without the brand. Nevertheless, given that the term 'back-of-house-ising' is ridiculously inelegant, the term "back-of-house franchising" will be adopted to identify this form of quasi-franchising.

2 Why Franchising Works: The System and the Brand

A convenient summary of the operation of and reasons for business format franchising is set out in Australia's 1997 *Fair Trading Report*:

> Franchising is an increasingly popular form of economic organisation providing an alternative means of expanding an existing business or an alternative means of entering an industry. Under the system, the franchisor, holding property rights over a marketing system, business service or product (identified by a brand name or trademark) enters a contract or agreement with the franchisee and grants, under certain conditions, the right to produce or distribute the franchisor's product or service.
>
> Substantial benefits exist for both franchisees and franchisors under the system. The franchisor derives income from any initial franchising fee and from access to a continuing cash flow through product sales and from licence fees without having to provide additional capital or to directly manage the franchisee. The franchisor gains from access to established business systems, developed products or services, training and business advice, group advertising and lower risk (House of Representatives Standing Committee on Industry, Science and Technology 1997).

The success of franchising is attributable to a number of factors, particularly the brand under which the business is conducted—the front-of-house elements—and the system underlying the external manifestation of the chain—the back-of-house elements. The debate—if indeed there is one—surrounding the importance of the brand versus the system in determining the respective contribution of each to franchising is a matter for after dinner conjecture. Franchising works because it synergistically combines these two key drivers. Determining the respective contributions of each is a sterile debate. But, in a paper that proposes a form of "franchising" which dispenses with the brand aspect—the standardised, public, external link to the consumer—it is necessary to address the contribution of each.

2.1 The Brand

The contemporary brand manager has a ready response to the suggestion of Shakespeare's Juliet that a 'rose by any other name would smell as sweet' (*Romeo and Juliet*) (Act II Scene ii). Trademarks 'encourage the confidence of investors, consumers and those licensed to use them, and accumulate value in an asset recognised by legal and accounting principles'(Terry and Giugni 2009). Brands are an entrenched feature of contemporary business and have a long heritage (Blackett 2004).[1] Since the beginning of commerce, entrepreneurs have sought to distinguish their products from those of competitors, but with the growth of the modern services sector branding has moved beyond the trademarks which simply identify and distinguish products and services to encompass entire consumption experiences. Sophisticated "servicescapes"—described as 'the environment in which the service is assembled and in which the seller and customer interact, combined with tangible commodities that facilitate performance or communication of the service' (Booms and Bitner 1981)—create the environment in which service is delivered. Corporations undertake branding exercises to 'differentiate the company [and] to project the unique personality of the corporation' (Diefenbach 1987). Today the brand 'is the most important and sustainable asset of any organisation [and] should be the central organising principle behind every decision and every action' (Clifton and Simmons 2003). Many companies are primarily engaged in the production and maintenance of brands as opposed to products (Willigan 1992).[2]

This reversal in the prominence of the roles of products and their signifiers has not happened overnight. How and why has branding become about more than the product itself? Scott Bedbury, then Starbucks' Vice President of marketing, stated that 'consumers don't truly believe there's a huge difference between products', which is why brands must 'establish emotional ties with their customers' (Haig 2003). This is a realistic assessment in an age of standardised mass production which necessitates the establishment of other than solely product related points of difference (Klein 2002). The establishment of points of difference from the myriad other business ideas, products and services in the marketplace is what franchising has seized upon, somewhat ironically attempting to replicate an original and unique success story in the search for greater growth and, with it, greater profits.

Branding takes on particular significance in a franchise context. Tangible and intangible elements of the consumption experience—the servicescape—contribute to customers' identification with a franchised brand. The significance of brands in society generally, and franchising in particular, is undoubted, immense and unchallengeable. That branding and franchising are inextricably intertwined is illustrated

---

[1] Branding is a by-product of the requirement of the mid-thirteenth century Assize of Bread and Ale requiring medieval traders to distinguish their goods by marks to enable the identification of the manufacturers of adulterated goods. The unintended consequence was the promotion of branding. Customers began to select particular bakers and brewers whose product they enjoyed on the basis of their direct experience of the distinguishing mark.

[2] One of the most successful brand merchants in the world, Nike founder Phil Knight, has expressed Nike's philosophy in these terms: "For years we thought of ourselves as a production-oriented company meaning we put all our emphasis on designing and manufacturing the product. But now we understand that the most important thing we do is market the product. We've come around to saying that Nike is a marketing-oriented company, and the product is our most important marketing tool" (Willigan 1992).

by the Interbrand trademark league tables in which franchise brands are prominent (www.interbrand.com). The brand drives franchising, and franchising drives the brand. Although there is no universal definition of franchising, every definition promulgated or proposed anywhere in the world includes the brand as an essential element (Blair and Lafontaine 2005).

2.2 The System

Despite the significance of branding, the single most important development in franchising was the realisation that entire business, operational and management systems could be cloned. The success of franchising is attributable to a number of factors, in particular the brand under which the business is conducted and the system under which the service/product is provided. Franchising in its contemporary business format mode indeed takes its name from the imposition of an entire business, operational and management system on a looser and more limited intellectual property licence. The development of business format franchising is attributable to the quantum leap that entire business systems could be cloned. The McDonald's story is instructive. The extraordinary success of Dick and Mac McDonald's San Bernardino outlet in the early 1950s attracted not only the custom of consumers but also the interest of entrepreneurs keen to share in its success. The original intellectual property licences granted by the McDonald brothers nevertheless did not lead to successful derivative businesses. It was Ray Kroc's realisation that system controls and obligations could be, and needed to be, engrafted onto the brand licence that led to the now legendary success of the chain and set in motion a retail revolution which continues to increase in influence (Love 1995).

Business format franchising is characterised by an ongoing business relationship between franchisor and franchisee which includes the product, service and trademark, as well as the entire business concept itself—a marketing strategy and plan, image, comprehensive operational standards, systems and formats, operating manuals, training, quality control, purchasing leverage, site selection, equipment and supplies and a continuing process of assistance, guidance and supervision. The 'system' is a complex equation which delivers uniformity and replicability. It promotes the reliability and familiarity of franchised goods and services and franchised brands. It encompasses both front-of-house and back-of-house features, both of which are essential in traditional business format franchising. This paper nevertheless suggests that back-of-house elements can be systematised and licensed independently of the brand and the tangible elements of it.

3 Challenges to Brand Sovereignty

The power of a good brand is 'simply staggering' (Murphy 1987) and franchising is a proven strategy for brand owners 'to seek to exploit more widely the equity in these valuable assets' (Blackett 1998). Franchising is, and will inevitably be, associated with the brand—with the "front-of-house" features which identify the outlet as part of a network which incorporates and exploits a range of synergistic benefits. Given the immense and undoubted significance of brands, what possible reason could there be for exploiting a form of "franchising" which does not incorporate, as an essential, entrenched and inviolate element, the brand and image? While branding is a massively significant driver of contemporary business, and of course of franchising, it is nevertheless not an exclusive or inevitable ingredient of business operation. Branding of particular products and services may be inevitable despite the rise of generic branding but branding of businesses, whether in the narrow sense of the name by which the business is known or the wider image and standards associated with it, is not essential.

3.1 The Brand and the Start Up Business

The brand is a characteristic which matures with time. The success of McDonald's original San Bernardino hamburger outlet in the early 1950s is a matter of record (Love 1995). But, in its early expansion was it the brand or the system which was influential? It may in fact have been neither: the original licensees displayed a cavalier disregard for both the brand and the rudimentary system which supported it (Love 1995). The McDonald brothers' breakthrough was nevertheless in relation to their back-of-house operations, not their front-of-house branding. McDonald's trademark attorney would no doubt have pointed out the problems with protecting a common family name. Similar advice in relation to geographic and generic names was no doubt given to Colonel Sanders in relation to Kentucky Fried Chicken. The point is that, at least in the early days of franchising, it is rarely the brand which is the most important ingredient. The concept, the recipes, the specifications, the system, the management and operational formats are the critical features. While some of these aspects may form part of the brand experience it is nevertheless possible to distinguish between back-of-house and front-of-house functions. Brands cannot exist in a vacuum—they must be attached to something, be it a product or a service—at least at the early stages of their development. The normal place for a start-up to begin developing awareness of their offerings is through product or service marketing. As noted by Lee Hower, 'start-up brands are by definition built upon the success and attributes of the products or services they provide. For start-ups, brands are less intentionally "built" than they are "derived" from successful products' (Hower 2007).

3.2 The Anti-brand Movement

Although the brand is subsidiary to the system in business start-ups, there are more fundamental reasons organisations may choose not to embrace branding apparatus. The unrestrained use of branding, which has been so effective in tapping into our souls and marrying our innermost desires to tangible embodiments of a brand's image in the form of phenomena for sale, may be on the wane. In the words of Saatchi and Saatchi CEO Kevin Roberts, 'brands are running out of juice' (Roberts 2004). Canning has suggested that

> [t]he world is changing, and companies have to change along with it. Groups that have spent decades building up equity in their brands are suddenly discovering that the solid foundations they thought they were creating – foundations that would anchor their businesses into the future through the good times and the bad – are now at the mercy of the combined forces of rapidly changing technology and consumer empowerment caused by the upsurge of social networks (Canning 2010).[3]

A growing resistance to transnational brands and corporate globalisation (Hollenbeck and Zinkhan 2006), documented in the literature and predicted by Naomi Klein (Klein 2002) in *No Logo*, has been played out on the big stage with anti-corporate and anti-brand sentiments being expressed in demonstrations against multinationals and their perceived ever growing power in which prominent international franchise systems are frequently the target. Klein points out that a franchise is 'technically owned by the franchisee, even if every detail of the outlet—from the sign that hangs out front to the precise temperature of the coffee—is controlled by a head office hundreds or even thousands of miles away' (Klein 2002). This reality may contribute to some consumers' growing sense of estrangement from branded franchise products and services. The phenomenon of globalised standardisation—termed the "McDonaldisation of society" by George Ritzer (Ritzer 2011)—perhaps not surprisingly co-opts the name of franchising's most prominent international star as its descriptor. Although concerns as to the 'potential excesses of overzealous franchising—particularly for the character and diversity of Australian cities and towns' and for the 'high streets in the UK [which] are virtually indistinguishable insofar as they are littered with franchises' (Bruce 2012)—are clearly overstated, such sentiments are nevertheless becoming more commonly vented.

3.3 Generation Y and Brand Indifference

The anti-brand movement identified by Klein, while not signalling the demise of branding as many of its more extreme disciples may have liked, has nevertheless

[3] There are many instances of brands with a soft human face which have touched hearts around the world being exposed for engaging in practices and activities anathema to the type of image built up through the branding exercise, and thus straining or even breaking the trust once established. As uncompromisingly expressed by Klein: "The travels of Nike sneakers have been traced back to the abusive sweatshops of Vietnam, Barbie's little outfits back to the child labourers of Sumatra, Starbucks' lattes to the sun-scorched coffee fields of Guatemala, and Shell's oil back to the polluted and impoverished villages of the Niger Delta" (Klein 2002). Beyonce and Lady Gaga are rarely cited in academic papers but their words resonate in this context: "Trust is like a mirror, you can fix it if its broke, but you can still see the crack in that mother fucker's reflection". Beyonce and Lady Gaga, lyrics from the song "Telephone".

resonated with modern consumers. Consumers have at the same time become aware of many of the techniques used by brands to infiltrate culture and penetrate their innermost spaces with rationalised techniques for product and service provision. George Gilder, who predated Klein, may have been more realistic in his modest assessment that 'the medium will change from a mass-produced and mass consumed commodity to an endless feast of niches and specialties... A new age of individualism is coming and it will bring an eruption of culture unprecedented in human history' (Gilder 1990).[4] Under the influence of Generation Y, and with more widespread awareness of branding techniques, traditional standardised transnational brands are not always the holy grail of consumption experiences. With the largest consumer group being Gen X and soon Gen Y, an important trait of both generational groups may be impacting upon the significance of the idea of the brand, at least in a franchising context. That trait is individuality, a trait which is hard to express as a franchisee in a branded franchise system where virtually all of the operations of the business are dictated by a franchisor eager to develop and protect its brand. That trait also makes it difficult to express one's individuality if consuming a franchised brand's products or services. This demographic will soon constitute the largest group of franchisees in Australia (Franchising Australia Survey 2010). This trait therefore impacts upon franchising from two angles: from the consumer side (a consumer who places a premium on individuality of consumption may prefer dining at an unbranded outlet rather than at an outlet of a branded chain) and from the business entrepreneur side (an opportunity for expressing individuality while benefiting from management and operational systems which facilitate this).

3.4 The Standardised Chain and the Brand

It follows that the romance of the brand is not compelling for all consumers, particularly in certain industry sectors. Commenting recently on Gloria Jean's Coffees—one of the fastest growing franchise systems in Australia with about 500 outlets in Australia and almost 1000 coffee houses in 40 countries—Toby Smith qualified his acknowledgment of Gloria Jean's impressive growth with the comment that

> [A] chain is a chain is a chain. And they are a franchise chain... The problem with a chain, its like a formula. Its generic. They all look the same. It's just a mish-mash of Starbucks, Costa Coffee and everyone else. It doesn't make an impact on the customer (Carruthers 2010).

[4] G Gilder, *Life After Television* cited in Klein 2002. As noted by Canning "New Age marketers and ad agencies can no longer afford to build pyramids that will withstand the onslaught of change. Now they must be kites, ready to go where the winds of social change take them" (Canning 2010).

Given that Toby Smith is a competitor, and the founder and proprietor of the boutique Toby's Estate café outlets, one may be sceptical of his criticism. It is nevertheless true that there are certain industry sectors, particularly in the hospitality industry, in which association with a national or international brand may not only not be an advantage but may actually constitute a disadvantage. While many agree with Howard Shultz's statement that 'it's the romance of the coffee experience, the feeling of warmth and community people get in Starbuck's stores' (Klein 2002) which draw customers to it, there is a growing number of people for whom the standardised brand experience is not optimal. Indeed, Toby Smith's comments may be more widely representative of a growing category of consumers who prefer their coffee, their food and their beer in an original, eclectic and idiosyncratic neighbourhood outlet. In some sectors—café, restaurant and pub are obvious examples—standardised branding can be a disadvantage. Lashley and Lincoln observe 'signs of consumer resistance to brands beyond a certain size' (Lashley and Lincoln 2000). They give, as an example, Scruffy Murphy's—a pub concept which after a rapid growth in the number of pub units suffered an 'equally rapid decline of the brand'. Australian cultural norms may be particularly influential in this area. Professor Geoffrey Garrett, Chief Executive of the US Studies Centre at the University of Sydney, has opined that

> Americans view chains as comforting. No matter where you are, Starbucks will always be there and taste the same. Australians prefer one-offs that are a bit quirkier. The fact that every shop is different is a fun part of the challenge. The Down Under demise of Krispy Kreme says more about our attitude to chains than our attitudes to the US' (Urban 2010).[5]

While pest control operators might benefit from a nationally recognised name, a bar, restaurant or café proprietor might not.

4 Quasi-Franchising and the Brand Continuum

Given the diminished significance of the brand in the contexts noted above it is possible to develop a "branding continuum" with traditional business format franchising at one extreme, and "back-of-house franchising" which eschews brand and its servicescape accoutrements at the other. Between these extremes there is a range of possibilities which challenge, to a greater or lesser extent, brand sovereignty.

[5] Lethlean comments that "Starbucks hasn't worked in Australia the way it has in the rest of the world? Who's surprised? With such a strong Italian coffee culture in nearly all our cities and towns, backed up by a rapidly emerging—so called third wave—specialist roaster-café scene it gives me a little comfort to know not everybody confuses free wi-fi with a quality coffee experience" (Lethlean 2010).

4.1 Back-of-House Franchising

The concept of replicating back-of-house functions without brand identification is of course not uncommon in business. Systematising back-of-house functions is a key factor in successful business operation and a necessary pre-requisite for replication of the business. Systems are an inevitable ingredient not only of business format franchising but all viable business operations. The company owned and managed chain operating under a brand—in effect a company owed franchise network—will of course have developed and documented a range of system issues equivalent to that of a franchise chain, including technology, purchasing arrangements, training, bookkeeping, and a range of other management services. However, even a company owned group which operates a series of individual businesses rather than under a network brand will improve profitability and management through applying back-of-house systems. Such arrangements are also common in various forms of groups of independent proprietors—from informal cooperative arrangements to more structured groups formed to obtain the benefits of proven systematised and efficient management systems. The rise of the 'celebrity chef' has been a catalyst for innovative business structures. While some, such as Gordon Ramsay and Jamie Oliver, operate models akin to franchising with consistent branding, an alternative model is the retention of idiosyncratic servicescape and branding at multiple sites but with centralised back of house functions. Prominent Australian chef George Calombaris comments 'I want to cook; I don't want to run the whole business. I don't know how to do that' (Cornell 2012). Calombaris acknowledges that 'if he was going to achieve what he wanted to on the food front, he needed business minds who could run balance sheets with the same skill' (Cornell 2012).

Company owned and managed networks, albeit unbranded, apply back-of-house technologies. Pub groups, where a company acquires a portfolio of hotels and deliberately retains the individuality and idiosyncracies of the pubs in the group, is an example. The business model is simple: the pub group acquires hotels where it believes the profits can increase under new management. Back-of-house functions are provided centrally for the individual units within the group (Reilly 2010). The individual pub is a company owned and managed outlet rather than a franchised outlet but the systems, management expertise and economies of scale applied to the individual units equate to those that would be found in a franchised network. Franchising the back-of-house functions could be particularly attractive in an environment where stringent banking conditions constrain growth through the purchase of further outlets. The hands-on proprietorship of the "back-of-house franchisee" would be expected to lead to it outperforming a manager in similar circumstances based on experience in conversion franchising (Giles et al. 2009). The back-of-house management systems provide the aspiring business proprietor with the opportunity to establish and operate a business with greater confidence, efficiency, and likelihood of success than that offered by solely independent proprietorship.

Associations to facilitate back-of-house activities are not uncommon and are provided to a greater or lesser extent by various forms of cooperation including trade associations and buying groups. A recent item from the Sydney Morning Herald provides an example of informal cooperation to achieve back-of-house efficiencies while retaining individual branding, image, and front-of-house architecture:

> In one of Australia's wealthiest suburbs its registered clubs are struggling under state taxes, shifting demographics and ever growing operational cost to survive and stay in the black. The solution is to form an alliance to pool resources and cut costs as 'together the group can create economies of scale by working together' through pooling casual labour, sharing courtesy buses, advertising as a group, bulk buying, and using the same cleaners, tradespeople, accountants and auditors. The strategy is not amalgamation, and membership will not be pooled. The plan being to keep all clubs viable and operating independently (Munro 2010).

In some cases relatively informal arrangements can mature into structured franchise systems. The progression from buying group, to marketing group with front-of-house branding and standards, to a franchise system with a complete repertoire of back-of-house and front-of-house architecture is not uncommon.[6]

The essence of 'back-of-house franchising' is simply franchising without the brand and associated trade dress, image and external indicia that symbolise membership of a standardised chain. It is a form of B2B franchising under which the business proprietor benefits from a range of back-of-house systems which remove many of the challenges in establishing a business—and without which business entry is difficult if not impossible—while retaining discretion in relation to front-of-house features. The underlying arrangements are imperceptible to consumers. Back-of-house franchising provides the opportunity for 'franchisees' to be able to express their entrepreneurial individuality, the scope for which is limited in a traditional business format franchise. This option may be particularly attractive to Generation Y who are believed to be more entrepreneurial than previous generations (Milman 2010) but who want to establish their own businesses which express their own individuality. McCrindle notes that Generation Y need a "trusted guide" which a back-of-house franchisor can provide (McCrindle).

The concept of providing back-of-house management services, economies of scale, and technology is not new. It operates in a range of contexts to enable the individual proprietor to focus on core business. Trade associations and outsourcing arrangements operate in this space. Outsourcing provides the mechanism for the contracting out of a business function to an external service provider. Business proprietors may utilise a range of outsourcers for the provision of a range of business services. Although back-of-house franchising may be thought of as a sophisticated form of outsourcing under which the back-of-house franchisor provides a complete range of business services, this analogy is limiting. A back-of-house franchisor

[6] Autobarn, a prominent Australian after-market automotive parts is a franchise system owned by the original members of the buying group. Mitre 10, a prominent hardware chain is a cooperative owned by its members and, apart from ownership, indistinguishable from a franchise system.

provides not only the complete package of back-of-house services, but also an integrated back-of-house system which outsourcing does not pretend to offer. The concept of B2B back-of-house services being provided in a systematised, structured and disciplined manner by a back-of-house service provider transcends traditional notions of outsourcing.

As with traditional franchising, back-of-house franchising is a practical strategy for which there is no one model. It is driven by a market for back-of-house services to enable an independent business entrepreneur to build a business free of the confines of brand and image but with the benefit of key back-of-house support services without which entirely independent business operation is difficult and risky. In its most derivative form the back-of-house service provider would secure premises and furnish them, or facilitate the licensed proprietor's eclectic furnishing of them. In its least derivative form, the provider would provide a range of back-of-house services commensurate with those in a traditional franchised operation. Given that there is no brand, the only image and standards are those of cleanliness, service and "appropriateness" for the particular concept and location. Back-of-house franchising is a deliberate top-down franchise model for providing back-of-house services independently of any brand and where the eclectic, idiosyncratic and individual appearance of the premises is an essential part of the concept. Under this model the services and the controls will vary with the particular concept. Back-of-house franchising provides the tools for an efficient and structured business operation through the licensing of business systems to empower local operators.

Despite the significance of the "system" in franchising, the franchisor's goodwill essentially resides in the brand. Trademarks 'differentiate an enterprise's goods or services from those of its competitors, indicate the source or origin of the goods or services, represent and secure the goodwill of the enterprise, serve as a symbol of the value or quality of the goods or services, and build brand loyalty' (Terry and Giugni 2009). The back-of-house "franchisor" without brand architecture of course faces a challenge in marketing and maintaining relationships simply on the basis of goodwill in back-of-house systems. This challenge can nevertheless be exaggerated. A business format franchisor provides an extensive range of back-of-house services which are no less significant to a back-of-house "franchisee" than to a typical business format franchisee. As in traditional business format franchising the franchisor's income stream is both driven by and dependent upon franchisee performance. A continuing back-of-house relationship is based on a contract between the parties but legal realities must be based on commercial viability and commensurate value. Because back-of-house franchising is based exclusively on back-of-house services the nature, range and quality of such services may even exceed that of the traditional model to attract and retain licensees. Although the nature, extent and quality of the back-of-house services must be commensurate with the value proposition in order to sustain an effective and continuing back-of-house relationship, the back-of-house contract supports the underlying arrangement. As with traditional business format franchising, ownership of the business premises and control over the look and feel of the business are key factors contributing to the integrity of a continuing back-of-house franchising relationship. A back-of-house "franchisee" who, in breach of contract, terminates a back-of-house relationship relinquishes occupancy rights and faces standard contractual consequences.

4.2 *Branded Back-of-House Franchising*

In a more conservative iteration of the back-of-house model, an "umbrella" brand may be employed by the back-of-house franchisor to signify the outlet is a member of the particular back-of-house system and in which goodwill accrues. Outlets would be identified by a generic system name but back-of-house franchisees would retain discretion in relation to servicescape aspects. The back-of-house franchisor would prescribe system standards but not branding standards and servicescape architecture.

Examples of innovative practices falling short of traditional business format franchising are increasingly being adopted. Modern developments in medical practice have resulted in private medical practice being a business endeavour as well as a clinical endeavour. A range of companies are now providers of practice management services to doctors. While it is usual for the medical practice to retain its own name there is a strong element of standardisation across a range of diagnostic and associated administrative services and in the "look and feel" of the premises. The "umbrella" brand of the practice management provider is nevertheless usually prominent. Similar arrangements exist in relation to certain hospitality groups. Hospitality groups such as "Leading Hotels of the World" (http://corp.lhw.com/) provide an umbrella brand and offer 'a variety of membership options to hotels starting from the planning and development stages, including brand licensing, technical and training consultation, revenue optimisation, sales, marketing and other services'. Such groups operate on membership which is restricted to hotels which can meet basic standards of service. An umbrella brand exists—and is prominent for marketing purposes—but individual hotels within the group retain their individuality subject to prescribed image and standards.

4.3 *Customised Business Format Franchising*

An iteration of the franchise model, owing more to traditional business format franchising than back-of-house franchising, is a more personalised form of business format franchising allowing, to some extent, franchisee flexibility in the provision of system services. While academics continue to debate whether franchisees are entrepreneurs (Torikka 2011), franchisors face the challenge of how to 'balance franchisee aspirations for entrepreneurial autonomy with the franchisor's efforts to enforce compliance to operational standards' (Davies et al. 2009). Pizanti and Lerner emphasise 'the need for a balance between control and autonomy, claiming that excessive levels of control or of autonomy can be counter-productive and negative' (Pizanti and Lerner 2003). Empirical research has demonstrated that franchisee commitment to core brand standards can increase in circumstances where franchisees are allowed the freedom to control other, more peripheral or non-core aspects of the business.

The uniformity versus customisation debate is most usually played out in the context of international business format franchising expansion where compromises to accommodate local conditions are usually necessary. Increasingly however, adaptation to local customer needs is finding application in domestic contexts. There is increasing recognition that, even in traditional business format franchising, while the core brand components such as brand name, logo and essential product features should be as consistent as possible across the network, peripheral attributes can be modified. Streed and Cliquet accept that '[s]ervice personalisation, and more specifically customised personalisation, presents... an effective opportunity for chains to adapt to local customer needs without jeopardizing brand integrity' (Streed and Cliquet 2008). An extreme form of 'customised' or 'personalised' business format franchising is that adopted by the "Great Harvest" franchise system documented by Streed and Cliquet (Streed and Cliquet 2008). The founder of the 200 outlet bakery system describes the concept as a "freedom franchise" because of the 'extreme freedom' given to franchisees. Franchisees trade under the Great Harvest banner but the franchisor 'welcomes and rewards entrepreneurial spirit', and although 'know how such as recipes and management processes are provided each store [can] build its own identity for a better fit in the local business landscape' (Streed and Cliquet 2008). The Great Harvest website states:

> Most franchises... require their owners to do things their way, with little or no variation. Cookie cutter-style. That's because the franchisor is trying to build a national brand, the foundation of which is consistency. The problem with this sort of franchise, if you're an entrepreneur-type, is that they aren't very much fun. All the good stuff about opening your own business – figuring out what you want to offer and what color the walls will be – aren't your decisions to make. They've already been made.
>
> At the other end of things is starting up and running your own Mom and Pop shop. There you have all the freedom in the world to create this thing just the way you want, but you're flying solo, with no one else to lean on. That's why so many start-ups fail.
>
> We provide you with middle ground between the advantages of a traditional franchise and the fun of a let's-do-it-all-ourselves start-up. Our philosophy is simple. Let's create unique neighborhood bakeries that are a reflection of the Great Harvest brand and the bakery owner. We are no cookie cutter franchise. We are a freedom-based franchise that encourages excellence and individuality, not to mention a spirit of fun and generosity.

The franchising literature's recognition of franchisee personalisation is nevertheless limited by the sovereignty of the brand architecture and servicescape. While brand integrity is critical in business format franchising, the extent to which franchisors can tolerate departure from system standards without concept infringement is a developing issue driven by practical commercial considerations. This paper suggests that franchise systems will have to allow franchisees the ability to localise beyond service personalisation and yet retain control over key aspects of operations. The determination of core and non-core aspects will be an interesting challenge.

5 Conclusion

Two decades ago William Davidow and Michael Malone in *The Virtual Corporation* (1992) identified driving forces that were transforming the marketplace and corporations. Contemporary commercial models needed to become adaptable, flexible and responsive. Franchising is not immune to such broader trends. While independence and individualism have traditionally been seen as the enemies of franchising (Terry 2011) the paradigm is shifting. There are increasing commercial pressures to allow greater franchisee autonomy. This can be achieved through a franchisor allowing franchisees operating under the system's brand to customise or personalise peripheral if not core aspects of the system (Streed and Cliquet 2008; Kaufmann and Eroglu 1999) or, in the *back-of-house* iteration suggested in this paper, by renouncing front-of-house elements including the system name and brand. Recent research suggests that '"pockets" of freedom may foster entrepreneurial activities by franchisees that in turn, might enhance the performance not only of the individual owners, but that of the entire franchise system [and that] allowing franchisees such freedoms strategically may offer a distinct competitive advantage to franchisors who capitalise on the benefits of such arrangements' (Grunhagen et al. 2012). This may be through allowing operational autonomy with respect to HR policies (Grunhagen et al. 2012) or through innovative service personalisation strategies (Streed and Cliquet 2011).

Branding is, and will inevitably remain, an integral and non-negotiable characteristic of business format franchising. There are however market niches in which outlets may be more attractive to consumers because they are not associated with the standardised and formulaic uniformity and the generic replicability which are the hallmarks of business format franchising. For most consumers this is not because they are consciously part of an anti-brand movement but simply because they value the idiosyncratic nature of customised rather than standardised ambience. The proprietors of such establishments would nevertheless benefit from back-of-house systems and support in operational and managerial aspects of running their businesses. Service providers of course exist to provide discrete services to independent business proprietors through outsourcing arrangements but the packaging of entire back-of-house services in a coherent and coordinated manner is a more sophisticated business model.

While franchising accommodates intrapreneurial franchisees prepared to work within the system, the opportunities for entrepreneurial franchisees who require an outlet for their individuality beyond the confines of the brand and system are of course limited. At the big end of town a national master franchisee acquiring the rights to sub-franchise an overseas system will inevitably have some flexibility in "customising" the system for local conditions but for the typical franchisee "standardisation" is a necessary reality. It would nevertheless be foolish to suggest that this is an immutable truth. Franchising's continual adaptation to accommodate changing circumstances is a major factor in its increasing influence. Under the influence of Gen X and Gen Y for whom the trait of individuality is apparently much stronger than for previous generations, customisation reflecting franchisee individuality may make inroads into the standardisation which is today the main characteristic of franchising. Back-of-house franchising provides a strategy which is attractive to consumers as well as to aspiring entrepreneurs for whom business entry may be intimidating if not practicably impossible. The "franchisee" acquires the right, and the obligation, to use the "franchisor's" back-of-house system but the front-of-house architecture is not prescribed, or at least not in any detail.

Given that every definition of franchising promulgated or proposed includes a 'brand' element,[7] it is inappropriate to use the term franchising to identify this most extreme iteration of business format franchising. James Whitcombe Riley is credited with the proposition that 'if it walks like a duck and swims like a duck and quacks like a duck, it is a duck'. Without the brand element, this creature is not a franchise. As Commander Spock may have said to Captain Kirk, "It's franchising, Captain, but not franchising as we know it".[8] This no-franchising form of franchising is nevertheless likely to assume increased prominence. For effective business operation—albeit at a more modest level than global domination—the back-of-house elements are essential but the brand and associated front-of-house elements aspects are not inevitable.

References

Blackett T (1998) Trademarks. Palgrave Macmillan, Basingstoke

Blackett T (2004) The case for brands. In: Clifton R, Simmons J (eds) Brands and branding. Bloomberg Press, New Jersey

Blair RD, Lafontaine F (2005) The economics of franchising. Cambridge University Press, New York

Booms B, Bitner J (1981) Marketing strategies and organization structures for service firms

Bruce K (2012) A forewarning on franchising. Business Review Weekly, 15 March 2012

Canning S (2010) Brands must go with the flow in a changing world. The Australian, 15 November 2010

Carruthers F (2010) Gloria Jean's brews up for Asia. Australian Financial Review, 11 Nov 2010

Clifton R, Simmons J (2003) Brands and branding. Bloomberg Press, New Jersey

Cornell A (2012) Masterclass. The Australian Financial Review Magazine, May 2012

Davies M, Lassar W, Manolis C, Prince M, Winsor R (2009) A model of trust and compliance in franchise relationships. J Bus Venturing 26(3):321

Diefenbach J (1987) The corporate identity as the brand. In: Murphy JM (ed) Branding: a key marketing tool. Macmillan, London

[7] Because "back-of-house franchising" involves no brand element it falls outside the scope of the definition of franchising in every regulated sector, but it is nevertheless a business venture which may be caught by the prior disclosure provisions of business opportunities regulation such as the US Federal Trade Commission *Disclosure Requirements and Prohibitions Concerning Business Opportunities* (2007).

[8] From the Star Trek TV series which debuted in 1966. The original quote was of course "It's life Captain, but not life as we know it".

Franchising Australia Survey (2010) Report by the Asia Pacific Centre for Franchising Excellence, Griffith University
Gilder G (1990) Life after television. Whittle Direct Books, Knoxville
Giles S, Redfern M, Terry A (2009) Franchising law and practice. LexisNexis, Sydney, Chap. 1
Grunhagen M, Dada O, Wollan M, Watson A (2012) The moderating role of HR operational autonomy on the entrepreneurial orientation – performance link in franchise systems. Proceedings of the 26th Annual International Society of Franchising Conference
Haig M (2003) Brand failures. Kogan Page Publishing, USA
Hollenbeck CR, Zinkhan GM (2006) Consumer activism on the internet: the role of anti-brand communities. Adv Consum Res 33:479–485
House of Representatives Standing Committee on Industry, Science and Technology. Finding a balance: towards fair trading in Australia (May 1997) para 3:4–5
Hower L (2007) http://www.agilevc.com/venturesome-archives/2007/1/9/brand-vs-product.html
Kaufmann P, Eroglu S (1999) Standardisation and adaptation in business format franchising. J Bus Venturing 14(1):69–85
Klein N (2002) No logo. Picador, USA
Kos S (1990) Franchisor liability for franchisee misconduct, Legal Forums. Franchisor's Association of Australasia
Lashley C, Lincoln G (2000) Licensed retail. In: Lashley C, Morrison A (eds) Franchising hospitality services. Butterworth-Heinemann, Oxford
Lethlean J (2010) Fat chance of losing our taste for franchises. Weekend Australian, 6 November 2010
Love JF (1995) Behind the golden arches. Bantam Books, New York
McCrindle M. Understanding Generation Y, The Australian Leadership Foundation. Available at http://www.learningtolearn.sa.edu.au/Colleagues/pages/default/mccrindle/?reFlag=1
Mendelsohn M (2004) The guide to franchising, 7th edn. Thomson Learning, London
Milman O (2010) Generation Y keen to start businesses: survey, 21 September. Available at http://www.smartcompany.com.au/start-up/20100921-generation-y-keen-to-start-new-businesses-survey.html
Munro K (2010) Foes turn friends as struggling clubs unite. Sydney Morning Herald, 19 November 2010
Murphy JM (ed) (1987) Branding: a key marketing tool. Macmillan, London
Parliamentary Joint Committee on Corporations and Financial Services, Parliament of Australia, Opportunity not opportunism: improving conduct in Australian franchising (December 2008), p xiii. Available at http://www.aph.gov.au/senate/committee/corporations_ctte/franchising/report/index.htm.
Pizanti I, Lerner M (2003) Examining control and autonomy in the franchisor-franchisee relationship. Int Small Bus J 21(2):131
Reilly T (2010) Singo's shout: taste's so good i bought the pubs. Sun Herald, Sydney 21 November 2010
Ritzer G (2011) The McDonaldization of society, 6th edn. Sage Publications, Thousand Oaks
Roberts K (2004) Lovemarks. Murdoch Books, Sydney
Streed O, Cliquet G (2008) Concept uniformity: control versus freedom in business format franchising. In: Hendrikse G, Tuunanen M, Windsperger J, Cliquet G (eds) Strategy and governance of networks: cooperatives, franchising and strategic alliances. Springer, Berlin, pp 205–220
Streed O, Cliquet G (2011) Service personalisation vs concept uniformity: the case of franchised quick service restaurants. Proceedings of the 25th Annual International Society of Franchising Conference
Terry A (2011) Franchising, togetherness and individualism. Franchising, March/April 2011
Terry A, Giugni D (2009) Business and the law, 5th edn. Cengage Learning, Australia & New Zealand

Torikka J (2011) Applying the general theory of entrepreneurship to franchising. Proceedings of the 25th Annual International Society of Franchising Conference

Urban R (2010) Changing tastes too fast for food chains. Weekend Australian 6 Nov, quoting Geoffrey Garrett

Willigan G (1992) High-performance marketing: an interview with Nike's Phil Knight. Harv Bus Rev 90:92

Entrepreneurial Processes of the Finnish Franchisee Training Program's Graduates

Jenni Torikka

Abstract The current study reports the results of the third and final phase of the longitudinal study on trainees of the Finnish franchisee training program. The study targeted the 46 respondents who in the previous phases indicated that they became either franchisees or self-employed in stand-alone businesses after graduation. The purpose of the study was to find out what factors influenced the respondents' entrepreneurial decision-making processes and what role the training program played in these processes. Trainees' entrepreneurial decision-making processes proved to be dissimilar. Some were pushed while others were pulled to entrepreneurship. The program was found to be a factor in the respondents' entrepreneurial decision-making process, but its role and significance was seen differently. Logistic regression analysis was performed to discover what factors could be used to predict respondents' likelihood of becoming franchisees or self-employed. Findings of the study provide implications for both franchisors and potential franchisees as well as for organisations planning and offering entrepreneurship training.

Keywords Entrepreneurial decision-making process • Entrepreneur • Franchisee • Opportunity recognition

1 Introduction

Since the early work on franchising at the end of the 1960s, studies on the question of whether franchisors and franchisees are to be considered entrepreneurs have from time to time been undertaken. However these have been relatively scarce

J. Torikka (✉)
University of Jyväskylä, Jyväskylä University School of Business and Economics, P.O. Box 35, FI-40014 University of Jyväskylä, Finland
e-mail: jenni.m.torikka@jyu.fi

T. Ehrmann et al. (eds.), *Network Governance*, Contributions to Management Science, DOI 10.1007/978-3-7908-2867-2_16, © Springer-Verlag Berlin Heidelberg 2013

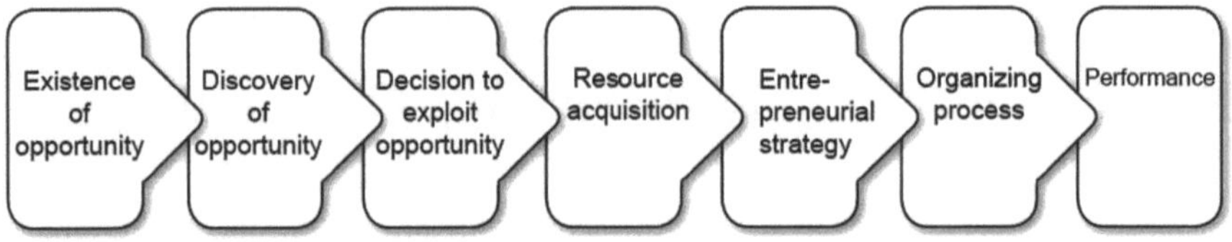

Fig. 1 The direction of the entrepreneurial process (Shane 2003)

compared to other topics and points of view discussed by franchising scholars (see e.g. Elango and Fried 1997; Young et al. 2000; Dant 2008; Young and McIntyre 2011). Some studies have covered topics closely related to entrepreneurship, for example independence and innovation in the franchising context, but have not specifically analysed franchising as a form of entrepreneurship. The current study takes an entrepreneurship viewpoint toward franchising and utilises Shane's (2003) framework of *a general theory of entrepreneurship*.

There are many investigations on motivations to franchise from the franchisor's perspective, but relatively few studies on the process of becoming a franchisee. Similarly, a variety of examinations on subjects related to the decision-making process of a potential franchisee have been conducted from the franchisor's viewpoint, but a small number have presented the franchisee's standpoint. In particular, investigations where franchisees are considered as entrepreneurs are limited in number. The current study contributes to filling this void in the literature.

Shane's (2003) framework modelled the entrepreneurial process and he identified franchising as one option for exploiting an entrepreneurial opportunity (cf. Shane and Hoy 1996; Tuunanen and Hoy 2007). Shane referred to people in the entrepreneurial process who will decide among other things, whether they will have a franchised or stand-alone business. Figure 1 presents Shane's view of the direction of the entrepreneurial process.

This investigation explores the factors that influenced the entrepreneurial decision-making processes of the respondents. Therefore, the focus is on the first three activities of Shane's entrepreneurial process model; *existence of opportunity*, *discovery of opportunity* and *decision to exploit opportunity*. Previous studies on the process of becoming a franchisee have related to these three activities. The subsequent four activities of Shane's model; *resource acquisition, entrepreneurial strategy, organising process and performance*, belong to a so-called execution stage. In a franchising context these execution-stage activities are carried out within the franchise relationship i.e. the franchisee executes them in cooperation with the franchisor. This is a specific feature of franchising. The four activities of the execution stage are not the focus of the current study.

Selected entrepreneurship, franchising and effectiveness of education studies form the theoretical background of the current study: Shane's (2003) model of the entrepreneurial process is complemented by few other entrepreneurship investigations which relate to the process of becoming an entrepreneur. In addition, studies on the process of becoming a franchisee are included to take into account the specific features of franchising that influence a franchisee's entrepreneurial process. Previous research has indicated that intentions, motivation and satisfaction connect to the entrepreneurial process and therefore, studies of entrepreneurial and franchisee intentions, motivation and (dis-)satisfaction were included in the theoretical background. The Finnish franchisee training program was the starting point of the longitudinal study, and assessing its effectiveness was one of the goals of the investigation. Consequently, measures of impact were included in this third phase, but no theoretical discussion of the effectiveness of entrepreneurship education is presented here.[1]

1.1 Focal Definitions of the Study

This section presents essential definitions to help frame readers' understanding of this study. First, in the current investigation, franchising is seen as a form of entrepreneurship, i.e. franchising is understood as a form of starting and growing new ventures and organisational forms, and as a mechanism for introducing new products and services to expanding markets (see Vesper 1980; Baumol 1986). Furthermore, regarding this study, it is essential to recognise the differences between two forms of franchising, namely trade name-product distribution franchising and business format franchising. The former is a supplier-dealer relationship in which the franchisee acts under the principal's brand and as a distributor of its products. In the latter case, cooperation between the parties is closer and more extensive, involving the transfer of the entire business operation concept to be used by the franchisee (Tuunanen 2005). Hence, in discussing franchising in this study, only business format franchising will be considered, because it provides the franchisee entrepreneur with an opportunity to set up and run an entire business.[2] This is congruent with the entrepreneurship approach applied in the study.

Second, in this study franchisees are considered as entrepreneurs (cf. Baumol 1986), despite the fact that contradictory views have been presented (see e.g. Rubin 1978; Knight 1984; Norton 1988; Anderson et al. 1992; Hoy 2008; Seawright et al. 2011; Ketchen et al. 2011).[3] In addition, the term entrepreneur does not refer here to

---

[1] Literature review on entrepreneurship education and training and its effectiveness is presented in the article (Torikka 2007) that reports the results of the second phase of the longitudinal study.

[2] The study was carried out in Finland, which is a full member state of the European Union. The EU recognises only business format franchising as a form of franchising.

[3] In actual fact the term franchisee means the company receiving the granted right i.e. franchise (Elango and Fried 1997). Frequently however, especially in the case of small and medium size companies, the owner-manager of the company or the entrepreneur is called a franchisee. Franchisee is then thought of as a person and the word franchisee is used synonymously with entrepreneur (/small business owner/self-employed). By saying this, it is recognised that franchisees can be companies that are larger in size than their franchisor and franchising may only be one line of business for these companies (see e.g. Hoy et al. 2000; Tuunanen 2005). Large franchisee companies are not the focus of the current study.

a highly creative venture based on a new and novel product and service. Instead, it is used in a more everyday sense, and is meant to be interchangeable with self-employed or small businessperson (cf. Baumol 1986; Stanworth 1995).[4] This is consistent with Shane (2003). Nevertheless, many entrepreneurship researchers have not seen the terms as interchangeable, and indeed, there has been a debate on the differences in the meaning of the terms.[5] Shane (2003) provided one solution by extending the meaning of the term entrepreneurship. By founding a new business Shane meant the formation of a business venture or not-for-profit organisation that had not previously existed. He defined the performance of work for personal profit rather than for wages paid by others as self-employment. Furthermore, he added that depending on the situation, a self-employed person may incorporate a business and employ others. Shane also assumed that being entrepreneurial does not require the creation of a new firm, and that an entrepreneur can use market mechanisms, such as licensing or franchising, to exploit entrepreneurial opportunities. In this study, a franchisee is regarded as a person and the word franchisee is used synonymously with franchise owner and entrepreneur. Likewise, self-employed is regarded as a self-employed person in a stand-alone business, and is used synonymously with the word entrepreneur.

Third, the definition of entrepreneurial decision-making process refers to the activities and time before a person starts to operate as an entrepreneur (regardless of the form). This period can be likened to the first three activities of Shane's entrepreneurial process model which the current study focuses on. However, note that the data collection was conducted after the respondents had started operating as franchisees and self-employed persons (this is the case with many entrepreneurship studies) which might have influenced their perceptions on their entrepreneurial decision-making process. Finally, in this study the expression entrepreneurship education and training program is used to refer to education targeted at people equally interested in becoming entrepreneurs, self-employed or small business owners.

---

[4] Entrepreneurship and entrepreneurial in the English language are often normative statements concerning individuals, often founder-owners, who manage their firms in a certain way. There are other languages e.g. Finnish where no other terms (with connotations of growth orientation or self-employment) for entrepreneur exist. International comparisons are difficult, because the core term entrepreneurship is very culturally oriented (Huuskonen 1992; Gibb 2002).

[5] Carland et al. (1984) raised a discussion on definitions of entrepreneur and small business owner and their differences. The discussion was commented on by Gartner (1988) and again by Carland et al. (1988), and has continued among entrepreneurship researchers (see e.g. Cunningham and Lischeron 1991; Gibb 2002; McKenzie et al. 2007). When the terms entrepreneur and entrepreneurial venture have been defined, the central issues in the discussion have been e.g. the personality traits and characteristics of the owner/founder of the company; the innovativeness, newness, and creativity of the owner/founder and the company; purpose of establishing and managing the company; and profitability and growth goals of the owner/founder and the company. However, many researchers have not provided any definition, which has also led to heterogeneous selection in sampling. Thus, the comparability of the studies has been problematic.

This article is organised as follows: to start with, the key features of *the general theory of entrepreneurship* introduced by Shane (2003) are described and focal findings of earlier studies of becoming a franchisee and studies of entrepreneurial intention, motivation and (dis-)satisfaction are summarised. Subsequent to the succinct literature review, the research design is explained and results of the empirical study are set out and discussed. Finally, conclusions and recommendations for future research are presented.

2 Literature

2.1 Shane's Model of the Entrepreneurial Process

The study of entrepreneurship spans a wide range of fields including decision sciences, economics, management, sociology, psychology, and history. Because of this, approaches from various disciplines have been applied to entrepreneurs, their behaviour, and the companies they operate. However, no consensus has been reached regarding definitions of entrepreneurship, the process of becoming an entrepreneur, or the factors influencing the process (see e.g. Cunningham and Lischeron 1991; Gibb 2002; McKenzie et al. 2007). Shane (2003) presented a conceptual framework for entrepreneurship in response to the failure of prior research to provide one.

The assumptions in Shane's *individual-opportunity nexus* framework that are adopted in this study are as follows: entrepreneurship is a process (see also Bird 1989; Huuskonen 1992; Hoy and Shane 1998; Bygrave 2004); entrepreneurial opportunities are objective and exist independently of the actors within a system;[6] specific individuals are required for the discovery and exploitation of entrepreneurial opportunities, since opportunities themselves lack agency. Shane found that all the activities of the process are influenced by individual (psychological and demographic) and environmental (industry and macro-environment) factors (cf. Price 1997; Williams 1999; Kaufmann 1999). The process is illustrated in Fig. 2. The framework proposed by Shane assumes that entrepreneurial activity is directional and ordered, but it accepts the possibility of feedback loops and non-linearity.[7]

According to Shane *entrepreneurship requires*: (1) the existence of opportunities or situations in which people believe that they can use new means-ends frameworks

[6] The view represented by Shane is called the *discovery view* and it is in marked contrast to an alternate *creative view*, according to which opportunities do not exist in any objective form, but are merely a social construction (Venkataraman 2003, xi; Alvarez and Barney 2007). These two views represent different ontological and epistemological starting points.

[7] Huuskonen (1992; cf. Bird) found that if a person abandons the intention to become an entrepreneur, the negative decision may not be permanent. The decision is linked to background, personal and environmental factors and the process may start again later.

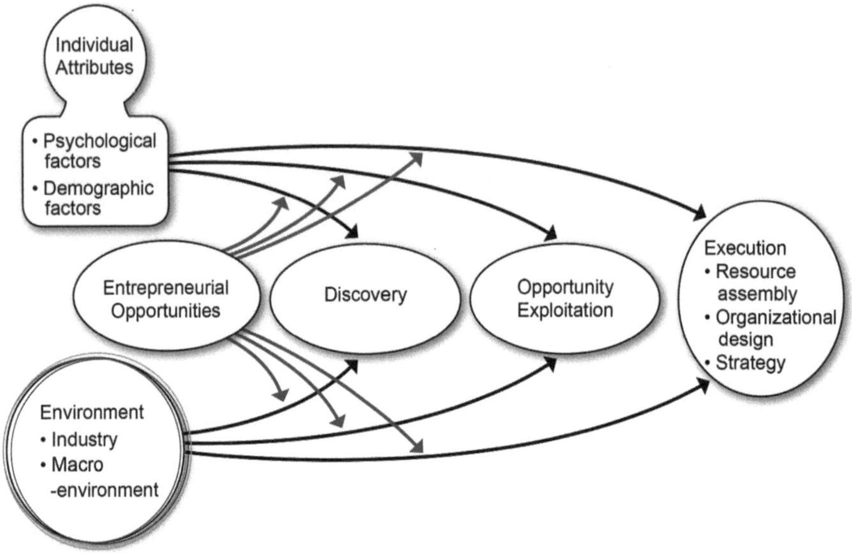

Fig. 2 Model of the entrepreneurial process (Shane 2003)

to recombine resources in order to generate profit; (2) differences between people, since people vary in their ability and willingness to recognise and act upon opportunities, and this influences the entrepreneurial process; (3) risk bearing, since the exploitation of opportunity is uncertain; (4) organising, i.e. creating a new way of exploiting the opportunity that did not previously exist; (5) some form of innovation, meaning the recombination of resources into a new form, according to the judgment of the entrepreneur.

Incorporated in Shane's model is the definition of *innovation*. In his view, the entrepreneurial process requires some form of innovation, but it can be much milder than Schumpeter's (1934) notion of innovation, i.e. something resulting in new combinations that will speed up creative destruction. What is needed is a recombination of resources into a new form. This type of milder innovation is often associated with Kirzner's (1997) perspective and can be applied to franchising, especially franchisees.

2.2 The Process of Becoming a Franchisee

Prior research has found that the decision to become a franchisee is an aspect of the more general process of becoming an entrepreneur (see e.g. Bradach and Kaufmann 1988; Spinelli 1994; Kaufmann and Stanworth 1995; Stanworth and Kaufmann 1996; Price 1997; Williams 1999; Kaufmann 1999; Guilloux et al. 2004; Bennett et al. 2010; see also Usbasaran et al. 2001; Shane 2003). However, the entrepreneurial decision-making process of a franchisee includes franchising-specific issues (see e.g. Stanworth and Kaufmann 1996; Price 1997; Kaufmann 1999; Gauzente 2002; Bennett et al. 2010) of which one is the role of the franchisor. A franchisor recruits franchisees and no franchisee can start operations without the approval of franchisor.

There is a need for further investigation of the entrepreneurial decision-making process of a franchisee conducted from the franchisee's perspective since the number of studies on the topic is scarce and previous research has presented some contradictory findings. Very few studies take a holistic view of the prospective franchisee's entrepreneurial decision-making process, i.e. that also consider the influence of background and environmental factors (cf. contingency view[8]). However, it can be said that there are numerous pull and push factors, i.e. positive and negative motivation factors[9] in franchising like in any other form of entrepreneurship that influence a person's entrepreneurial decision-making process. Further, various factors motivate different people depending on their background and the context they are in (cf. Stanworth and Kaufmann 1996; Vesalainen and Pihkala 1999; Tuunanen and Hyrsky 2001; McMullen and Shepherd 2002; Shane 2003; Bennett et al. 2010).

Several studies on the process of becoming a franchisee have been conducted at the time when respondents were about to decide whether or not to become franchisees. It should be borne in mind that intentions are not always the best predictor of action. On the other hand, in cases where respondents are studied after making the decision, one must remember, that respondents' perceptions of their motivations and of the factors influencing their decisions may have been influenced by e.g. the amount of time that has passed and the extent to which their expectations have been met and goals have been realised.

2.3 Entrepreneurial Process and Intention, Motivation and (Dis-)Satisfaction

Previous research on entrepreneurship, franchising, management and marketing indicates that intentions, motivation and satisfaction closely relate to the entrepreneurial process of both the self-employed and franchisees (see e.g. Brockhaus 1980;

[8] The view of different situational factors that influence the entrepreneurial process is called contingency view (see e.g. Gilad and Levine 1986; Littunen 2001).

[9] Examples of push factors (negative motivations to entrepreneurship): job dissatisfaction, prospect of unemployment, unemployment, local identity/willingness to stay in certain area. Examples of pull factors (positive motivations to entrepreneurship): desire to work for oneself, existence of role models, knowledge of market opportunity, being able to raise capital e.g. receiving an inheritance. Classifying factors to push or pull factors is difficult since factors might be strongly connected and their influence might be subjective and case-specific: what is negative to one person might be neutral or positive to another.

Gilad and Levine 1986; Bird 1988, 1989; Cooper and Artz 1993; Hing 1995; Vesalainen and Pihkala 1999; Krueger et al. 2000; Tuunanen 2002; Shane 2003; Davies et al. 2011). In addition, entrepreneurial intentions,[10] motivation and satisfaction are intertwined throughout entrepreneurial process.

Research has also found that factors that motivate people to become self-employed or franchisees may not be the same as those that motivate retention or subsequent exit from either mode (see e.g. Stanworth et al. 1995; Stanworth and Kaufmann 1996; Morrison 1997; Tuunanen and Hyrsky 2001; Grünhagen and Dorsch 2003; Davies et al. 2011). It has also been indicated that (dis-)satisfaction experienced in operating as self-employed or a franchisee connects to intentions to continue or quit as well as to intentions to set up a new stand-alone business, enter through franchising or to join another franchise system (post-engagement intention) (see e.g. Hing 1995; Morrison 1997; Gauzente 2003; Harmon and Griffits 2008; Huang and Phau 2009).

The empirical part of the study will be presented next. The research process is depicted first, after which results are described and discussed.

3 Research Design

The study on trainees of the Finnish franchisee training program and their entrepreneurial processes is a longitudinal study consisting of three phases. Table 1 describes the research process in more detail.[11]

The longitudinal study applies an entrepreneurship framework and focuses on examining the entrepreneurial paths of the respondents. For that reason, respondents who in the previous phases of the study indicated that they became franchisees or self-employed after the training program were chosen as targets of the third phase. The purpose of the third phase was to find out what factors influenced respondents' entrepreneurial decision-making processes and the extent to which the training program played a role in the process. The training program was the starting point of the longitudinal study and assessing its effectiveness was one of the goals of the investigation. Consequently, some measures of impact were included also in the third phase.

Measures employed in the empirical study are founded in a range of literature. No single model served the purpose of the investigation and thus frameworks from entrepreneurship, franchising and effectiveness of education studies were adapted

[10] Bird (1988) described entrepreneurial intention as a state of mind, directing attention, experience and action toward a specific object (goal) or pathway to its achievement (means) (see also e.g. Bird and Jelinek 1988; Bird 1989; Krueger et al. 2000).

[11] A broader description of the franchisee training program including its background is given and results of the first phase of the longitudinal study are reported in two prior articles: Torikka and Tuunanen (2003) and Torikka (2004). Results of the second phase are reported in the article by Torikka (2007).

Table 1 Description of the research process

Phase of the study	Study method	Individual phases	Respondents
Phase 1	Postal survey	• Targets: trainees of the training programs 1 and 2, held in 1999 in Helsinki • Data collection completed in fall 2000 • Targets: trainees of the training programs 3–5, held in 2000 in Turku, Tampere and Helsinki • Data collection completed in spring 2001 • Targets: trainees of the training programs 6–10, held in 2001 in Helsinki (3), Turku and Tampere • Data collection completed in spring 2002	$N = 214$ $n = 176$
Phase 2	Phone interviews	• Targets: trainees of the training programs 1–5, who responded to the survey • Data collection completed in fall 2003 • Targets: trainees of the training programs 6–10, who responded to the survey • Data collection completed in spring 2004	$N = 176$ $n = 143$
Phase 3	Phone interviews	• Targets: 46 franchisees and self-employed, who responded to phone interviews in phase 2 • Data collection completed in fall 2007	$N = 46$ $n = 39$

to compose the measures. The theoretical background of the third phase of the longitudinal study is based on: Shane's (2003) model of entrepreneurial process which is complemented with selected other well-known measures of entrepreneurship literature; studies of the process of becoming a franchisee; studies of entrepreneurial and franchisee intentions, motivation and (dis-)satisfaction; and studies of effectiveness of education.[12]

Data collection was conducted at a time when respondents had operated as franchisees and as self-employed for several years. This could be considered a limitation of the study as e.g. respondents' entrepreneurial experience may have influenced their perceptions of the motivations and factors that affected their entrepreneurial decision-making processes (hindsight bias[13]). Yet, it is noteworthy that investigating the motivations and factors that influence respondents' decisions to become entrepreneurs is very challenging before the start-up because intentions

[12] No theoretical discussion on effectiveness of entrepreneurship education and training will be presented in this article since it was not the focus of third phase (see Torikka 2007).

[13] Delmar and Davidsson (2000) described hindsight bias as "the risk of incorrect reporting due to memory loss or re-interpretation of facts as a consequence of events that have occurred after the time of start-up." Hindsight bias has received considerable attention in cognitive psychology over the last two-three decades. It has been found to be a robust phenomenon occurring in a variety of settings and judgements. Recent research has conceived of hindsight bias as consisting of three separable and partially independent components: memory distortions, impressions of foreseeability and impressions of necessity (see e.g. Blank et al. 2008; Nestler et al. 2010; Bernstein et al. 2011).

do not necessarily predict action. Furthermore, it is difficult to foresee or estimate the occurrence and influence of triggering events[14] beforehand.

The third phase was carried out to complete the longitudinal study. More information about the respondents' entrepreneurial processes as well as confirmation for the information received in the earlier phases of the longitudinal study was needed. The possibility of hindsight bias was acknowledged and an effort was made to take that into consideration in planning and conducting the current study. The longitudinal design acted as a confirming factor limiting bias and enhancing the quality of the study.

4 Results

4.1 Sample and Sub-group Descriptions

The present study concentrates on those 46 respondents of the phase 2 of the longitudinal investigation who after graduation from the franchisee training program became franchisees or self-employed. Telephone interviews made in the fall of 2007 showed that among the 39 respondents there were 3 people who had already established and operated their businesses for a few years before participating in the program. When the three enrolled in the program they were interested in becoming franchisees and their companies were either being run by others or in a resting state. However, after the program they returned to running their businesses. Since the circumstances of these 3 and the other 36 respondents were not comparable, the 3 were excluded from the analyses.

Some of the remaining 36 respondents had closed or sold their businesses between the second and third phase of the longitudinal study and were salaried employees or unemployed at the time of the interviews. The goal was to compare the entrepreneurial decision-making processes of franchisees and the self-employed. Therefore, possible differences between the groups of operating and former franchisees as well as between operating and former self-employed people were examined. Selection of demographic, background, and entrepreneurial process, intention, motivation and satisfaction variables were examined employing binomial test, t-test, Man-Whitney U-test and cross-tabulation in the analysis. Since no significant biases were found between the groups of operating and former entrepreneurs, the respondents who had closed their businesses were included in the groups of operating entrepreneurs. Consequently, data analysis was carried out with 20 franchisees and 16 self-employed individuals.

---

[14] Triggering event is described as an event or circumstance (such as lay-off, divorce, winning or inheriting money) that enables or gives a person a final push or pulls him/her to entrepreneurship (see e.g. Bygrave 1989, 2004; Shindehutte et al. 2000; Davidsson 2005). The notion of triggering event is banded with push and pull factors of entrepreneurship and entrepreneurial intention and motivation (see e.g. Shapero and Sokol 1982; Bygrave 1989, 2004).

4.2 Sub-group Comparisons

Research has identified personal and environmental, i.e. contextual, factors that influence the probability that a person will discover and exploit entrepreneurial opportunities. The empirical measures employed in the study were extracted from the literature. Measures regarding respondents' entrepreneurial decision-making processes, demographic and background information, career experience, entrepreneurial satisfaction and intention, and impact of the training program on their decisions to select franchising or self-employment were analysed. Before logistic regression analysis (LRA) preliminary analysis was conducted to compare the groups of franchisees and self-employed. The groups were small, and thus, the statistical tests were chosen carefully. It was found that some tests were sensitive to the small number of respondents and therefore, diverse tests were utilised to enhance the reliability of the analysis: binomial test, t-test, Man-Whitney U-test and cross-tabulation were employed. No statistically significant differences appeared between the groups but the comparisons gave an indication of the factors worthy of being included in LRA models.

Shane (2003) found that entrepreneurial opportunities can take the form of new products and services, new ways of organising, new raw materials, new markets and new production processes and that one way to exploit an entrepreneurial opportunity is franchising. The results of the current study supported Shane's view. Two thirds (67 %) of the analysed 36 respondents discovered an entrepreneurial opportunity via franchising. Four of those who first operated as franchisees became self-employed after the term of their franchise contract period ended, and continued in the same industry where they had operated as franchisees. Respondents also found business opportunities via prior job or position and via other social networks such as family, friends, acquaintances and hobbies. Participation in the franchisee training program proved to be a form of interaction with other people and a way to gain access to information about entrepreneurial opportunities: respondents spoke of social interaction and discussions with franchisors' representatives, other trainees and instructors of the program. Additionally, the training program provided the trainees with information about markets: franchisors operating in Finnish markets, available franchisee positions, how to run a franchised business and also general information on issues related to owning and running a business in Finland.

Entrepreneurship literature has identified demographic and background factors that influence the likelihood of exploiting entrepreneurial opportunities. Such factors are gender, age, marital status/working spouse, education and unemployment (see e.g. Shane 2003). No statistically significant differences were found regarding these factors between the groups. Moreover, none of these appeared as a strong predictor of becoming a franchisee or self-employed among the respondents. Along with the background factors, entrepreneurship and franchising studies have presented career factors that often are closely linked to the decision to become an entrepreneur. Such factors are unemployment, general business experience, functional experience, industry experience, entrepreneurial experience and vicarious experience (see. e.g. Cooper and Dunkelberg 1986; Tervo and Niittykangas 1994; Kaufmann and Stanworth 1995; Stanworth and Kaufmann 1996; Price 1997; Delmar and Davidsson 2000; Littunen 2000; Shane 2003; Bennett et al. 2010). When the two entrepreneur groups were compared no considerable differences were found regarding the career factors.

According to the franchising literature the prospective franchisee needs to address the same issues that a prospective stand-alone entrepreneur does, plus issues that are specific to franchising. The current study aimed to take the specific features of franchising into account and hence, few questions presented to franchisees differed from the questions posed to self-employed. Franchising studies have shown that the decision to become a franchisee is one facet of the more general decision process of becoming an entrepreneur. Findings of this investigation were in line with the previous studies. Entrepreneurship was a long-time dream to nearly all respondents and the majority (90 %) of the franchisees first decided to become entrepreneurs. For 70 % of the franchisees the form of entrepreneurship,[15] i.e. franchising, was the second and industry the third decision. Closely related to the decisions on the form of entrepreneurship and industry was selection of the franchise. Six out of ten franchisees chose the franchise they did because the business format of that particular franchise was suitable for them. Nearly half (45 %) chose the franchise because the industry felt interesting and suitable.

Motivations to become an entrepreneur differed to some extent between franchisees and the self-employed which was expected based on previous research. However, there were factors that motivated respondents in both groups. The most essential factors that motivated respondents to become franchisees related to franchising as a form of business and entrepreneurship: 75 % of franchisees mentioned “ready-made and proven business idea and formula”, 70 % mentioned “scale economies provided by the franchise system” and 60 % mentioned “support provided by the franchisor”. Many reported that franchising provided them with an opportunity to fulfil a long-time goal of becoming an entrepreneur. The special feature of franchising is that education and/or work experience in the same field or self-employment experience are seldom expected from a prospective franchisee since the franchisor will provide training for the franchisee (see e.g. Curran and Stanworth 1983; Emerson 1998; Kaufmann 1999; Tuunanen 2005; Clarkin and Swavely 2006). However, findings of prior studies on franchisors’ preferences regarding prospective franchisees’ entrepreneurial experience (see e.g. Ozanne and Hunt 1971; Peterson and Dant 1990; Kaufmann and Stanworth 1995; Stanworth and Kaufmann 1996; Kaufmann 1999; Williams 1999; Tuunanen and Hyrsky 2001; Bennett et al. 2010) and work experience in the same business sector (see e.g. Tatham et al. 1972; Edens et al. 1976; Knight 1984, 1986; Owen 1989; Fenwick and Strombom 1998; Jambulingam and Nevin 1999) are equivocal. Some franchisors seem to prefer prior relevant experience whereas others do not. Results of the current study showed that 35 % of the franchisees had work experience in the

[15] Management studies on franchising recognise form of entrepreneurship as organisational form.

same business sector as the franchise they purchased and similarly, 35 % had prior entrepreneurial experience. Consistent with prior entrepreneurship literature (see e.g. Cooper and Dunkelberg 1986; Tervo and Niittykangas 1994; Delmar and Davidsson 2000; Littunen 2000; Shane 2003), the results of this investigation showed that many franchisees had career experience that may have provided them with the knowledge and skills to aid them in recognising entrepreneurial opportunities and assist in running their own businesses: 75 % had worked in a micro-size company, 55 % had work experience in marketing and sales and 50 % in management.

The following were the most noteworthy factors that motivated the self-employed to become entrepreneurs. Interestingly, many franchisees were motivated by the same factors: "opportunity to employ oneself and others" motivated 63 % of the self-employed and 40 % of the franchisees. "Opportunity to act out one's ideas, goals and dreams" motivated 44 % of the self-employed and interestingly, few (20 %) franchisees mentioned the same factor. "Other motivating factors" was mentioned by 44 % of the self-employed. Examples of these were "higher incomes reached via self-employment", "opportunity to work for myself" and "willingness to develop the business and grow the company that I purchased". Like franchisees, many of the self-employed had gained career experience that entrepreneurship research has identified as beneficial in recognising entrepreneurial opportunities as well as in running a business: 50 % had work experience in a micro-size company, 69 % had work experience in marketing and sales and 56 % in management. Also 50 % had work experience in the same business sector as the business they started or bought whereas 31 % had prior entrepreneurial experience.

Logistic regression analysis was performed to find out whether the factors presented in the literature and specific for this study could be used to predict respondents' likelihood of becoming either franchisees or self-employed. Various LRA models were built for both groups but only some models were found to have good predictive power and very few variables appeared statistically significant. Diverse factors were discovered to predict becoming a franchisee and a self-employed. The results for franchisees are presented in Table 2. The Chi-square value for the model was statistically significant, Nagelkerke R^2 value was 0.73 and 89 % of the total observations were classified correctly by the model. Variable "scale economies provided by the franchise system" was statistically significant and "ready-made and proven business idea and formula" was weakly significant.

The results for the self-employed are reported in Table 3. The Chi-square value of the model for self-employed was statistically significant, Nagelkerke R^2 value was 0.50 and 81 % of the total observations were classified correctly by the model. There were two statistically significant variables "work experience in a micro-size company" and "triggering event: other". Examples of the "other" triggering events mentioned by the respondents were long-time willingness to become an entrepreneur, finding and/or purchasing business premises, and willingness to find something new to enrich one's life. In addition, "opportunity to employ oneself and others" and "work experience in the same business sector as the business started" were weakly significant.

Table 2 LRA model on factors that predict becoming a franchisee

Variables		Coefficient	S.E.	Sig.
Career experience upon becoming a franchisee				
Work experience in other areas/tasks (e.g. customer service)		−2.945	1.764	0.095
Work experience in the same business sector as the business started		−1.334	1.490	0.371
Motivations to become a franchisee				
Ready-made and proven business idea and formula		2.714	1.476	0.066
Scale economies provided by the franchise system		4.025	1.814	0.026
The most important factors influencing the decision to become a franchisee				
Environmental/contextual factors		−0.809	1.547	0.601
Constant		−2.601	1.739	0.135
Chi-square	28.113			
Sig.	0.000			
Nagelkerke R^2	0.726			

Independent variables used in the analyses were dichotomous and had values of 0 = yes and 1 = no

	Predicted		
Observed	*Franchisees*	*Self-employed*	Percentage correct
Franchisees (n = 20)	17	3	85.0
Self-employed (n = 16)	1	15	93.8
Overall percentage			88.9

Table 3 LRA model on factors that predict becoming a self-employed

Variables		Coefficient	S.E.	Sig.
Career experience upon becoming a self-employed				
Work experience in a micro-size company		−2.557	1.148	0.026
Work experience in the same business sector as the business started		2.004	1.145	0.080
Work experience in other areas/tasks (e.g. customer service)		1.710	1.128	0.130
Motivations to become a self-employed				
Opportunity to employ oneself and others		1.893	1.033	0.067
Triggering event				
Other		1.865	0.934	0.046
Constant		−2.853	1.329	0.032
Chi-square	17.007			
Sig.	0.004			
Nagelkerke R^2	0.504			

Independent variables used in the analyses were dichotomous and had values of 0 = yes and 1 = no

	Predicted		
Observed	*Self-employed*	*Franchisees*	Percentage correct
Self-employed (n = 16)	11	5	68.8
Franchisees (n = 20)	2	18	90.0
Overall percentage			80.6

Results of the LRA models indicate that factors related to franchising as a form of business and entrepreneurship, i.e. "scale economies provided by the franchise system" and "ready-made and proven business idea and formula" were emphasised in the entrepreneurial decision making processes of the franchisees (Table 2). Becoming an entrepreneur was a long-time dream for many respondents and franchising provided a few of them with the opportunity that they may have not otherwise had been brave enough to pursue. The franchisees' comments supported the results of the logistic regression analysis: "It (becoming an entrepreneur) was easier via franchising", "The concept was ready", "I wanted to become an entrepreneur, but I didn't have the courage to do it by myself", "I had no special skills or knowledge to base a business idea on".

Previous entrepreneurship research has shown that working in a small business increases a person's awareness of the opportunities for entrepreneurship, develops person's entrepreneurial qualifications and helps to develop an overall picture of running a small business (see e.g. Cooper and Dunkelberg 1986; Tervo and Niittykangas 1994; Delmar and Davidsson 2000). Results of the LRA models can be interpreted to indicate that work experience in a micro-size company was not emphasised in the decision making process of the self-employed but was linked to becoming a franchisee (Table 3). This finding was somewhat contradictory to prior entrepreneurship studies but it is in line with franchising studies which have shown that many franchisees have prior self-employment experience (see e.g. Peterson and Dant 1990; Kaufmann and Stanworth 1995; Stanworth and Kaufmann 1996; Tuunanen and Hyrsky 2001; Welsh et al. 2011). Entrepreneurship research has also found that new entrepreneurs often rely on their work experience when starting a company (see e.g. Littunen 2000; Shane 2003). The results of the current study were consistent with these prior findings. Work experience in the same business sector as the business started appeared to be linked to the decision making process of the self-employed (Table 3).

Results of the LRA models further indicated that factors closely related to a desire to become self-employed and to find something new in one's life were emphasised in the decision-making processes of the self-employed (Table 3). Noteworthy is that unemployment did not appear as a strong predictor of becoming an entrepreneur even though one third of the respondents (30 % of franchisees and 38 % of self-employed) were unemployed at the time they entered the program. On the other hand, "opportunity to employ oneself" appeared to explain becoming self-employed which may have reflected the influence of unemployment on the decision to become self-employed. In addition, 40 % of the franchisees identified "opportunity to employ oneself" as a motivating factor in becoming a franchisee. This shows that the same factor can be seen negatively and/or positively, i.e. as a push and/or a pull factor to entrepreneurship and consequently, the perception of the interpreter plays a key role.

Based on the results of the various LRA models built for the two entrepreneur groups the following findings can be summarised: the franchising specific factors (scale economies, ready-made and proven business idea and formula and franchisor's support), and work experience in small businesses (work experience in a micro-size business and work experience in family business) were emphasised in the decision-making processes of the franchisees. In comparison with the franchisees, strong desire to become self-employed, willingness to find something new to one's life as well as relevant work experience (in the same business sector as the business started and in other areas like customer service) were emphasised in the decision-making processes of the self-employed.

4.3 Impact of the Training Program, Intentions and Satisfaction on the Respondents' Entrepreneurial Processes

An essential goal of the study was to examine the role of the franchisee training program in respondents' entrepreneurial decision-making processes. Over 80 % responded that the program had an effect on their decision but only 25 % indicated that the effect was strong. Nearly half (47 %) believed that they would have made the decision to become franchisees or self-employed without completing the training. However, 22 % could not say how they would have decided in such case. The majority of respondents (89 %) believed that entrepreneurship training may influence a person in such a way that s/he will become an entrepreneur. Note that the question referred to entrepreneurship training in general. Yet, all highlighted that at the same time the person needs to have a strong desire to become an entrepreneur and no training can make someone yearn to become an entrepreneur. Many told that the franchisee training program provided them the time and place for information gathering, analysing, networking and career decision-making. Some saw that the program supported the decision they had already made whereas some indicated that during the program they realised that franchise ownership was not a suitable option for them. Several expressed that via the program they found out about franchise opportunities and made contact with the representatives of the franchisor they chose. A small number pointed out that the program gave them general information on issues related to owning and running a business in Finland. Based on these findings, it can be said that the program was valuable both for those who became franchisees and those who became self-employed, even though the program did not appear as a statistically significant predictor of becoming a franchisee or becoming a self-employed.

Entrepreneurship and franchising research has indicated that entrepreneurial (dis-)satisfaction and intention closely relate to entrepreneurial process and intertwine throughout the process (see e.g. Bird 1988; Cooper and Artz 1993; Hing 1995; Gassenheimer et al. 1996; Littunen 2001; McMullen and Shepherd 2002; Tuunanen 2002; Gauzente 2003; Davies et al. 2011). Thus, measures of entrepreneurial (dis-)satisfaction and intention were included in the study. Few of these measures were also utilised in assessing effectiveness of the training program. Entrepreneurship seemed to be a positive experience for most of the respondents. The majority in both entrepreneur groups and in total (86 %) were satisfied with their decision to become entrepreneurs. Similarly, the majority in both groups and in total (78 %) were satisfied with the time they had operated as entrepreneurs. Most respondents (86 %) were also willing to recommend franchise ownership or self-employment based on their experience.

Seventy percent of the respondents who were operating as franchisees or self-employed at the time of the interviews intended to continue operations. Some also said that they would like to try something else in the future, even though they were quite satisfied with their status as franchisees or self-employed. The most often mentioned reasons for willingness to exit operations were "boredom or willingness to do something else" (mentioned by 11 respondents), and "ageing or retirement" (mentioned by 7 respondents). Six respondents mentioned "long-term unprofitable business" and "other reasons related to business". A small number were dissatisfied with their decision to become entrepreneurs ($n = 4$) as well as with the time they had operated as entrepreneurs ($n = 7$). One fourth (25 %) expressed their disappointment and had no intention of becoming entrepreneurs in future. Even though all respondents had not succeeded as entrepreneurs and had closed operations or were planning to exit, many were satisfied with their decision to become entrepreneurs and some were looking for new entrepreneurial opportunities. Several respondents said that entrepreneurship is a way of life for them.

5 Conclusions

The current study reports the results of the third and final phase of the longitudinal study on trainees of the Finnish franchisee training program. The study focused on the entrepreneurial decision-making processes of the 20 franchisees and 16 self-employed. The purpose was to find out what factors influenced respondents' entrepreneurial decision-making processes and what was the role of the franchisee training program in their processes.

The investigation was exploratory in nature. Recent developments of entrepreneurship theory were applied to franchising. Moreover, franchising was seen as a form of entrepreneurship and franchisees as entrepreneurs, which is not a widely accepted view in entrepreneurship or franchising research. Studies on the entrepreneurial process of franchisees are scarce but previous research has found that the decision-making processes of becoming either a franchisee or self-employed in a stand-alone business are connected. To examine the entrepreneurial decision-making processes of the targeted franchisees and self-employed, frameworks from entrepreneurship and franchising studies were adapted since no single model served the purpose of the investigation.

Shane's (2003) *general theory of entrepreneurship* breaks away from earlier, rather rigid, detailed definitions of entrepreneurship and trait theory perspectives, and provides a holistic, process-based view. For example, diverse definitions for entrepreneurs, i.e. people who own and run businesses (such as self-employed person, small businessman/-woman/-owner) are seen as being essentially interchangeable; they do not have connotations of being "more" or "less" entrepreneurial in nature. Furthermore, entrepreneurship is not bound to a certain period of time. Instead, it is seen as a process that can have feedback loops and be non-linear. Moreover, although the entrepreneurial process requires innovation, a recombination of resources into a new form is sufficient to fulfil this requirement (Kirznerian perspective).

Shane's (2003) framework is well-suited to explaining new forms of market and interorganisational relations such as franchising. In addition, franchisees fit well with the requirements set for entrepreneurs in Shane's model. For instance, the entrepreneurial opportunity does not have to be based on a novel product or service, it may take the form of new ways of organising, new raw materials, new markets and new production processes and becoming an entrepreneur may occur through market mechanisms such as franchising (cf. Vesper 1980; Baumol 1986). An example could be the following: instead of establishing a stand-alone business that is based on a novel product or service a person may recognise a business opportunity in a franchisor's supply, the franchise concept, which can be utilised in the local market. A franchisee will use new means-ends frameworks, and his/her knowledge and skills, to introduce the franchisor's concept to local markets. Furthermore, even though the franchisee receives a ready-made business concept and a known brand, s/he will bear the risk consequences of operating the business in a particular local market, plus the risk derived from the operations of fellow franchisees and the franchisor. As the owner of the venture, the franchisee will eventually bear the risk of failure alone. To be able to operate the business in the particular local market successfully, the franchisee needs to exercise entrepreneurial decision-making and creativity. This implies that franchisees have innovative potential which is important for the competitiveness and development of the whole franchise and thus useful for the franchisor (see. e.g. Jambulingam and Nevin 1999).

Many theories that have been used to explain franchising e.g. transaction cost theory, agency theory, and resource-based theory represent mainly the franchisor's perspective. Applying Shane's model of entrepreneurial process to franchisees' entrepreneurial process will aid in outlining the franchisees' perspective and in broadening the picture of the franchise relationship. Therefore, Shane's model will complement the previously used theories.

Shane's model is meant to be applicable to all forms of entrepreneurship. Nevertheless, there are forms of entrepreneurship such as franchising where the entrepreneurial process is not as straightforward as in more conventional forms and Shane's model does not take that into account. For the purpose of the current study, frameworks from franchising studies were applied to complement Shane's model. The essential, specific features of franchising that influence the franchisee's entrepreneurial process are the following: (1) Franchisor has an active role in the entrepreneurial process of a prospective franchisee. Franchisor selects franchisees and acceptance by the franchisor is the precondition for anyone to become a franchisee. (2) Franchising suits some prospective franchisees better than others. A franchisee is a collaborative partner in a franchise relationship and a member of the franchise system and this background affects all the decisions and actions of a franchisee (cf. Fenwick and Strombom 1998; Bennett et al. 2010). (3) The franchise relationship is contract-based and builds on an efficient division of labour and profit sharing between the franchisor and the franchisee. In addition to the above mentioned features, there are other factors that may influence the decision to become an entrepreneur and which might differ between franchisees and self-employed as studies of e.g. Bradach and Kaufmann (1988), Spinelli (1994), Kaufmann and Stanworth (1995), Stanworth and Kaufmann (1996), Williams (1999) and Kaufmann (1999) suggested. An example of these factors is the influence of the prospective entrepreneur's perception of risk.

The findings of the current study showed that factors related to franchising as a form of business had a significant role in the decision-making process of franchisees. Scale economies provided by the franchise system, a ready-made and proven business idea and formula as well as franchisor's support were shown to be important factors for the franchisees. Thus, the old saying "being in business for yourself but not by yourself" seemed to be applicable to the examined franchisees. In comparison with the franchisees, factors closely related to the desire to become self-employed were emphasised in the entrepreneurial decision-making processes of the self-employed. Career experience found by prior research as beneficial in recognising entrepreneurial opportunities and in running a business as well as experience related to small businesses seemed to have a role in the entrepreneurial decision-making processes of the respondents. For instance, work experience in a micro-size company appeared to be linked to becoming a franchisee whereas work experience in the same business sector as the business started was emphasised in the decision-making process of the self-employed. These findings support the conclusion that Shane's entrepreneurial process model can be applied to franchising and that the processes of becoming a franchisee and self-employed in stand-alone business are connected. At the same time the findings showed that although prospective franchisees and self-employed in stand-alone business address many of the same issues in their entrepreneurial decision-making process, the point of view may differ because franchising has features that distinguish it from many other forms of entrepreneurship. For example, instead of focusing on finding a distinguishable and viable business idea, prospective franchisees need to put effort into finding a suitable business partner (franchisor) and concept (franchise). The findings further illustrated that entrepreneurial processes can be very case-specific.

Research has indicated that entrepreneurial intentions, motivation and satisfaction intertwine throughout entrepreneurial process. The findings of this study were consistent with prior studies. For instance, entrepreneurship was a long-time dream for nearly all respondents, i.e. they had a firm intention of becoming entrepreneurs. Moreover, most respondents were satisfied with their decision to become franchisees or self-employed and intended to continue operations. On the other hand, despite that some had not succeeded as entrepreneurs, they were satisfied with their decision to become entrepreneurs and few were looking for new entrepreneurial opportunities. This is in accordance with Shane's (2003) notion that entrepreneurial processes can have feedback loops and be non-linear. In addition to Shane's notion, the view of entrepreneurship as a career (see e.g. Bird 1989; Katz 1994;

Dyer 1994; Feldman and Bolino 2000; Carter et al. 2003) as well as the concept of habitual entrepreneurship (see e.g. Carland et al. 2000; Usbasaran et al. 2003) can be applied to many of the respondents. For example, few who started their entrepreneurial career as franchisees continued later as self-employed in stand-alone businesses. Another example is that a small number of the operating franchisees and self-employed had new business ideas in mind and they intended to exploit those ideas in future.

An essential goal of the longitudinal study was to examine the impact of the Finnish franchisee training program on the entrepreneurial decision-making processes. Results of the logistic regression analysis indicate that the program was not a statistically significant predictor of becoming a franchisee or becoming a self-employed among the respondents. However, even though the impact of the program was seen and expressed in various ways by the respondents, the program influenced their decisions to become entrepreneurs: the respondents received support and tools for career decision-making, and they obtained focal information on franchise ownership and general issues related to owning and running a business. Furthermore, via the program many franchisees found out about available franchisee opportunities and made contact with representatives of the franchisors who introduced their franchise offerings in the program. In addition, many of the self-employed viewed the program as beneficial entrepreneurship training. Based on these findings, it can be said that the program was instrumentally valuable both for those who became franchisees and those who became self-employed.

Generally speaking, the franchisee training program served as a franchising information channel in Finland, where franchising is still relatively unknown and a poorly understood form of business and entrepreneurship. It was also a get-together venue for franchisors and prospective franchisees in a market that lacks franchise fairs and exhibitions. In addition, the program had significant indirect effects such as the franchised and stand-alone businesses established as well as the revenue and jobs created by those businesses. To conclude, the Finnish franchisee training program was a worthwhile government-financed small business policy activity. It met the need of locating new franchisees and benefited the Finnish franchising sector and the economy indirectly through the resulting established and continued businesses.

The third phase was carried out to complete the longitudinal study. The response rate remained relatively high in this final phase although the number of respondents was small. Comparisons between franchisees and the self-employed added to the challenge as respondents were divided into even smaller groups. The study also had a retrospective viewpoint and possibility of hindsight bias was acknowledged. Taking both the small number of respondents and the retrospective point of the investigation into consideration, effort was made to improve the quality of the study: The empirical measures employed were based on literature and selected complementary items were used. Both quantitative and qualitative techniques were applied in data gathering. Information given by the respondents was compared to their responses in earlier phases of the longitudinal study as well as the data obtained from third party resources.[16] Statistical tests used were chosen carefully and diverse tests were utilised to enhance the reliability of the analyses. Overall, the longitudinal design provided an opportunity to compare the results obtained in the three phases and to gain a more extensive picture of the entrepreneurial process of the respondents and the factors influencing it. In that way the longitudinal design acted as a confirming factor enhancing the quality of the study.

Even though steps were taken to improve the quality of the study, the small number of respondents and the possibility of hindsight bias can be seen as limitations of the study. If hindsight bias was present, it could probably be seen in respondents' expressions of the impact of the franchisee training program on their decision to become franchisees or self-employed: In retrospect, respondents might have seen the role of the program different. The outcome, i.e. respondents' entrepreneurship (experience) may also have influenced their retrospective perception of the decision-making process and the factors relating to it. Especially, how successful the respondents saw that they were as entrepreneurs and whether operating as a franchisee or self-employed met their initial expectations. Moreover, in hindsight the respondents may have highlighted the impact of the most memorable or outcome consistent factor in their decision-making process. Research has shown that hindsight bias effects can be individual and hindsight bias may be caused by e.g. memory distortions and even unintentional cognitive processes. Additionally, diverse sense-making processes, self-presentation and motivational issues can play a significant role in the way people perceive and express prior events in hindsight (see Musch and Wagner 2007; Pezzo and Pezzo 2007; Blank et al. 2008; Ash 2009; Nestler et al. 2010; Bernstein et al. 2011).

Due to the limitations of the study the results need to be interpreted with caution. Nevertheless, the study has merit: it is the concluding part of a three phase longitudinal investigation. Moreover, it contributes to entrepreneurship and franchising literature and provides new knowledge on entrepreneurial processes that vary among the diverse forms of entrepreneurship. The study also illuminates the numerous ways that entrepreneurship training may influence the entrepreneurial process.

5.1 Proposals for Future Studies

This study invites franchising and entrepreneurship scholars to apply recent views and definitions of entrepreneurship to franchising research. It also calls for more studies on franchisees as entrepreneurs. In addition, more comparative studies on

[16] Information given by the respondents concerning e.g. the contents and the timing of the training program was compared with the information received from the organisers of the program. Additionally, information concerning e.g. the company form and the time of establishment and closure of the company were compared with the data obtained in the national business information system which is a joint service for businesses and organisations that are clients of the Finnish Tax Administration and Trade Register.

franchisees and the self-employed in stand-alone businesses are needed to obtain a clearer picture of these dissimilar types of entrepreneurs. Franchisor's role in the entrepreneurial process of a franchisee would also be a fruitful subject to future investigations. The numerous particular features of franchise relationships distinguish franchising from other forms of entrepreneurship and provide an intriguing and inspiring canvas for entrepreneurship and franchising researchers.

Acknowledgements The author would like to express her gratitude to professor, Dr. Hannu Littunen, University of Eastern Finland, docent, Dr. Mika Tuunanen and Dr. Jenny Buchan, University of New South Wales for their insightful comments on the study and earlier versions of this article.

References

Alvarez SA, Barney JB (2007) Discovery and creation: alternative theories of entrepreneurial action. Strateg Entrepren J 1(1–2):11–26
Anderson RL, Condon C, Dunkelberg J (1992) Are franchisees "Real" entrepreneurs? J Bus Entrepren 4(1):97–105
Ash IK (2009) Surprise, memory, and retrospective judgment making: testing cognitive reconstruction theories of the hindsight bias effect. J Exp Psychol Learn Mem Cogn 35(4):916–933
Baumol WJ (1986) Entrepreneurship and a century of growth. J Bus Venturing 1(2):141–145
Bennett S, Frazer L, Weaven S (2010) What prospective franchisees are seeking. J Mark Channels 17(1):69–87
Bernstein DM, Erdfelder E, Meltzoff AN, Peria W, Loftus GR (2011) Hindsight bias from 3 to 95 years of age. J Exp Psychol Learn Mem Cogn 37(2):378–391
Bird B (1988) Implementing entrepreneurial ideas: the case for intention. Acad Manage Rev 13(3):442–453
Bird B (1989) Entrepreneurial behavior. Scott, Foresman and Company, Glenview
Bird B, Jelinek M (1988) The operation of entrepreneurial intentions. Entrepren Theory Prac 13(2):21–29
Blank H, Nestler S, von Collani G, Fischer V (2008) How many hindsight biases are there? Cognition 106(3):1408–1440
Bradach J, Kaufmann P (1988) Franchisee or independent businessperson: some observations on the decision process. In: Hills GE, La Forge W (eds) Research at the marketing/entrepreneurship interface. University of Illinois at Chicago, Chicago, pp 38–48
Brockhaus RH (1980) The effect of job dissatisfaction in the decision to start a business. J Small Bus Manage 18(1):37–43
Bygrave WD (1989) The entrepreneurship paradigm (I): a philosophical look at its research methodologies. Entrepren Theory Prac 14(1):7–26
Bygrave WD (2004) The entrepreneurial process. In: Bygrave WD, Zacharis A (eds) The portable MBA in entrepreneurship, 3rd edn. Wiley, Hoboken, pp 1–27
Carland JC, Carland JW, Stewart WH Jr (2000) The indefatigable entrepreneur: a study of the dispositions of multiple venture founders. J Bus Entrepren 12(1):1–18
Carland J, Hoy F, Boulton WR, Carland JAC (1984) Differentiating entrepreneurs from small business owners: a conceptualization. Acad Manage Rev 9(2):354–359
Carland J, Hoy F, Carland JAC (1988) "Who is an Entrepreneur?" is a question worth asking. Am J Small Bus 12(4):33–39
Carter NM, Gartner WB, Shaver KG, Gatewood EJ (2003) The career reasons of nascent entrepreneurs. J Bus Venturing 18(1):13–39

Clarkin JE, Swavely SM (2006) The importance of personal characteristics in franchisee selection. J Retailing Consum Serv 13(2):133–142
Cooper AC, Artz K (1993) Determinants of satisfaction for entrepreneurs. In: Frontiers of entrepreneurship research 1993, Proceedings of the Babson College entrepreneurship research conference. University of Houston, Small Business Development Center, Houston
Cooper AC, Dunkelberg WC (1986) Entrepreneurship and paths to business ownership. Strateg Manage J 7(1):53–68
Cunningham JB, Lischeron J (1991) Defining entrepreneurship. J Small Bus Manage 29(1):45–61
Curran J, Stanworth J (1983) Franchising in the modern economy – towards a theoretical understanding. Int Small Bus J 2(1):8–37
Dant RP (2008) A futuristic research agenda for the field of franchising. J Small Bus Manage 46(1):91–98
Davidsson P (2005) Researching entrepreneurship. Springer Science + Business Media, New York
Davies MAP, Lassar W, Manolis C, Prince M, Winsor RD (2011) A model of trust and compliance in franchise relationships. J Bus Venturing 26(3):321–340
Delmar F, Davidsson P (2000) Prevalence and characteristics of nascent entrepreneurs. Entrepren Reg Dev 12(1):1–23
Dyer WG Jr (1994) Toward a theory of entrepreneurial careers. Entrepren Theory Prac 19(2):7–21
Edens FN, Self DR, Grider DR Jr (1976) Franchisors describe ideal franchisee. J Small Bus Manage 14(July):39–47
Elango B, Fried VH (1997) Franchising research: a literature review and synthesis. J Small Bus Manage 35(3):68–81
Emerson RW (1998) Franchise terminations: legal rights and practical effects when franchisees claim the franchisor discriminates. Am Bus Law J 35(4):559–646
Feldman DC, Bolino MC (2000) Career patterns of the self-employed: career motivations and career outcomes. J Small Bus Manage 38(3):53–67
Fenwick GD, Strombom M (1998) The determinants of franchisee performance: an empirical investigation. Int Small Bus J 16(4):28–45
Gartner WB (1988) "Who is an Entrepreneur?" is the wrong question. Am J Small Bus 12(4):11–32
Gassenheimer JB, Baucus DB, Baucus MS (1996) Cooperative arrangements among entrepreneurs: an analysis of opportunism and communication in franchise structures. J Bus Res 36(1):67–79
Gauzente C (2002) Using qualitative methods in franchise research – an application in understanding the franchised entrepreneurs' motivations. Forum Qual Soc Res 3(1), Art. 20. Retrieved November, 10, 2009 from: http://www.qualitative-research.net/index.php/fqs/article/view/884
Gauzente C (2003) Measuring franchisees' satisfaction: theoretical considerations and empirical testing. J Retail Distribution Manage 31(10):508–517
Gibb A (2002) In pursuit of a new 'Enterprise' and 'Entrepreneurship' paradigm for learning: creative destruction, new values new ways of doing things and new combinations of knowledge. Int J Manage Rev 4(3):233–269
Gilad B, Levine P (1986) A behavioral model of entrepreneurial supply. J Small Bus Manage 24(4):45–53
Grünhagen M, Dorsch MJ (2003) Does the franchisor provide value to franchisees? Past, current and future value assessments of two franchisee types. J Small Bus Manage 41(4):366–384
Guilloux V, Gauzente C, Kalika M, Dubost N (2004) How France's potential franchisees reach their decisions: a comparison with franchisers' perceptions. J Small Bus Manage 42(2):218–224
Harmon TR, Griffits MA (2008) Franchisee perceived relationship value. J Bus Ind Mark 23(4):256–263

Hing N (1995) Franchisee satisfaction: contributors and consequences. J Small Bus Manage 33(2):12–25
Hoy F (2008) Organizational learning at the marketing/entrepreneurship interface. J Small Bus Manage 46(1):152–158
Hoy F, Shane S (1998) Franchising as an entrepreneurial venture form. J Bus Venturing 13(2):91–94
Hoy F, Stanworth J, Purdy D (2000) An entrepreneurial slant to franchising research. In: Sexton DL, Landström H (eds) The Blackwell Handbook of Entrepreneurship. Blackwell Business, Oxford, pp 408–432
Huang Y, Phau I (2009) Mapping the profiles of franchisees: getting to know the black sheep, rough diamonds, whingers and best buddies. Direct Mark Int J 2(4):221–238
Huuskonen V (1992) Yrittäjäksi ryhtyminen – Teoreettinen viitekehys ja sen koettelu. (The process of becoming an entrepreneur – a theoretical framework with empirical experiments) Publications of the Turku School of Economics and Business Administration. Series A-2-1992
Jambulingam T, Nevin JR (1999) Influence of franchisee selection criteria on outcomes desired by the franchisor. J Bus Venturing 14(4):363–395
Katz JA (1994) Modeling entrepreneurial career progressions: concepts and considerations. Entrepren Theory Prac 19(2):23–39
Kaufmann PJ (1999) Franchising and the choice of self-employment. J Bus Venturing 14(4):345–362
Kaufmann PJ, Stanworth J (1995) The decision to purchase a franchise: a study of prospective franchisees. J Small Bus Manage 33(4):22–33
Ketchen DJ Jr, Short JC, Combs JG (2011) Is franchising entrepreneurship? Yes, no, and maybe so. Entrepren Theory Prac 35(3):583–593
Kirzner I (1997) Entrepreneurial discovery and the competitive market process: an Austrian approach. J Econ Lit 35(1):60–85
Knight RM (1984) The independence of the franchise entrepreneurs. J Small Bus Manage 22(2):53–61
Knight RM (1986) Franchising from the franchisor and franchisee points of view. J Small Bus Manage 24(3):8–15
Krueger NF, Reilly MD, Carsrud AL (2000) Competing models of entrepreneurial intentions. J Bus Venturing 15(5–6):411–433
Littunen H (2000) Networks and local environmental characteristics in the survival of new firms. Small Bus Econ 15(1):59–71
Littunen H (2001) The birth and success of new firms in a changing environment. Jyväskylä Studies in Business and Economics no. 8, Academic Dissertation. University of Jyväskylä, Jyväskylä
McKenzie B, Ugbah SD, Smothers N (2007) "Who is an entrepreneur?" Is it still the wrong question? Acad Entrepren J 13:23–43, Retrieved October 13, 2009 from http://findarticles.com/p/articles/mi_hb6005/is_1_13/ai_n29411855/
McMullen JS, Shepherd DA (2002) Regulatory focus and entrepreneurial intention: action bias in the recognition and evaluation of opportunities. In: Frontiers of entrepreneurship research 2002, Proceedings of the Babson College entrepreneurship research conference. University of Colorado at Boulder, Robert H. and Beverly A. Deming Center for Entrepreneurship, Boulder
Morrison KA (1997) How franchise job satisfaction and personality affects performance, organizational commitment, franchisor relations and intention to remain. J Small Bus Manage 34(3):39–67
Musch J, Wagner T (2007) Did everybody know it all along? a review of individual differences in hindsight bias. Soc Cogn 25(1):64–82
Nestler S, Blank H, Egloff B (2010) Hindsight $\neq$ hindsight: experimentally induced dissociation between hindsight components. J Exp Psychol Learn Mem Cogn 36(6):1399–1413
Norton SW (1988) An empirical look at franchising as an organizational form. J Bus 61 (2):197–217

Owen T (1989) Franchising for future growth. Small Bus Rep 14(2):30–34
Ozanne UB, Hunt S (1971) The economic effects of franchising. Senate Select Committee on Small Business, Washington
Peterson A, Dant RP (1990) Perceived advantages of the franchise option from the franchisee perspective: empirical insights from a service franchise. J Small Bus Manage 28(3):46–61
Pezzo MV, Pezzo SP (2007) Making sense of failure: a motivated model of hindsight bias. Soc Cogn 25(1):147–164
Price S (1997) The franchise paradox: new directions, different strategies. Cassell, London
Rubin PH (1978) The theory of the firm and the structure of the franchise contract. J Law Econ 21:223–233
Schumpeter JA (1934) The theory of economic development: an inquiry into profits, capital credit, interest, and the business cycle. Harvard University Press, Cambridge
Seawright KW, Smith IH, Mitchell RK, McClendon R (2011) Exploring entrepreneurial cognition in franchisees: a knowledge-structure approach. Entrepren Theory Prac. doi:10.1111/j.1540-6520.2011.00467.x (Article first published online: 25 April 2011)
Shane S (2003) A general theory of entrepreneurship. The individual-opportunity nexus. Edward Elgar, Cheltenham
Shane S, Hoy F (1996) Franchising: a gateway to cooperative entrepreneurship. J Bus Venturing 11(5):325–327
Shapero A, Sokol L (1982) The social dimensions of entrepreneurship. In: Kent C, Sexton L, Vesper K (eds) Encyclopedia of entrepreneurship. Prentice-Hall, Engelwood Cliffs, pp 72–90
Shindehutte M, Morris MH, Kuratko DF (2000) Triggering events, corporate entrepreneurship and the marketing function. J Mark Theory Prac 8(2):18–30
Spinelli S (1994) Franchising. In: Bygrave WD (ed) The portable MBA in entrepreneurship. Wiley, New York, pp 353–376
Stanworth J (1995) The franchise relationship: entrepreneurship or independence? J Mark Channels 4(1/2):161–176
Stanworth J, Kaufmann PJ (1996) Similarities and differences in U.K. and U.S. franchise research data: towards a dynamic model of franchisee motivation. Int Small Bus J 14(3):57–70
Stanworth J, Kaufmann PJ, Purdy D (1995) The Blenheim/University of Westminster franchise survey: a comparison of UK and USA data. International Franchise Research Center, Special Studies, No 5, March. University of Westminster Press, London
Tatham R, Bush R, Douglas R (1972) An analysis of decision criteria in franchisor/franchisee selection process. J Retailing 48(1):16–21, 94
Tervo H, Niittykangas H (1994) The impact of unemployment on new firm formation in Finland. Int Small Bus J 13(1):38–53
Torikka J (2004) Finnish franchisee training program – findings from the first ten programs 1999–2001. In: Windsperger J, Hendrikse G, Cliquet G, Tuunanen M (eds) Economics and management of franchising networks. Physica, Heidelberg, pp 320–340
Torikka J (2007) Franchisees can be made – empirical evidence from a follow-up study. Int J Entrepren Small Bus 4(1):68–96
Torikka J, Tuunanen M (2003) Finnish franchisee training program – an exploratory study. In: Alon I, Welsh DHB (eds) International franchising in industrialized markets. Western and Northern Europe. CCH, Chicago, pp 165–187
Tuunanen M (2002) An ounce of prevention is a pound of cure: findings from national franchisee (dis-)satisfaction study. J Mark Channels 10(2):57–89
Tuunanen M (2005) Essays on franchising in Finland, empirical findings on franchisors and franchisees, and their relationships. Jyväskylä Studies in Business and Economics no. 37, Academic Dissertation. University of Jyväskylä, Jyväskylä
Tuunanen M, Hoy F (2007) Franchising – multifaceted form of entrepreneurship. Int J Entrepren Small Bus 4(1):52–67
Tuunanen M, Hyrsky K (2001) Entrepreneurial paradoxes in business format franchising: an empirical survey of Finnish franchisees. Int Small Bus J 19(4):47–62

Usbasaran D, Westhead P, Wright M (2001) The focus of entrepreneurial research: contextual and process issues. Entrepren Theory Prac 25(4):57–80
Usbasaran D, Wright M, Westhead P (2003) A longitudinal study of habitual entrepreneurs: starters and acquirers. Entrepren Reg Dev 15(3):207–228
Venkataraman S (2003) Foreword. In: Shane S A general theory of entrepreneurship. The individual-opportunity nexus. Edward Elgar, Cheltenham, pp xi–xii
Vesalainen J, Pihkala T (1999) Motivation structure and entrepreneurial intentions. In: Frontiers of entrepreneurship research 1999, Proceedings of the Babson College entrepreneurship research conference. University of South Carolina, the Darla Moore School of Business, Columbia
Vesper K (1980) New venture strategies. Prentice Hall, Englewood Cliffs
Welsh DHB, Desplaces DE, Davis AE (2011) A comparison of retail franchises, independent businesses and purchased existing independent business startups: lessons from the Kaufmann firm survey. J Mark Channels 18(1):3–18
Williams DL (1999) Why do entrepreneurs become franchisees? J Bus Venturing 14(1):103–124
Young JA, McIntyre FS (2011) The International Society of Franchising: a review of conference papers across its first 25 years. In: Proceedings of the International Society of Franchising conference, Boston
Young JA, McIntyre FS, Green RD (2000) The International Society of Franchising proceedings: a thirteen year review. In: Proceedings of the International Society of Franchising conference, San Diego

The Failure of Pre-purchase Disclosure to Protect Franchisees of a Franchisor in Administration

Jenny Buchan

Abstract Pre-purchase disclosure is an important aspect of the due diligence process for business format franchisees. It focuses on the financial fitness of the franchisor entity, and on the specific franchise opportunity the franchisee is evaluating. Equipped with disclosed information a diligent franchisee theoretically has the opportunity, *ex ante*, to identify and evaluate risks and protect itself from the consequences of making a bad investment decision. This chapter examines the efficacy of disclosure for franchisees whose insolvent franchisor enters administration. Problems arise out of the content and timing of disclosure, the difficulty of verifying the disclosed information and the conflicting requirements of the legislation protecting franchisees and that regulating administrators. Pre-purchase disclosure cannot empower franchisees to anticipate or address the consequences of franchisor administration. It fails to protect franchisees. Suggested avenues for redressing this situation are identified.

Keywords Angus & Robertson • Australia • Business format franchise • Due diligence • Franchisee • Franchising code of conduct • Franchisor bankruptcy • Franchisor in administration • Insolvent • Pre-purchase disclosure • Risk

1 Introduction

Business format franchising ('franchising') is adopted by thousands of businesses selling products and services in about two-thirds of the approximately 192 countries in the world.[1] We deal with franchisees in our daily lives: to buy travel, a meal, a

[1] See Spencer (2010).

J. Buchan (✉)
School of Taxation and Business Law, Australian School of Business, University of New South Wales, Kensington, NSW 2052, Australia
e-mail: jm.buchan@unsw.edu.au

T. Ehrmann et al. (eds.), *Network Governance*, Contributions to Management Science,
DOI 10.1007/978-3-7908-2867-2_17,

printer cartridge or a haircut. Whilst the key players in a franchise network are the franchisor and its franchisees, today's franchisors operate within complex, inter-connected webs of contractual relationships. Alan Felstead observed that '[o]nly by examining the inter-connections between firms can one fully appreciate that the ability to exercise control over production may stretch beyond a firm's legal borders'.[2] By examining the interaction between firms and the law one can appreciate the far-reaching consequences of adverse events that involve the franchisor. A poor business decision by the franchisor, a member of the group of companies of which it forms part or one of its other stakeholders can damage its franchisees and leave them with limited options.[3]

The franchise is marketed as a proven concept. Believing they are investing in a proven business concept, franchisees invest confidently. Notwithstanding the mantra of 'proven business' policy makers have recognised that franchisees are vulnerable in the franchise relationship and have implemented policies to address that vulnerability. One widely adopted strategy is to require franchisors to provide pre-purchase disclosure information to intending franchisees. This identifies some of the major commercial, financial and legal responsibilities and risks for the incoming franchisee. It focuses on the relationship between the franchisor and franchisee, with little information about the franchisor's related entities, those beyond the firm's legal borders.

However, franchises are not always proven before advertising for franchisees and in some cases franchisors continue selling franchises when the franchisor is already insolvent.[4] Many franchisors do not survive 10 years.[5] In light of the number of franchisors that enter administration, one variable which should be evaluated for its capacity to protect a franchisee from the potential adverse consequences of its franchisor's business failing is the disclosure document. This chapter identifies some consequences of franchisor failure for franchisees. Specifically, it demonstrates the ineffectiveness of disclosure in reducing the vulnerability of franchisees whose franchisor enters administration.

The chapter is structured in the following way. First the twenty-first century franchise model is described. Secondly, a case study of a franchisor in administration[6] is presented. The role, scope and effectiveness of pre-purchase disclosure in Australia are then reviewed in the light of the franchisor's administration. The inability of pre-purchase disclosure to provide a solution to franchisor failure is discussed. Implications for policy are then considered. Finally potential solutions are identified.

---

[2] Felstead (1993), p. 30.

[3] Croonen and Brand (2010), pp. 1021–1038.

[4] For example, Australian franchisor *Beach House Group* (BHG) had 22 franchisee-owned gyms @ setup fee of $300,000 + equipment & working capital. The administrator was appointed in November 2008. Cor Cordis Liquidators Report (19 October 2009) 13 states: in the opinion of BHG's liquidator the franchisor was *insolvent in 2005 and remained so* from that time. BHG was still accepting franchise fees from new franchisees in the third quarter of 2006.

[5] Perrigot and Cliquet (2004).

[6] The equivalent to Chap. 11 bankruptcy in the United States.

2 Business Format Franchising in the Twenty-First Century

Through franchising franchisors can expand branded businesses very rapidly and uniformly by creating a distribution channel that harnesses the commitment, energy, local knowledge, equity and access to debt of individual owner-operators, the franchisees. A business operator, the franchisor, grants the right to legally independent but contractually dependent franchisees, to clone a facet of its business. Franchising can also be explained as

> a form of industrial organisation . . . where the techniques of mass-production are mastered and replicated across time and space by successfully separating conceptual work by a small core of managers and planners [franchisors] from the performance of standardised work tasks by a largely unskilled contingent of easily replicable workers [franchisees].[7]

In addition to funding growth in the usual ways available to a stand-alone business (shareholder equity or debt) a franchisor is able to finance growth by selling franchises. Franchisees consequently outnumber franchisors by many times.[8] Franchising is now a significant part of the commercial landscape in developed[9] and developing economies.[10]

A fundamental aspect of franchising is the separation that is achieved through franchise agreements of ownership (by franchisees), from control (which remains with the franchisor). This can also be expressed as a separation of risk bearing, a function allocated to franchisees, and decision functions, which rest with the franchisor.[11] Franchisees are exposed to significant risk and this poses interesting new challenges for policy makers and regulators. The effectiveness of their responses is tested when a franchisor's business fails.

Challenges arise for franchisees through a combination of three features that uniquely come together in franchising. Firstly, there is a fundamental disconnect between the corporate governance of the franchisor and any legal principle that the franchisor or its administrator should factor the interests of franchisees into decisions; secondly, the nature of the franchise agreement; and thirdly, asymmetrical features of the law and the franchise agreement. None of these features are satisfactorily addressed by disclosure.

Twenty-first century franchisors operate in ways that rival the complexity of sophisticated multinationals. The law responds to potential exploitation and abuse of power within corporations by imposing a range of checks and balances, such as directors' duties, shareholder remedies, disclosure, audits, independent directors,

---

[7] Taylor (1997).

[8] For example Australia's 1,025 franchisors in 2010 had 62,000 franchisees, an average of 60 franchisees per franchisor. Frazer et al. (2010), p. 11.

[9] [One] out of every 12 businesses [in the USA] is a franchised business' Franchise Consultants Inc 2011.

[10] For example, South Africa's *Consumer Protection Act*, 2010 specifically addresses franchisees as consumers.

[11] See Fama and Jensen (1983), p. 301, 304.

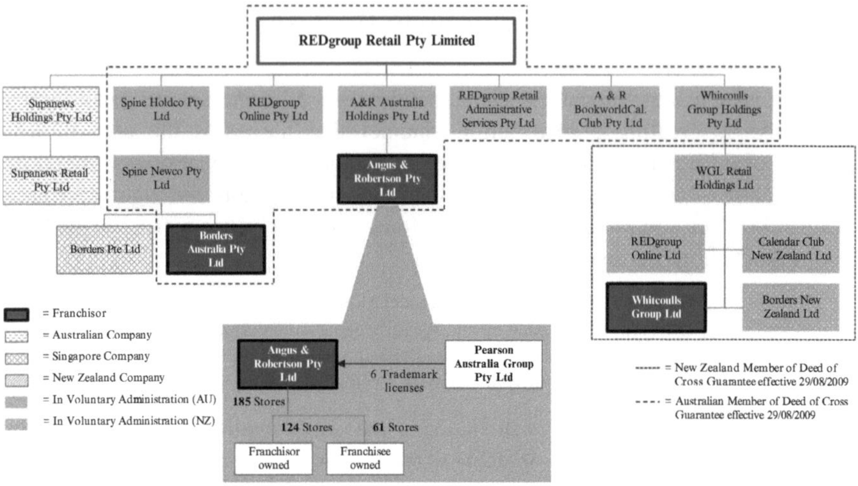

Fig. 1 Franchisors within the REDgroup: Borders Australia, Whitcoulls and Angus & Robertson

securities laws, insolvency procedures and corporate governance principles.[12] However, these checks and balances imposed on the non-franchised players within a corporation or a corporate group are absent in the regulation of the relationship between a franchisor and its franchisees. Denied access to the traditional mitigation mechanisms, franchisees rely heavily on pre-purchase disclosure and contractual rights. Franchising thus poses exploitation-management challenges for policy makers.

The franchise relationship is primarily regulated through the franchise agreement, a contract drafted by a franchisor to preserve and protect its interests. Policy makers worldwide treat business format franchising as a simple, single contract-based franchisor/franchisee relationship. Policy initiatives start from the position that franchisors and franchisees are both business people and franchisees can thus negotiate terms that will protect them. Franchisees contract with the franchisor that is often a lightly capitalized member of a corporate group. For example, as can be seen in Fig. 1[13] 'the franchisor' Angus & Robertson Pty Ltd ('A & R'), a proprietary company is one of 18 companies and trusts, including three franchisors, operating as a corporate group in three countries. The group is controlled by parent REDgroup Retail Pty Ltd.

[12] Lessing (2009); employees' rights, for example, are protected through a wide range of laws such as the Fair Entitlements Guarantee Act 2012 (Cth); and the OECD Principles of Corporate Governance, 2004, http://www.oecd.org/dataoecd/32/18/31557724.pdf.

[13] Based on diagram in Ferrier Hodgson, REDgroup Retail Pty Limited and associated companies (Administrators appointed), Report by Administrators Pursuant to Section 439A4(a) of the Corporations Act 2001, Appendix C, 11.

Through standard form, and thus non-negotiable franchise agreements franchisors' shift risks that would normally be borne by an employer or a financier to the franchisee. Such risks might include the risk that the franchisee's venture might fail as a result of poor site selection by the franchisor. In a negotiated contract risks would normally be measured and costed, then borne or rejected knowingly by contracting parties. This level of evaluation is impossible or prohibitively expensive in franchising. No amount of due diligence would make franchise agreements negotiable. A franchisee, like a consumer, does its best to evaluate the offering then signs the agreement or walks away from the opportunity.

Asymmetries of risk, information, adviser, reward, regulation and rights over franchisees' interests are pronounced.[14] They favour franchisors. Pre-entry to the franchise system the asymmetry is exacerbated by barriers to the conduct of effective due diligence. Barriers include the difficulty and cost of gaining an understanding of the complex networks such as the REDgroup in Fig. 1, within which franchisors like A & R operate. Before being allowed to sign the contract the franchisee has disclosed every detail of its personal finances. The franchisor on the other hand is able to withhold information about all aspect of its businesses except what has to be disclosed. The franchise agreement itself formally embeds the power and risk imbalance in the relationship.

Not all franchisors succeed in business. 'The Franchising Australia 2010 survey revealed that 56 franchise systems ceased operating and a further 88 of the 1100 trading in 2008 ceased franchising in the two year period to 2010'.[15] This was 12.36 % in 2 years. Whilst 'ceased' does not necessarily equate to failure, it does sometimes. '[A]necdotal Australian data provides a [further] insight into the size of the problem. The 1999 Australian Franchising Yearbook and Directory listed 347 franchisors. Of these, 251 (72 per cent) were no longer franchising by 2011'.[16] The exits included franchisors that had exited franchising but possibly remained in business and others like Kleenmaid with 15 franchisees,[17] Kleins Jewellery with 134 franchisees,[18] Traveland with 270 franchisees, Beach House Group with 22 franchisees, Healthzone Limited with 80 franchisees, Refund Home Loans with 320 franchisees and Tyrecorp with 33 franchisees that entered administration and/or were wound up.

These numbers give cause for concern as they continue despite pre-purchase disclosure being mandatory under Australian law since 1998. Accepting that franchisors may fail, it is important to ask what happens to franchisees when their

---

[14] Buchan (2008), p. 407.

[15] Buchan et al. (2011a).

[16] Buchan et al. (2011a); and see also Blair and Lafontaine (2005), p. 44 and 272; Lafontaine and Shaw (1998), p. 95; Perrigot and Cliquet (2004).

[17] The Kleenmaid franchisees paid initial fees of $60,000–120,000 each plus set-up costs. They subleased premises from the franchisor's associate, Kleenmaid Property. Deloitte (2009), p. 12.

[18] "Kleins collapse turns nasty" available at http://www.smartcompany.com.au/retail/cleins-collapse-turns-nasty.html. See also "What happens if a franchisor fails" at http://www.birkettlong.co.uk/site/library/legalnews/what_happens_if_a_franchisor_fails.html.

franchisor fails and whether pre-purchase disclosure could have helped franchisees protect their investments. Before looking at the disclosure in more detail I will demonstrate some consequences of an administrator being appointed to a franchisor through a case study of a recent franchisor failure.

3 Case Study: Angus & Robertson Booksellers

Franchisees starting a new A&R store paid up to $380,000 in total to establish their business. This investment becomes sunk costs. As part of their commitment to become franchisees they signed franchise agreements and premises leases and invested in stock. After paying for the shop fit out, stock is an A&R franchisee's second major cash investment and costs $900–1,000 per square metre.[19] Franchisees hire staff whose wages and entitlements are an ongoing cost. A & R franchise agreements coordinated the terms with lease term but aimed at providing 5 years with a further 5 years renewal, thus newly signed franchisees could look forward to establishing a business over a period of up to 10 years.

As Fig. 2 shows, A & R traded for 91 years before it started franchising and over 100 years before its first merger in 1990. A & R knew the business of buying and selling books. Thereafter it experienced continued development until being purchased by a venture capitalist Pacific Equity Partners in 2004. The venture capitalist formed REDgroup Retail Pty Ltd ('REDgroup') in 2009 and loaned it $138 m, secured over the assets of five bookselling brands including franchisors Borders, Whitcoulls and A & R. By August 2009 REDgroup had 20 % of the Australian book market.[20]

By sometime in 2010 it was becoming clear to those in the know that all was not well in the book retailing world. Borders in the US filed for Chap. 11 bankruptcy protection in February 2011 and Borders Australia quickly followed. Within 24 h, 'REDgroup Retail was placed in administration … owing an estimated $170 million'.[21] Whilst three franchise systems were affected the focus for this chapter is A & R.

3.1 Angus & Robertson 'Administrators Appointed'

The administration process is regulated under corporations law. In Australia this is the *Corporations Act* 2001 (Cth) ('CA'). Of the three possible paths through the insolvency process[22] the secured creditors, PEP, selected was voluntary administration. This process enabled PEP as:

[19] Angus & Robertson website. http://www.angusrobertson.com.au/franchise-terms-of-agreement viewed 13 October 2011.

[20] Zappone (2011a).

[21] http://www.franchiseadvice.com.au/index.php?option=com_content&task=view&id=247&Itemid=1 viewed 25 October 2011.

[22] See http://www.asic.gov.au/asic/asic.nsf/byheadline/Resources?openDocument.

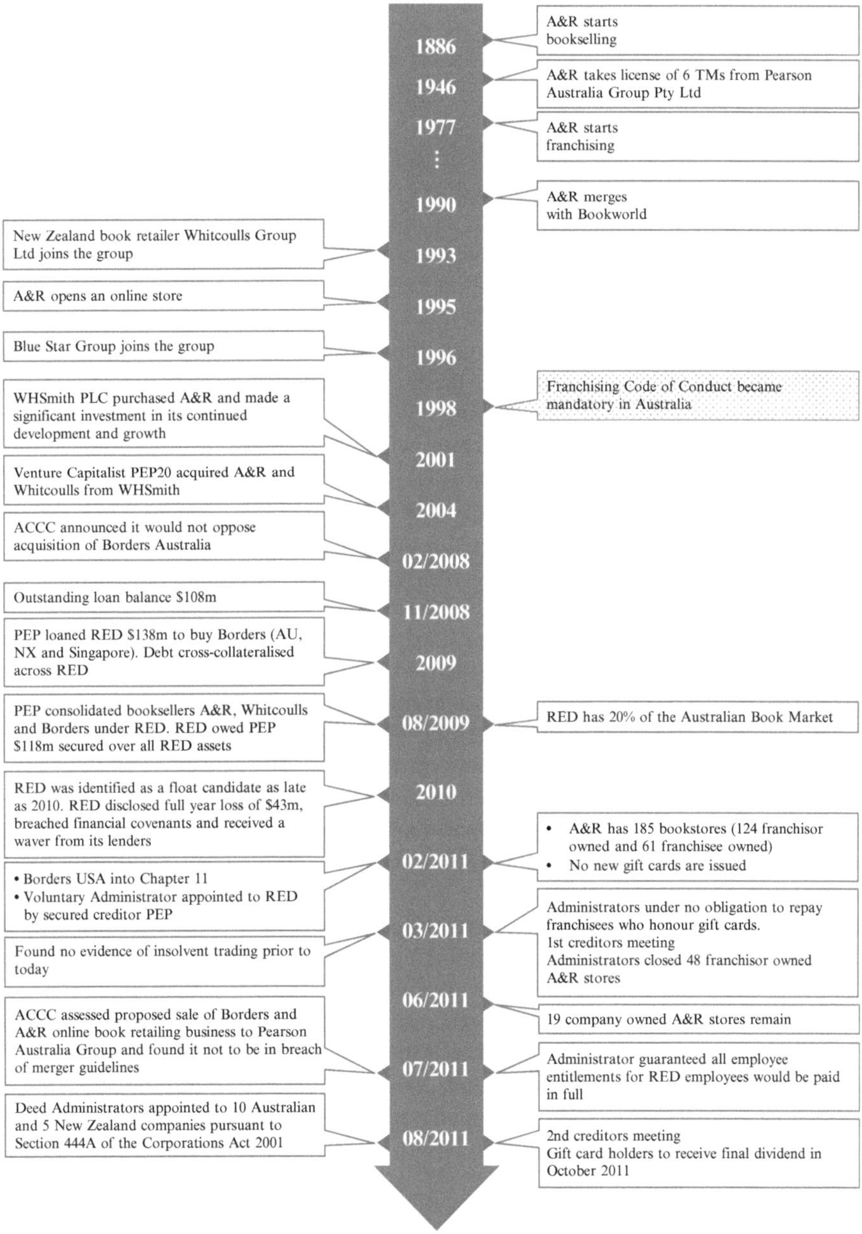

Fig. 2 Timeline of Angus & Robertson ascent and decline

> a secured creditor with a charge over most of the company's assets [to] appoint ... a 'voluntary administrator'. The role of the voluntary administrator is to investigate the company's affairs, to report to creditors and to recommend to creditors whether the company should enter into a deed of company arrangement, go into liquidation or be returned to the directors.[23]

The provisions of the *Corporations Act*, including time limits and order of priority of payments, must be adhered to strictly by administrators. This includes a requirement that a second creditors meeting must be held within 21 days of the appointment of the administrator. The court has discretion to delay this meeting if the administrators provide compelling reasons. In the REDgroup case the administrators were granted additional time to hold the second meeting of creditors. This enabled them to identify and negotiate with parties possibly interested in purchasing parts of the troubled business. The extended time frame placed the franchisees in limbo for several months.

Focusing here on the A & R franchise, within a month of their appointment the administrators closed '48 franchisor-owned A & R stores'.[24] They identified the A & R franchise agreements as saleable assets. The franchisees were required to continue trading whilst the administrators sought a buyer for the businesses or any of their component parts. A & R's Terms of Agreement and Financial Commitment for franchisees state 'Payment of stock and all other expenses is managed by the Franchise Owner'.[25] Accordingly franchisees appear to have held the premises leases in their own names and to have dealt directly with the book suppliers. In this case the franchisor's failure would not result in the franchisee losing the right to trade from their shops.

By 17 June 2011, four months after the administrators had been appointed they had closed a further '42 A&R [company-owned] stores [leaving only] 19 company-owned stores alongside the 48-strong franchise network'.[26] The 48 franchisees were included in the discussions with two potential buyers of their franchise agreements. Options were presented to the 48 A&R franchisees on 17 June. Ultimately most decided to join the Collins bookseller group[27] and the remainder joined an independent buying group called Leading Edge. From franchisee, Mrs Appleby's perspective 'Collins have innovative thinking and plans for the future. I'll probably do better under the Collins brand than I would have under the A&R brand'.[28]

Early in the administration period a circular sent by the administrators to franchisees stated that the appointment of the administrators did not automatically terminate the franchise and that royalties should continue to be paid by direct debit from franchisees' banks, as usual. From the administrators' perspective, franchisees

---

[23] Australian Securities and Investments Commission 2011.

[24] Zappone (2011b).

[25] Angus & Robertson website. http://www.angusrobertson.com.au/franchise-terms-of-agreement viewed 13 October 2011.

[26] Meeting on potential sale of Angus & Robertson underway, Smartcompany 2011.

[27] Collins itself had been formed as a buyers group after the Collins Booksellers franchisor failed in 2005 after selling books since 1929.

[28] Good staff and family support see stores write another chapter: Light at end of the tunnel, 13 September 2011, Sunshine Coast Daily, 20.

'continue[d] to trade as normal'[29] during the administration. Despite the lack of ongoing support by the franchisor, post administration royalties and marketing contributions paid by franchisees contributed $226,518 to the administrators' pool of funds. Before it was placed into voluntary administration the REDgroup network had over 2,500 employees.[30] Ultimately 'stock realisations [by the administrators were] sufficient to pay in full all [of the franchisor's] employee entitlements, totalling approximately $11.7 million'.[31]

Two features of the administration specific to A & R, gift cards, and consumer warranties on merchandise purchased before the administration and normally returnable if faulty, merit attention. Each will now be addressed.

3.2 *Gift Cards*

The February timing of the appointment of the administrator would mean that a high proportion of gift cards issued prior to Christmas and redeemable at any of the 185 A & R company-owned or franchisee owned stores, would not yet have been redeemed. Following their appointment the administrators '[c]hang[ed] the terms on which franchisees could honour customers' gift cards issued up to 2 April, halting all gift card redemptions on 3 April'.[32] This operational change placed franchisees in an unenviable position. Some such as Mr and Mrs Appleby 'honoured A&R gift cards for as long as they could, despite the fact that they were not legally bound to and despite making a loss on them'.[33] Their decision matrix is represented in Fig. 3.

Mr and Mrs Appleby bought two A&R franchised bookstores on Queensland's Gold Coast in November 2010. It is reasonable to predict that they had invested over $500,000. This timing enabled them to capitalise on the peak pre-Christmas trading time for booksellers. In the words of A & R,

> The book industry ... relies heavily on the Christmas trade. [A] large percentage of our sales, and our positive cash flow comes from the Christmas season. Franchisees must be prepared for the increase in stock required to meet Christmas demand and for the subsequent bills that must be paid.[34]

Little could the Applebys guess that at 3 pm on 17 February 2011, within 3 months of buying the stores and only 2 h after completing their franchisee induction at

---

[29] Ferrier Hodgson Press Release, 'Negotiations underway for sale of Angus & Robertson bookstores' 15 June 2011.

[30] Troubled bookstores face closure this week, ABC Melbourne, 1 March 2011. http://www.abc.net.au/news/stories/2011/03/01/3151898.htm?site=melbourne.

[31] Ferrier Hodgson press release 'Second Meeting of REDgroup creditors', 27 July 2011.

[32] Zappone (2011b).

[33] Light at end of the tunnel: Good staff and family support see stores write another chapter, 13 September 2011, Sunshine Coast Daily, 20.

[34] Angus & Robertson website. http://www.angusrobertson.com.au/franchise-partners.

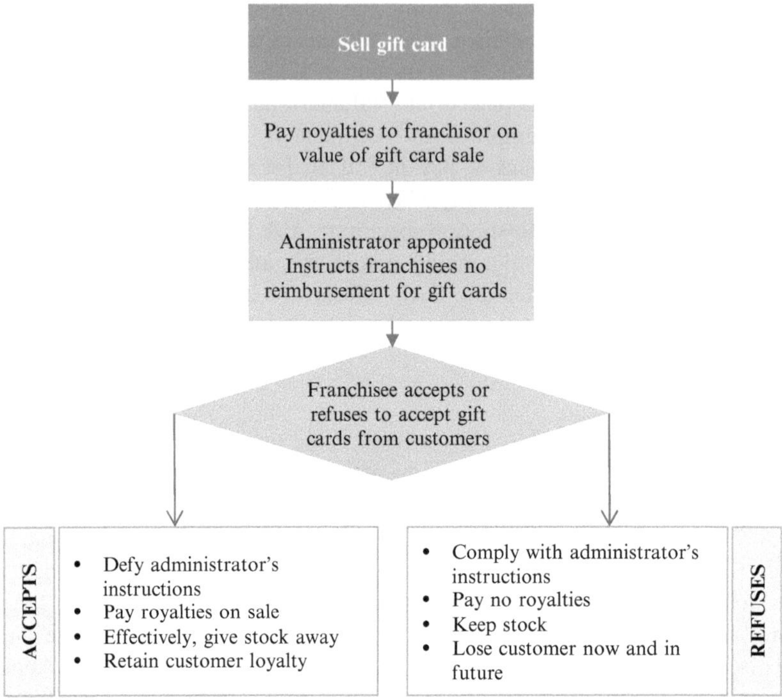

Fig. 3 Gift card choices

REDgroup Retail network's head office in Melbourne, their bank manager called to tell them their franchisor, A & R … had entered administration.[35] Franchisees had to decide whether to comply with the administrators' instructions or protect their investment by honouring gift cards to retain customer loyalty. As Mrs Appleby pointed out

> [n]ot many people were aware we were privately-owned and they didn't understand the gift card situation. When we sold a gift card, A&R took that money off us straight away. We didn't get that money back until someone used that gift card and then we had to claim it back.[36]

The Applebys had every incentive to protect their new investments.

3.3 *Consumer Warranties for Faulty Goods*

The administrator advised franchisees not to replace faulty purchases by consumers. Although this refusal would constitute a breach of the statutory

[35] Light at end of the tunnel: Good staff and family support see stores write another chapter, 13 September 2011, Sunshine Coast Daily, 20.

[36] Good staff and family support see stores write another chapter Light at end of the tunnel, 13 September 2011, Sunshine Coast Daily, 20.

warranties that provide consumer protection under Australia's *Competition and Consumer Act* 2010 (Cth), Schedule 2, Australian Consumer Law, ('ACL'). No claimant could take action whilst the moratorium against creditors initiating or pursuing litigation was effective. In this regard the administrators' powers under the CA effectively 'trump' the consumer's rights under the ACL.

3.4 What Would Make an A & R Franchisee a Creditor or Debtor?

Any franchisee owed money by a franchisor in administration would be a creditor and could attend creditors meetings. In all other respects franchisees had to continue to meet their contractual obligations with no right to participate in the administration process.

Franchisees became unsecured creditors for the face value of any unredeemed gift cards. If they had engaged in mediation with the franchisor any consequential agreement that the franchisor would pay the franchisee money would make the franchisee a creditor for any outstanding sum. There is no evidence of mediations occurring within A & R and no public record of mediation in Australia. Franchisees would also be creditors if they had purchased stock through the franchisor and items had been returned under warranty. Most of the franchisee's financial investment is in the form of the initial franchise fee, sunk costs and ongoing liabilities to third parties. Thus, for the bulk of the franchisee's investment, it is not a creditor of the franchisor.

Franchisees are all debtors of the franchisor in administration for any unpaid ongoing component of their royalties. If the franchisees had sub-leased premises from the franchisor they would also be debtors for rent owed after the appointment of the administrator. This appears not to have been the case with A & R.

3.5 Why Did A & R Fail?

The REDgroup's directors attributed A & R's failure to external events, including consumer spending patterns, the fact that Australian laws are seen to favour overseas online sellers (because of the application of Australia's Goods and Services Tax to domestic sales and the laws regulating parallel importation) the strong Australian dollar 'which had appreciated against the US dollar and the pound sterling by 20% since September 2009'[37] giving Australians particularly strong overseas buying power.

The administrators agreed with the directors and added a further four internal factors that contributed to the failure of the REDgroup, namely management buying decisions did not meet market demands. There was

[37] Ferrier Hodgson, REDgroup Retail Pty Limited and Associated Companies (Administrators Appointed) – Report by Administrators Pursuant to Section 439A(4)(a) of the *Corporations Act 2001*, 25 July 2011, 13–14.

> more emphasis on 'buying' than 'selecting' stock resulting in overstocking with aged, poor stock; failure to recognize and promptly address loss making stores; under-utilisation of space in stores and poor organisation with no logical grouping and general lack of consistent business processes with little use and reference to signed off critical paths and event management cycles.[38]

There may be a more fundamental cause of this particular administration that a different perspective would reveal. Insolvency practitioners acknowledge that

> [b]ankruptcy provides a useful business tool for a company to reorganize its operations, deleverage its balance sheet, accomplish a sale of assets, obtain new financing or improve its capital structure. For example, bankruptcy may assist a franchisor in addressing ... overexpansion in the market and the need to eliminate units, an unworkable equity structure, desire to sell or merge with another entity, threat of franchisee litigation, desire to refinance but the lender has expressed concern about financial or other issues.[39]

The view expressed by a New Zealand commentator was that for PEP:

> [v]oluntary administration was seen as a cost-effective way... to exit its ill-fated foray into book retailing. [PEP] knew what they were doing and they used the law to the maximum possible extent to extract everything they could out of it. ...Effectively this was a staged exit. ... REDgroup's total secured debt was $A118 million ($NZ147 million), most of it owed to [secured creditor] Pacific Equity Partners.[40]

The likelihood that REDgroup's voluntary administration was a strategic move is given weight by the REDgroup administrators' conclusion that there was no evidence of trading while insolvent. The administration was voluntary, triggered by a secured creditor Pacific Equity Partners ('PEP') that owned the majority of shares in the REDgroup. This administration can arguably be categorized as part of a considered business strategy. Voluntary administration thus provided an opportunity for the venture capitalist PEP to exit its investment when the financial climate did not permit it to exit by floating the REDgroup. Significantly, from the perspective of where the losses should fairly fall, this failure could not be attributed to anything the franchisees had done or failed to do.

3.6 Could A & R Franchisees Have Anticipated the Administration?

Australia's Franchising Task Force concluded that '...the most vulnerable franchise systems are those that have recently commenced franchising and have less than, say, 12-15 units'.[41] Awareness of this dated information would have instilled confidence in aspiring A & R franchisees, and their financiers and advisers.

---

[38] Ferrier Hodgson, REDgroup Retail Pty Limited and Associated Companies (Administrators Appointed) – Report by Administrators Pursuant to Section 439A(4)(a) of the *Corporations Act 2001*, 25 July 2011, 13–14.

[39] Foster and Johnsen (2005).

[40] Gray (2011).

[41] Franchising Task Force Final Report to the Minister for Small Business and Customs, December 1991. 2.7.

Some franchisees would have entered the A & R system before PEP became the owner in 2004[42] or before PEP created the REDgroup in 2009. Those that knew of the involvement of the PEP might have been reassured to read that PEP was described as 'a leading Australasian private equity firm focusing on buyouts and late stage expansion capital in Australia and New Zealand' that assists businesses move closer to their 'full potential'.[43]

If payment for premises rental and stock had normally been made to the franchisor and thence to suppliers the franchisor might have become a bad credit risk with its suppliers and the franchisees might have then been alerted by changed trading terms. This does not appear to have happened. In their report to creditors the administrators wrote 'it is difficult to maintain an argument that the Group was insolvent for any material period prior to 17 February 2011'.[44] Whilst Rod Sims, makes the well founded observation that 'many [franchisees] enter the franchising relationship without a great deal of business experience but nevertheless invest large sums of money - and sometimes their nest egg ... without doing sufficient due-diligence checks',[45] no amount of due diligence could anticipate the events that cause some franchisor administrations.

REDgroup took out a loan with Pacific Equity Partners Fund IV, LP for $138 million to complete the acquisition of Borders. 'The debt due to PEP [was] cross-collateralized across the group. PEP has lodged a Proof of Debt in the Administration of each company for $118,547,419'.[46] The franchisees could have had no knowledge of this debt, or the manner in which it was secured. It is possible that all franchise agreements were included in this security. If they were then this compromises their value as security for loans taken out by franchisees to purchase their businesses. Even if they had known of the debt to PEP, franchisees would not have the ability or the resources to evaluate its significance. At a personal level, the Applebys

> had done a tour of the office [the same day the administrator was appointed], [the REDgroup] had welcomed us, the CEO had been in to talk to us the day before," ... "And [Mrs Appleby said] call me naive, but I can't believe that any of the staff in that building on that day knew. Not the day-to-day workers'.[47]

---

[42] See Fig. 2.

[43] https://www.pep.com.au/pages/default.asp (accessed 3 November 2011).

[44] Report by Administrators pursuant to Section 439A(4)(a) of the Corporations Act 2001, prepared by the administrator of the REDgroup Retail Pty Limited and Associated Companies (Administrators Appointed), S. Sherman, J Meluish and J Lindholm, Ferrier Hodgson, 25 July 2011, 2.

[45] Parker (2011).

[46] Report by Administrators pursuant to Section 439A(4)(a) of the Corporations Act 2001, prepared by the administrator of the REDgroup Retail Pty Limited and Associated Companies (Administrators Appointed), S. Sherman, J Meluish and J Lindholm, Ferrier Hodgson, 25 July 2011, 7.

[47] Good staff and family support see stores write another chapter. Light at end of the tunnel, 13 September 2011, Sunshine Coast Daily, 20.

Neither the Applebys nor any of the other franchisees could have anticipated the financial state of the owner of the franchisor, or the appointment of the administrator. However, the A & R franchisees were lucky: they contracted independently with book suppliers and landlords so would not lose their credibility as buyers when the franchisor itself became a bad credit risk. The administrators found buyers for the franchise agreements. This does not always happen. Other franchisees lose their investment if a buyer is not found, or if the franchisor holds the head lease and the landlord has a more credible substitute tenant to replace the licensee/sub-lessee franchisee.

Against this background the role, scope and effectiveness of Australia's pre-purchase disclosure requirement in the face of a franchisor's administration will now be considered.

4 Role, Scope and Effectiveness of Pre-purchase Disclosure

One globally widely adopted mechanism intended to reduce the imbalance of contracting power between franchisor and franchisees and mitigate some aspects of asymmetry is pre-purchase disclosure. To comply with Australia's Franchising Code of Conduct ('Code')[48] all franchisors must maintain a disclosure document that adheres strictly to the format set out in Annexure 1 or 2. They must supply the appropriate version to prospective and current franchisees in prescribed situations. This disclosure packages much of the information that a franchisee needs to be aware of before signing a franchise agreement into one document.

There is no requirement in Australia for disclosure documents to be registered, or for the franchisor to be identified as a franchisor in any public register. All details of mediation conducted to satisfy the requirements of the Code remain confidential including the names of the parties. Thus, information about franchisors other than information they supply in disclosure or on their own websites is difficult and/or expensive to obtain, and breaches of the disclosure regime are usually impossible to detect unless they lead to litigation.

The express purposes of the Disclosure are twofold:

> 6A (a) to give to a prospective franchisee, or a franchisee proposing to enter into, renew, extend or extend the scope of a franchise agreement, information from the franchisor to help the franchisee to make a reasonably informed decision about the franchise; and (b) to give a franchisee current information from the franchisor that is material to the running of the franchised business.[49]

The front page of the disclosure every franchisee receives before signing a franchise agreement in Australia warns:

[48] *Trade Practices (Industry Codes—Franchising) Regulations* 1998 (Cth), Statutory Rules 1998 No. 162 as amended made under the *Trade Practices Act 1998*.

[49] *Trade Practices (Industry Codes—Franchising) Regulations 1998* (Cth), Clause 6A.

DISCLOSURE DOCUMENT FOR FRANCHISEE OR PROSPECTIVE FRANCHISEE;

> This disclosure document contains some of the information you need in order to make an informed decision about whether to enter into a franchise agreement.
>
> Entering into a franchise agreement is a serious undertaking. Franchising is a business and, like any business, the franchise (or franchisor) could fail during the franchise term.[50]

In a masterful understatement the notification continues,

> This could have consequences for the franchisee.[51]

A short warning at the point in the relationship when the intending franchisee has invested money and energy in examining a range of different options and is psychologically committed to proceeding with 'the one' is unlikely to ring alarm bells. The range of commonly experienced consequences is not spelt out, so the 'consequences' remain an abstract concept. Whilst this cursory warning may encourage a franchisee to shelter some personal assets from the business debts, a franchisee's ability to do so would depend on the amount of security other than personal assets it had available to fund the purchase of the franchise.

Franchisors are required to disclose information about themselves, their associates and the specific franchise opportunity. Franchisees are encouraged seek legal and financial advice in relation to the material disclosed. A & R's disclosure in Australia would thus have provided franchisees with over 250 items of information under 23 major headings. The items with the potential to reveal information about the franchisor, its plans and its attitude to risk are: Items 2 (Franchisor details), 3 (Business experience), 4 (Litigation), 7 (Intellectual property), 9 (Supply of goods or services to a franchisee), 15 (Franchisor's obligations) and 18 (Obligation to sign related agreements).

Under Clause 2, 'franchisor' and 'associate' are defined. The franchisor was A & R. Under the heading of associate the franchisor must disclose a person

> (a) who:
>
> (i) is a director or related body corporate, or a director of a related body corporate, of the franchisor; or
> (ii) for a franchisor that is a proprietary company — directly or indirectly owns, controls, or holds with power to vote, at least 15% of the issued voting shares in the franchisor; or
> (iii) is a partner of the franchisor; and whose relationship with the franchisor is relevant to the franchise system, including supplying goods, real property or services to a franchisee.[52]

Without access to the A & R disclosure document provided to the A & R franchisees it is not clear whether the existence of controlling shareholder PEP

---

[50] *Trade Practices (Industry Codes—Franchising) Regulations 1998* (Cth), Annexure 1, 1.1(e) and Annexure 2, 1.1(e).

[51] *Trade Practices (Industry Codes—Franchising) Regulations 1998* (Cth), Annexure 1, 1.1(e) and Annexure 2, 1.1(e).

[52] *Trade Practices (Industry Codes—Franchising) Regulations 1998* (Cth), Clause 3, 1(b).

would have been disclosed. PEP would not have been supplying goods, real property or services to the franchisees. It can be seen from Fig. 1 that it was distant from, and had a finance role superficially unrelated to supporting the franchisees. A search of the Australian Securities and Investments Commission records reveals eight PEP entities.[53] Even if it had been aware of PEP's involvement it is fanciful to suggest that an individual prospective franchisee could or should be required to conduct due diligence on a set of complex entities. Additionally, the administrator's conclusion that there was no evidence of the REDgroup having traded whilst insolvent suggests that even the most diligent franchisee would have found nothing to cause alarm if it had conducted searches of the corporations' records.

Under Item 3, as previously noted A & R had been in the bookselling business for over 100 years and franchising for 33 years when the Applebys signed on. REDgroup had the appearance of being a very well organized, well capitalized, geographically diversified, robust and focused player in the retail bookselling world. Any franchisee joining the group prior to PEP's involvement in 2004 would have received a disclosure document that could not have mentioned PEP. Any disclosure provided before 2009 would have preceded the existence of the REDgroup.

Item 4 would not have revealed any litigation current against the franchisor or its associates. As disputes resolved by mediation are confidential their existence is not disclosed.[54] Franchisors are not obliged to disclose the debt load that they, or their ultimate owner, is carrying unless that debt has triggered litigation that needs to be disclosed under the 'Litigation' item in the disclosure. One of the sub-items under Litigation information to be provided to an incoming franchisee is Item 4.1(a)(iii) contravention of the *Corporations Act 2001* (Cth). Trading while insolvent would be a contravention of the *Corporations Act* but as already noted the franchisor A & R was not thought to have been trading while insolvent in the opinion of the administrator. The nearest the parent appears to have come to trading while insolvent is, as identified by a journalist after the administrator had been appointed, that 'REDgroup . . . was also forced to get a waiver from its lenders after breaching some of its financial covenants'.[55] Even though a breach of a contract with a third party may be a warning bell about impending insolvency it does not need to be disclosed under 'Litigation'. In this case the breach was not by the franchisor but by the parent; and thus more remote and not requiring disclosure.

[53] Pacific Equity Partners Fund 1 Pty Limited ACN 083 026 822, Pacific Equity Partners Pty Limited ACN 082 283 949, Pacific Equity Partners Fund III (Australasia) Pty Limited ACN 117 565 410, Pacific Equity Partners Fund II (Australasia) Pty Limited ACN106 318 370, Pacific Equity Partners Fund 111 GP (Jersey) Limited 126745686, Pacific Equity Partners Fund III SPV Pty Limited ACN 119 059 040, Pacific Equity Partners Fund IV (Australasia) Pty Limited ACN 124 839 989 and Pacific Equity Partners Fund IV L.P. 150258165.

[54] For a full discussion of mediation and the problems the confidentiality of the process creates see Buchan et al. (2011b).

[55] Stafford (2011).

Under item 7 A & R would have revealed the existence of six registered trade marks[56] protecting the names 'Angus & Robertson' and 'Angus & Robertson where books come to life', all owned by a company called Pearson Australia Group Pty Ltd ('Pearson') that does not appear on any of the administrators' records. Pearson would be a creditor for any unpaid licence fee payable by A & R for the right to use the marks. The appointment of the administrators to the licensee is likely to have been an event allowing Pearson to terminate the trade mark licences with A & R. Pearson Australia Group Pty Ltd eventually emerged as the buyer of the REDgroup's online retailing business after gaining mergers approval from the ACCC.[57]

Item 9 would reveal items that the franchisee was obliged to buy from the franchisor. In the case of Angus & Robertson this may have been very little.

Item 15 requires disclosure of the franchisor's specific obligations. As these are typically couched in terms such as 'in the franchisor's discretion'[58] it poses no threat to franchisors and enables administrators to keep franchise agreements on foot and require the ongoing performance by franchisees so long as the franchisor is not in breach of a head-lease or other essential supply line contract. Where franchisees are selling instant use items like books and not future experiences like travel, the administrator is in a good position. By requiring franchisees to keep trading the administrator can be assured of a steady revenue stream.

The 'Obligation to sign related agreements' identified in Item 18 would direct the franchisees' attention to its premises lease. In some franchise systems the franchisor might require franchisees to lease fitout or to sign loan agreements and personal guarantees.

5 Implications

The organisation chart (Fig. 1) shows the position of the franchisor, Angus & Robertson Pty Limited within the corporate REDgroup of companies. The 'franchisor' makes disclosure as a single legal entity, but franchisors seldom stand alone. The REDgroup operated three franchised brands (A & R, Whitcoulls and Borders) through 18 separate corporations and trusts in three countries (Australia, New Zealand and Singapore). It is relatively easy and inexpensive for a prospective franchisee to conduct a search of one proprietary company. The more entities there are, the more expensive and difficult it becomes for franchisees to conduct meaningful due diligence or to make sense of what they see. The confidential nature of trusts, such as those that formed part of the ownership structure of the failed

[56] Numbered 299489 (Class 16), 343650 (Class 16), 637633 (Class 42), 861016 (Class 16, 35, 41), 1025323 (Class 16, 35, 41) and 1073382 (Class 16, 35, 41).

[57] http://www.accc.gov.au/content/index.phtml/itemId/1002218/fromItemId/751043 (viewed 20 June 2012).

[58] Spencer (2006).

Kleenmaid and Beach House Group franchises, means that conducting due diligence on a trust is impossible.

Of more value to franchisees than disclosure about 'the franchisor' would be a simple straightforward set of financial ratios with benchmarks from low through to high risks that could assist franchisees and their advisors. So also would be a comprehensive credit report of the franchisor and the group of companies it formed part of.

The front cover of the disclosure document, as already noted, warns a franchisee that the franchisor might fail. The franchisor is not required to reinforce this caution by identifying the possible consequences of its entering administration or becoming insolvent for the individual franchisee. Those consequences differ for each franchise network depending on the role of the franchisor in supplying premises, stock and on whether the franchisor/franchisee relationship is based on the franchisee paying royalties to or receiving commission from the franchisor. The variables extend to the type of product or service sold by the franchisee. A franchisee selling suites of kitchen equipment (ovens, refrigerators and the like) which the customer buys and pays a deposit on now and then pays the balance and takes delivery of in the future when the builder is ready for them is in a very different situation from a newsagent who sells magazines with a shelf life of one week. In the former case the customer does not get their kitchen suite as it is part of the security of the franchisor's lender. The customer becomes an unsecured creditor for the value of the deposit paid months ago. In the latter case the purchase is instant, so the franchisees customers are not creditors in the franchisor's insolvent estate.

Ultimately, each franchisor is the only party that knows how it structures its own business, which risks it passes to franchisees and what the consequences of its failure might be for each franchisee. Specific risks cannot be foreseen by an inexperienced franchisee. Nor can they satisfactorily be explained in a standard government-produced information booklet.

A further issue devaluing the pre-purchase disclosure as a franchisee protection mechanism is that administrators generally treat the Code, and its requirement that disputes be mediated, as not applying to them. In some cases, such as the Kleenmaid insolvency where the franchisees supply chain for both stock and premises was directly through the franchisor's insolvent group, mediation may have been pointless, but in other situations where the franchisees and their franchisor's businesses are not so intermeshed, it may be a valuable way for the franchisees to protect their interests during the administration.

The disclosure discloses a considerable amount of information about the franchise but the risk is that the franchisor will come from a bad family. In the case of A & R the 'bad family' element was not the franchisor that made disclosure. It was the venture capitalist that assembled the network and then failed to nurture it. The prior disclosure for A & R franchisees was wholly ineffective as a mechanism to protect franchisees from the consequences of their franchisor's administration.

6 Potential Solutions

Administration is a complex legal procedure. Any protection for franchisees must address all three possible outcomes of administration. It must also acknowledge that other stakeholders, such as employees and creditors may have to surrender some of their 'stake' in the administration if franchisees are to be given rights. Potential responses include contract-based or insurance-based solutions, or regulation. They must enable franchisees to set themselves up more securely *ex ante*, and position them better *ex post* the appointment of the administrator.

Contract based solutions include the possibility of inserting *ipso facto* clauses in franchise agreements; these would permit the franchisee to terminate the agreement if an administrator was appointed to the franchisor. Unless the process by which the right was exercised was refined beyond being a bald right this is not an ideal solution. It could make it difficult for the administrator to sell the franchise as a going concern, as was successfully achieved in A & R.

As noted earlier, if the franchise agreements form part of the franchisor's security they are less valuable as security for the franchisee's creditors. As it is the franchisee that paid for its business it may be appropriate to forbid franchisors from including the franchise agreements in their security.

Once an administrator recommends that the business cannot be saved and should be wound up, control of the company passes from administrators to the liquidators. Whilst administrators are notionally bound by the Code, they are regulated by the *Corporations Act 2001* (Cth). The liquidators' duties and liabilities are found only in the *Corporations Act*. Under section 568(1) *Corporations Act 2001* (Cth) liquidators have the power to disclaim onerous contracts. This enables them to walk away from head leases, trade mark licences, franchise agreements, obligations to pay commissions and any other contracts that are seen as unsaleable or a drain on resources. Thus any contract-based solutions to the problems franchisor administration creates for franchisees would be of limited value as they would not endure throughout the winding up option.

Given the number of franchisors and franchisees it is inconceivable that any contract based improvements to the franchisees' position would occur without statutory intervention. Thus, what remains are regulatory solutions. These range from specific legislation addressing franchisee rights and implying concepts such as fitness for purpose into all franchise agreements, to amendments to the legislation that regulates the conduct of the administration and insolvency processes.[59]

[59] Solutions are explored in detail in Buchan (2010) and Buchan (2013).

7 Conclusion

> Q: How do you start a small business?
> A: Sell a big business to a private equity company.[60]

The franchise model evolved only after policy and regulation to regulate governance of corporate groups and provide rights for employees, consumers and creditors was long-settled. When pre-purchase disclosure is tested by exposing it to a franchisor entering administration, the disclosure that is so widely relied on to protect franchisees, fails. We expect too much of this regulatory tool.

In A & R significant evolutions of the ownership and governance of the business model occurred in 2004 and 2009. Franchisees whose entry pre-dated 2004 or 2009 were not supplied with a crystal ball along with the disclosure document. For a franchisee, signing the franchise agreement after receiving disclosure of the franchisor entity is comparable to getting married having met only ones intended, but not any of their family, friends or business associates. There is nothing to stop the patriarch or matriarch remarrying and in doing so taking the family to a new future. Similarly, the franchisor might be taken over, or bought by a venture capitalist, introducing an unforseen dynamic into the franchise 'family'. As with our personal relationships, there is a limit to how much pre-purchase disclosure at a moment in time can reveal.

Changes of franchisor ownership and structure that accompany the entry of private equity, or the listing of a franchisor, are expected. They will arguably become more common as entrepreneur franchisors seek exit opportunities. It is time to recognise we are living in more complex times than when franchising started. Some of those changes lead to franchisor failure. Once the franchise network is handed to the administrator it is too late for the franchisees to react in any meaningful way. *Ex ante* warnings, such as that in the Code's disclosure, of the possibility of the franchisor failing are of no benefit if the franchisee is unable to accurately evaluate the risks to itself. Contract-based actions that franchisees may otherwise have the right to pursue are thwarted by the existence of the stay on proceedings under the *Corporations Act* once an entity is in administration. In a conflict between statutory requirements and any contract-based rights contained in franchise and ancillary agreements, the statute-based requirements of insolvency prevail.

In the realm of business failure, attitudes and expectations have reached a point where business failures are seen by many in the commercial world as

> a productive mechanism. ... part of a process in which inefficient and unprofitable businesses are replaced by efficient and profitable ones. ... Economies get better through a process of experimentation and natural selection.[61]

Franchisor administration and insolvency are key events where governance, contracts and asymmetry place franchisees and their investment in a very weak

---

[60] Stockdill (2011).

[61] Bickerdyke et al. (2000).

legal position vis à vis the franchisor's other stakeholders. Pre-purchase disclosure can do little to alleviate this situation. It is time the law accepted that franchisors will fail, and that the franchisees' failure to prepare for this eventuality is not a failure to conduct due diligence. Further, the blame for their franchisor's failure cannot be laid at the feet of the franchisees.

Failure to acknowledge the inability of pre-purchase disclosure to protect franchisees consigns franchisees permanently to economic externality status in its franchisor's insolvency. Franchisees should not be an externality in a franchisor's insolvency any more than carbon should be an externality to manufacturing.

References

Angus & Robertson Agreement (2011) http://www.angusrobertson.com.au/franchise-terms-of-agreement. Accessed 13 Oct 2011

Australian Securities and Investments Commission (2011) http://www.asic.gov.au/insolvency-resources. Accessed 25 Oct 2011

Bickerdyke I, Lattimore R, Madge A (2000) Business failure and change: an Australian Perspective Productivity Commission Staff Research Paper, Canberra, p 3

Blair RD, Lafontaine F (2005) The economics of franchising. Cambridge University Press, Cambridge, p 44 and 272

Buchan J (2008) Ex ante information and ex post reality for franchisees: the case of franchisor failure. Australian Bus Law Rev 36:407

Buchan JM (2010) Franchisor failure: an assessment of the adequacy of regulatory response. PhD Thesis, Queensland University of Technology

Buchan J (2013) Franchisees as consumers: Benchmarks, Perspectives and Consequences. Springer, New York

Buchan J, Qu C, Frazer L (2011a) Protecting franchisees from their franchisor's impending failure: a way forward for consumer protection regulators using indicators? Paper presented at the first international conference on comparative law and global common law 'Exchange and Integration of the Contemporary Legal Systems,' College of Comparative Law, China University of Political Science and Law, 24–25 September 2011, Beijing

Buchan J, Harris J, Gunasekara G (2011b) Franchise mediation: confidentiality or disclosure: a consumer protection conundrum. Paper presented at the 25th Annual International Society of Franchising Conference, Boston, 16–18 June 2011

Competition and Consumer Act 2010 (Cth), Schedule 2, The Australian Consumer Law

Consumer Protection Act (2010) (South Africa)

Corporations Act 2001 (Cth) (Australia)

Croonen EPM, Brand MJ (2010) Dutch druggists in distress: franchisees facing the complex decision of how to react to their franchisor's strategic plans. Entrepren Theory Prac 34 (5):1021–1038

Deloitte (2009) The Kleenmaid Group (Administrators Appointed) Administrators' Report to Creditors Pursuant to Section 439A Corporations Act 2001, 14 May 2009

Fair Entitlements Guarantee Act 2012 (Cth) No 159 of 2012

Fama EF, Jensen MC (1983) Separation of ownership and control. J Law Econ XXVI(2):301, 304, who do not include franchise networks in the spectrum of organizations discussed

Felstead A (1993) The corporate paradox: power and control in the business franchise. Routledge, London, 30

Ferrier Hodgson (2011) Report by Administrators pursuant to Section 439A(4)(a) of the Corporations Act 2001, prepared by the administrator of the REDgroup Retail Pty Limited and Associated Companies (Administrators Appointed), S. Sherman, J Meluish and J Lindholm, 25 July 2011

Ferrier Hodgson Press Release: (15 June 2011) Negotiations underway for sale of Angus & Robertson bookstores

Ferrier Hodgson Press Release: (27 July 2011): Second Meeting of REDgroup creditors

Foster SB, Johnsen C (2005) The war of the worlds: bankruptcy versus .... Paper presented at American Bar Association, 28th Annual Forum on Franchising, 2005

Franchising Task Force (1991) Final Report to the Minister for Small Business and Customs

Frazer L, Weaven S, Bodey K (2010) Franchising Australia 2010. Griffith University, Brisbane, p 11

Gray J (2011) REDgroup creditors set to vote on slim payout. The New Zealand Herald August 4, 2011 quoting David Joel, managing director at Macmillan Publishers NZ

Lafontaine F, Shaw KL (1998) Franchising growth and franchisor entry and exit in the US market: myth and reality. J Bus Venturing 13(2):95, in: Hoy F, Stanworth J (eds) (2003) Franchising: an international perspective. Routledge, London and New York, pp 163–164

Lessing J (2009) The checks and balances of good corporate governance. Corporate Governance ejournal. http://works.bepress.com/john_lessing/9. Accessed 10 Dec 2011

Parker L (2011) Do the numbers stack up? Sydney Morning Herald November 2, 2011 quoting Rod Sims, Chair of the Australian Competition and Consumer Commission

Pacific Equity Partners website (2011) https://www.pep.com.au/pages/default.asp. Accessed 3 Nov 2011

Perrigot R, Cliquet G (2004) Survival of franchising networks in France from 1992 to 2002. Paper presented at the 18th Annual International Society of Franchising Conference, Las Vegas, Nevada, 6–7 March 2004

Smartcompany (2011) Meeting on potential sale of Angus & Robertson underway, 17 June 2011

Spencer EC (2006) Standard form and relational aspects of franchise contracts. Paper presented at the 20th International Society of Franchising Conference, Palm Springs, California, 24–26 February, p 21

Spencer EC (2010) The regulation of franchising in the new global economy. Edward Elgar, London

Stafford P (2011) Borders Australia and Angus & Robertson Chains Collapse, Thursday, 17 February 2011 15:47. http://Www.Smartcompany.Com.Au/Retail/20110217-Borders-Australia-And-Angus-Robertson-Chains-Collapse.Html. Accessed 6 Oct 2011

Stockdill R (2011) No one to blame but themselves. Inside Retailing. 25 February 2011

Sunshine Coast Daily Good staff and family support see stores write another chapter: Light at end of the tunnel, 13 September 2011, 20

Taylor VL (1997) Contracts with the lot: franchises, good faith and contract regulation. New Zealand Law Review 459 at 461 citing Cox R (1987) Production, poser and world order: social forces in the making of history, Columbia University Press, New York, p 328

Trade Practices (Industry Codes—Franchising) Regulations 1998 (Cth), Statutory Rules 1998 No. 162 as amended made under the Trade Practices Act 1998

Zappone C (2011a) Borders, A&R call time on book vouchers. Sydney Morning Herald 23 March 2011

Zappone C (2011b) Booksellers' woes worsen as franchisees seek to defect. Sydney Morning Herald, 5 April 2011